HISTORY

OF THE

GREAT FIRES

IN

CHICAGO AND THE WEST.

A PROUD CAREER ARRESTED BY SUDDEN AND AWFUL CALAMITY; TOWNS AND COUNTIES LAID WASTE BY THE DEVASTATING ELEMENT.

SCENES AND INCIDENTS,

LOSSES AND SUFFERINGS,

BENEVOLENCE OF THE NATIONS, Etc., Etc.

WITH A

History of the Rise and Progress of Chicago, the "Young Giant."

To which is appended a Record of the Great Fires in the past.

By Rev. E. J. GOODSPEED, D.D.,
OF CHICAGO.

ILLUSTRATED.

SOLD ONLY BY SUBSCRIPTION.

New York:
H. S. GOODSPEED & CO., 37 Park Row, New York.
J. W. GOODSPEED, Chicago, Cincinnati, St. Louis, and New Orleans.
D. L. GUERNSEY, Concord, N. H.
SCHUYLER SMITH, London and Prescott, Ontario.

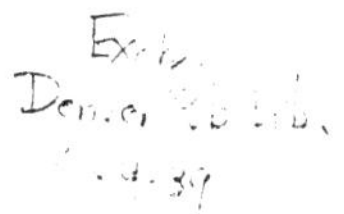

CONTENTS.

CHAPTER XV.

RIVER TUNNELS.

CHAPTER XVI.

MORALS AND RELIGION.

CHAPTER XVII.

ADVANTAGES OF CHICAGO.

IV.—THROUGH FIRE.

CHAPTER XVIII.

THE GREAT CONFLAGRATION.

CHAPTER XIX.

REMOTE CAUSE OF WESTERN FIRES.

CHAPTER XX.

PROGRESS OF THE FIRE.

CHAPTER XXI.

ORIGIN OF THE FIRE.

CHAPTER XXII.

THE BURNING.

CHAPTER XXIII.

AFTER THE FIRE.

CHAPTER XXIV.

INCIDENTS AND EXPERIENCES.

V.—MINISTERED TO BY CHRISTIAN CHARITY.

CHAPTER XXIX.

SUPPLEMENTAL WORK.

CHAPTER XXX.

AID FROM THE STATE.

CHAPTER XXXI.

APPRECIATION OF THE BOUNTY.

VI.—CONVALESCENCE.

CHAPTER XXXII.

GENERAL EXPECTATION OF CHICAGO'S RESURRECTION.

CHAPTER XXXIII.

REBUILDING.

VII.—THE FUTURE.

CHAPTER XXXIV.

GLORIOUS ANTICIPATIONS.

CHAPTER XXXV.

FAITH AND HOPE.

CHAPTER XXXVI.

THE STRONG MEN REMAIN.

THE FIRES IN WISCONSIN AND MICHIGAN.

CHAPTER XLIII.

FIRES IN DAKOTA.

HISTORY OF THE GREAT FIRES IN THE PAST.

CHAPTER XLIV.

VIRGIL'S DESCRIPTION OF THE BURNING OF TROY.

CHAPTER XLV.

ANCIENT ROME IN FLAMES.

CHAPTER XLVI.

MOSCOW.

CHAPTER XLVII.

LONDON.

CHAPTER XLVIII.

NEW YORK'S EXPERIENCE OF FIRE.

CHAPTER XLIX.

OTHER CITIES VISITED.

CHAPTER L.

APPENDIX.

ILLUSTRATIONS.

PREFACE.

Among the remarkable phenomena of modern times, Chicago occupies a leading place. Richard Cobden, the English statesman, charged Goldwin Smith on the eve of his departure for America: "See two things in the United States, if nothing else—Niagara and Chicago,"—intimating thus that these were the two principal wonders of the New World to a stranger. Since our Great Conflagration, it has occurred simultaneously to many that the ambitious young city, always aspiring to lead, wished also to surpass the world in the way of a fire. And now, certainly, her fortunes attract and interest millions of mankind as never before. To satisfy this interest in part, many have undertaken to write up the city and its vicissitudes. Believing that the story of its changes, prosperity and calamity, of its help and hope, will be eagerly read by millions, we offer this contribution, gathered from many sources and carefully prepared, to the generous public, who have already signalized their interest in our welfare by the most magnificent bounty to our suffering thousands. Let the poet no longer sing—

"Oh, the rarity
Of Christian charity!"

but rather celebrate

"The quality of mercy,
Which droppeth like the gentle rain from Heaven."

Our parched soil, after fourteen weeks of drought, did not rejoice in the showers that fell from God, as we exulted in the beneficence that poured forth upon us in our extremity of need. "The Lord loveth a cheerful giver!"

Chicago is great in its ruins and hopeful in its prostration. The record of its herculean energy and manly heroism, and the outlook for its future, must animate and encourage the world, now smitten in every part by our misfortune.

We send forth this venture in humble gratitude to the Almighty for such a past, in submission to His providence, confidence for the future, and trust in the charitable generosity of the people, to whom it is boldly submitted for their patronage.

We have faithfully sought to arrange all the lights needed for a complete illustration of the stupendous events recorded. In the full illumination afforded by these various torch-bearers, many of them brilliant and glowing, the reader may expect to see and appreciate, as no one eye-witness could, what must ever be considered marvellous among the marvels of time!

E. J. G.

CHICAGO.

“Hear the loud alarum bells—
Brazen bells!
What a tale of terror, now, their turbulency tells!
In the startled ear of night
How they scream out their affright!
Too much horrified to speak,
They can only shriek, shriek,
Out of tune,
In a clamorous appealing to the mercy of the fire,
In a mad expostulation with the deaf and frantic fire
Leaping higher, higher, higher,
With a desperate desire
And a resolute endeavor,
Now—now to sit or never,
By the side of the pale-faced moon.
Oh, the bells, bells, bells,
What a tale their terror tells
Of despair!
How they clang, and clash, and roar,
What a horror they outpour
On the bosom of the palpitating air!”

NOTICE.

THE PUBLISHERS purpose, after defraying the expenses of publishing and selling this book, to devote a portion of the profits which may arise from the sale thereof to the aid of deserving mechanics, working-women, etc., who have suffered by the fires. Having already made several instalments, the following letters are appended to show the manner in which aid is proposed to be rendered.

THE PUBLISHERS.

CHICAGO, *November* 25, 1871.

H. S. GOODSPEED & CO.:

Gentlemen—I have received the elegant sewing-machine sent by you to me, to be given to the most deserving person of my acquaintance who suffered in the late terrible fire here. May God bless you in your endeavors to help our suffering people, so many of whom will have a hard struggle to live through the cold winter.

I am very truly yours,
MRS. LIZZIE AIKEN, *Missionary.*

$100. CHICAGO, *27th November*, 1871.

Received from J. W. GOODSPEED, of Chicago, One Hundred Dollars, for the Chicago Relief and Aid Society.

GEORGE M. PULLMAN, *Treasurer.*
Per W. C. NICHOLS, *Cashier.*

FREE READING-ROOMS AND LIBRARY OF THE YOUNG MEN'S CHRISTIAN ASSOCIATION,
97 W. RANDOLPH ST., CHICAGO, *Nov.* 28, 1871.

Mr. J. W. GOODSPEED, Publisher,
51 S. Carpenter street, Chicago:

Dear Sir—On behalf of the Young Men's Christian Association of Chicago, I would gratefully acknowledge the receipt of your order upon Lyon & Healy for a Burdett Organ for the use of the devotional meetings, upon account of Dr. Goodspeed's "History of Chicago and the Great Fire." May all the other results of that wonderful visitation in like manner tend to promote the praise of God and the edification of his Church.

Yours in Christ,
ROBERT PATERSON.

ROOMS LADIES' CHRISTIAN UNION,
COR. PEORIA AND JACKSON STREETS.

Mr. J. W. GOODSPEED, Publisher:

Dear Sir—The Ladies' Christian Union do most gratefully acknowledge the receipt of a *Home Shuttle* Sewing-Machine from you, as publisher of Dr. Goodspeed's "History of Chicago and Great Fires." It is a most timely and acceptable gift, and our prayers are that "He who loveth a cheerful giver" may reward you, who, in giving to the poor in time of their utmost need, but lend to Him.

Yours truly,
MRS. O. P. KNOX, *Pres't Ladies' Christian Union.*

A SCENE AT THE TAKING OF FORT DEARBORN, 1812.

HISTORY OF THE GREAT FIRES

IN

CHICAGO AND THE WEST.

I.—THE INFANT.

CHAPTER I.

CITIES, which are but an aggregation of individuals, have their periods of development, changes, growth, checks, prosperity and adversity, sickness and recovery, and alas! of decline and dissolution, like men. The proudest of earth's great gatherings of human beings had their origin in some accident, or, we may better say, some providential circumstance or course of events, and their progress from humble beginnings has been slow. As rivers rise in some small obscure fountain in the depth of the forest, or upon the mountain side, and wind onward for long distances, fed by other streams till they become like the foaming Rhine or the majestic Father of Waters, so the metropolis now teeming with vast multitudes of busy men, began in a group of lowly huts or cabins, and increased by degrees from within and from without, by births and immigration, till it reached

2

greatness and became a power in the earth. We may compare it to the snowball which boys roll along the whitened field till it becomes an immense mass. It was at first just a handful of white crystals massed together; it ends by assuming gigantic proportions. Our Londons and other capitals grew up in this manner, and had in their history all the elements of crudeness and feebleness which marked Chicago's infancy.

The age of fable has passed, and in telling the story of Chicago we have no Romulus and Remus suckled by a wolf to adorn our tale. Yet if all that was experienced by the first white people who settled the shores of this magnificent lake could be described with graphic pen, the story would be full of romance. We cannot point to such an origin as Venice had, which was the retreat of robber bands who built among the shallow waters and upon the mud a nest for themselves, to which they might bring their plunder. Yet upon these sands, and beside the river that winds along the prairie as if loth to leave the Lake, savages roamed or built their wigwams for temporary residence. And these waters echoed to the war-whoop, and the shriek of the despairing was heard in unison with the moan of the waves along the beach. But the white people who came to this Far West were men of adventurous, but not bloodthirsty natures, who sought for themselves a fortune in these untrodden virgin regions of the New World. These hardy pioneers were tired of restraint in older countries, and pined for the freedom of the wild prairies, where the winds were no freer than the spirits of the hunter. Woman, ever clinging fondly to man, accompanied the bold adventurer to cheer and bless him in his wanderings, and to help him sustain the hardships of frontier life.

In the fearful Indian massacre which early stained these shores with blood, there shone forth the heroism and fidelity of the female character—even as sixty years afterwards, in the horrors of the furious assault of murderous flames, woman exhibited heroism and nobleness, and proved herself worthy to be termed man's "help meet."

> O woman, in our hours of ease
> Uncertain, coy, and hard to please,
> And variable as the shade
> By the light, quivering aspen made;
> When pain and anguish wring the brow,
> A ministering angel thou!

CHAPTER II.

In the alternation of victory and defeat during the wars of France and England, the native people, the aborigines, were sometimes on the side of the colonists and sometimes against them. It was natural for them to incline to the dominant party, and they became the prey of intriguers who bought their treacherous aid with presents. It was needful to protect outlying settlements, where trade was carried on by adventurous white men, by means of forts and garrisons. Chicago, a term said to have denoted a king or deity, a skunk or a wild onion, was much haunted by the Indians, and a fort there arose to give the shelter of its guns to the whites. Often had it been marked for assault, but always escaped, till the period of the last war with Great Britain, when certain circumstances conspired to prepare the way for the great tragedy described by the historian Brown.

INDIAN MASSACRE.

When war was declared in 1812, the little garrison at Chicago, consisting of a single company, was commanded by Captain Heald; Lieutenant Helm and Ensign Ronan were officers under him, and Dr. Van Voorhes its surgeon.

On the 7th of August, 1812, in the afternoon, Winnemeg, or Catfish, a friendly Indian of the Pottawatomie tribe, arrived at Chicago, and brought dispatches from General Hull, containing the first intelligence of the declaration of war. General Hull's letter announced the capture of Mackinaw, and directed Captain Heald "to evacuate the fort at Chicago if practicable, and in that event to distribute all of the United States property contained in the fort, and the United States factory, or agency, among the Indians in the neighborhood, and repair to Fort Wayne." Winnemeg urged upon Captain Heald the policy of remaining in the fort, being supplied as they were with ammunition and provisions for a considerable time. In case, however, Captain Heald thought proper to evacuate the place, he urged upon him the propriety of doing so immediately, before the Pottawatomies (through whose country they must pass, and who were as yet ignorant of the subject of his mission) could collect a force sufficient to oppose them. This advice, though given in great earnestness, was not sufficiently regarded by Captain Heald; who observed, that he should evacuate the fort, but having received orders to distribute the public property among the Indians, he did not feel justified in leaving it until he had collected the Pottawatomies in its vicinity, and made an equitable distribution among them. Winnemeg then suggested the expediency of marching out, and leaving everything standing; "while the Indians," said he, "are dividing the spoils, the troops will be able to retreat without molestation." This advice was also unheeded, and an order for evacuating the fort was read next morning on parade. Captain Heald, in issuing it, had neglected to consult his junior officers, as it would have been natural for him to do in such an emergency, and as he probably would have done, had there not been some coolness between him and Ensign Ronan.

The lieutenant and ensign waited on Captain Heald to learn his intentions, and being apprised for the first time of the course he intended to pursue, they remonstrated against it. "We do not," said they to Captain Heald, "believe that our troops can pass in safety through the country of the Pottawatomies to Fort Wayne. Although a part of their chiefs were opposed to an attack upon us last autumn, they were actuated by motives of private friendship for some particular individuals, and not from a regard to the Americans in general; and it can hardly be supposed that, in the present excited state of feeling among the Indians, those chiefs will be able to influence the whole tribe, now thirsting for vengeance. Besides," said they, "our march must be slow, on account of the women and children. Our force, too, is small. Some of our soldiers are superannuated, and some of them are invalids. We think, therefore, as your orders are discretionary, that we had better fortify ourselves as strongly as possible, and remain where we are. Succor may reach us before we shall be attacked from Mackinaw; and, in case of such an event, we had better fall into the hands of the English than become victims of the savages."

Captain Heald replied that his force was inadequate to contend with the Indians, and that he should be censured were he to continue in garrison when the prospect of a safe retreat to Fort Wayne was so apparent. He therefore deemed it advisable to assemble the Indians and distribute the public property among them, and ask of them an escort thither, with the promise of a considerable sum of money to be paid on their safe arrival; adding that he had perfect confidence in the friendly professions of the Indians, from whom, as well as from the soldiers, the capture of Mackinaw had studiously been concealed.

From this time forward the junior officers stood aloof from their commander, and, considering his project as little short of madness, conversed as little upon the subject as possible. Dissatisfaction, however, soon filled the camp; the soldiers began to murmur, and insubordination assumed a threatening aspect.

The savages, in the mean time, became more and more troublesome, entered the fort occasionally in defiance of the sentinels, and even made their way, without ceremony, into the quarters of its commanding officer. On one occasion an Indian, taking up a rifle, fired it in the parlor of Captain Heald. Some were of opinion that this was intended as the signal for an attack. The old chiefs at this time passed back and forth among the assembled groups, apparently agitated; and the squaws seemed much excited, as though some terrible calamity was impending. No further manifestations, however, of ill feeling were exhibited, and the day passed without bloodshed. So infatuated, at this time, was Captain Heald, that he supposed he had wrought a favorable impression upon the savages, and that the little garrison could now march forth in safety.

From the 8th to the 12th of August the hostility of the Indians was more and more apparent; and the feelings of the garrison, and of those connected with, and dependent upon it for their safety, more and more intense. Distrust everywhere at length prevailed, and the want of unanimity among the officers was appalling. Every inmate retired to rest, expecting to be roused by the war-whoop; and each returning day was regarded by all as another step on the road to massacre.

The Indians from the adjacent villages having at length arrived, a council was held on the 12th of August. It was attended only by Captain Heald on the part of the military—the other officers refused to attend, having previously learned that a massacre was intended. This fact was communicated to Captain Heald; he insisted, however, on their going, and they resolutely persisted in their refusal. When Captain Heald left the fort they repaired to the blockhouse which overlooked the ground where the council was in session, and opening the port-holes, pointed their cannon in its direction. This circumstance, and their absence, it is supposed, saved the whites from massacre.

Captain Heald informed the Indians in council, that he would, next day, distribute among them all the goods in the United

States factory, together with the ammunition and provisions with which the garrison was supplied; and desired of them an escort to Fort Wayne, promising them a reward on their arrival thither, in addition to the presents they were about to receive. The savages assented, with professions of friendship, to all he proposed, and promised all he required.

The council was no sooner dismissed, than several waited on Captain Heald in order to open his eyes, if possible, to their condition.

The impolicy of furnishing the Indians with arms and ammunition, to be used against themselves, struck Captain Heald with so much force that he resolved, without consulting his officers, to destroy all not required for immediate use.

On the next day (August 13th), the goods in the factory store were distributed among the Indians; and in the evening the ammunition, and also the liquors belonging to the garrison, were carried, the former into the sally-port and thrown into the well, and the latter through the south gate, as silently as possible, to the river bank, where the heads of the barrels were knocked in, and their contents discharged into the stream.

The Indians, suspecting the game, approached as near as possible, and witnessed the whole scene. The spare muskets were broken up, and thrown into the well, together with bags of shot, flints, and gun-screws, and other things; all of little value.

On the 14th the despondency of the garrison was for a while dispelled by the arrival of Captain Wells and fifteen friendly Miamies. Having heard at Fort Wayne of the order to evacuate Chicago, and knowing the hostile intentions of the Pottawatomies, he hastened thither in order to save, if possible, the little garrison from its doom. He was the brother of Mrs. Heald, and having been reared from childhood among the savages, knew their character; and something whispered him "that all was not well." He was the son of General Wells of Kentucky, who, in the defeat of St. Clair, commanded three hundred savage warriors posted in front of the artillery, who caused extraordinary carnage among those who served it; and, uninjured himself, picked off the artillerists, until "their bodies were heaped up almost to the height of their pieces."

Supposing that the whites, roused by their reverses, would eventually prevail, he resolved to abandon the savages and rejoin his countrymen.

This intrepid warrior of the woods, hearing that his friends at Chicago were in danger, and chagrined at the obstinacy of Captain Heald, who was thus hazarding their safety, came thither to save his friends, or participate in their fate. He arrived, however, too late to effect the former, but just in time to effect the latter. Having, on his arrival, learned that the ammunition had been destroyed, and the provisions distributed among the Indians, he saw there was no alternative. Preparations were therefore made for marching on the morrow.

In the afternoon a second council was held with the Indians, at which they expressed their resentment at the destruction of the ammunition and liquor in the severest terms. Notwithstanding the precautions which had been observed, the knocking in of the heads of the whiskey-barrels had been heard by the Indians, and the river next morning tasted, as some of them expressed it, "like strong grog." Murmurs and threats were everywhere heard; and nothing, apparently, was wanting but an opportunity for some public manifestation of their resentment.

Among the chiefs there were several who participated in the general hostility of their tribe, and retained, at the same time, a regard for the few white inhabitants of the place. It was impossible, however, even for them to allay the angry feelings of the savage warriors, when provocation after provocation had thus been given; and their exertions, therefore, were futile.

Among this class was Black Partridge, a chief of some renown

Soon after the council had adjourned, this magnanimous warrior repaired to the quarters of Captain Heald, and taking off a meda. he had long worn, said: "Father, I have come to deliver up to you the medal I wear. It was given me by your countryman, and I have long worn it as a token of our friendship. Our young men are resolved to imbrue their hands in the blood of the whites. I cannot restrain them, and will not wear a token of peace when compelled to act as an enemy."

Had doubts previously existed, they were now at an end. The devoted garrison continued, however, their preparations as before; and amid the surrounding gloom a few gallant spirits still cheered their companions with hopes of security.

The ammunition reserved, twenty-five rounds to each soldier, was now distributed. The baggage-wagons designed for the sick, the women, and the children, containing also a box of cartridges, were now made ready, and the whole party, anticipating a fatiguing, if not a disastrous, march on the morrow, retired to enjoy a few moments of precarious repose.

On the morning of the 15th the sun rose with uncommon splendor, and Lake Michigan "was a sheet of burnished gold."

Early in the day a message was received in the American camp, from To-pee-na-bee, a chief of the St. Joseph's band, informing them that mischief was brewing among the Pottawatomies, who had promised them protection.

About nine o'clock the troops left the fort with martial music, and in military array. Captain Wells, at the head of the Miamies, led the van, his face blackened after the manner of the Indians. The garrison, with loaded arms, followed, and the wagons with the baggage, the women and children, the sick and the lame, closed the rear. The Pottawatomies, about five hundred in number, who had promised to escort them in safety to Fort Wayne, leaving a little space, afterward followed. The party in advance took the beach road. They had no sooner arrived at the sand-hills, which separate the prairie from the beach, about a mile and a half from the fort, than the Pottawatomies, instead of continuing in rear of the Americans, left the beach and took to the prairie. The sand-hills, of course, intervened, and presented a barrier between the Pottawatomies and the American and Miami line of march. This divergence had scarcely been effected, when Captain Wells, who with the Miamies was considerably in advance, rode back and exclaimed: "They are about to attack us; form instantly and charge upon them." The word had scarcely been uttered, before a volley of musketry from behind the sand-hills was poured in upon them. The troops were brought immediately into a line, and charged up the bank. One man, a veteran of seventy, fell as they ascended. The battle at once became general. The Miamies fled at the outset; their chief rode up to the Pottawatomies, charged them with duplicity, and brandishing his tomahawk, said, "he would be the first to head a party of Americans, and return to punish them for their treachery." He then turned his horse and galloped off in pursuit of his companions, who were then scouring across the prairie, and nothing was seen or heard of them more.

The American troops behaved gallantly. Though few in number, they sold their lives as dearly as possible. They felt, however, as if their time had come, and sought to forget all that was dear on earth.

While the battle was raging the surgeon, Doctor Voorhes, who was badly wounded, and whose horse had been shot from under him, approaching Mrs. Helm, the wife of Lieutenant Helm (who was in action, participating in all its vicissitudes), observed: "Do you think," said he, "they will take our lives? I am badly wounded, but I think not mortally. Perhaps we can purchase safety by offering a large reward. Do you think," continued he, "there is any chance?" "Doctor Voorhes," replied Mrs. Helm, "let us not waste the few moments, which yet remain, in idle

BLACK PARTRIDGE HOLDING MRS. HELM IN THE WATER.

or ill-founded hopes. Our fate is inevitable. We must soon appear at the bar of God. Let us make such preparations as are yet in our power." "Oh!" said he, "I cannot die. I am unfit to die! If I had a short time to prepare! Death!—oh, how awful!"

At this moment Ensign Ronan was fighting at a little distance with a tall and portly Indian; the former, mortally wounded, was nearly down, and struggling desperately upon one knee. Mrs. Helm, pointing her finger, and directing the attention of Dr. Voorhes thither, observed: "Look," said she, "at that young man; he dies like a soldier."

"Yes," said Doctor Voorhes, "but he has no terrors of the future; he is an unbeliever."

A young savage immediately raised his tomahawk to strike Mrs. Helm. She sprang instantly aside, and the blow intended for her head fell upon her shoulder. She thereupon seized him around his neck, and while exerting all her efforts to get possession of his scalping-knife, was seized by another Indian, and dragged forcibly from his grasp.

The latter bore her, struggling and resisting, towards the lake. Notwithstanding, however, the rapidity with which she was hurried along, she recognized, as she passed, the remains of the unfortunate surgeon stretched lifeless on the prairie. She was plunged immediately into the water and held there, notwithstanding her resistance, with a forcible hand. She shortly, however, perceived that the intention of her captor was not to drown her, as he held her in a position to keep her head above the water. Thus reassured she looked at him attentively, and in spite of his disguise recognized the "white man's friend." It was Black Partridge.

When the firing had ceased, her preserver bore her from the water and conducted her up the sand-bank. It was a beautiful day in August. The heat, however, of the sun was oppressive; and walking through the sand, exposed to its burning rays in her drenched condition, weary, and exhausted by efforts beyond her strength, anxious beyond measure to learn the fate of her friends, and alarmed for her own, her situation was one of agony.

The troops having fought with desperation till two-thirds of their number were slain, the remainder, twenty-seven in all, borne down by an overwhelming force, and exhausted by efforts hitherto unequalled, at length surrendered. They stipulated, however, for their own safety and for the safety of their remaining women and children. The wounded prisoners, however, in the hurry of the moment, were unfortunately omitted, or rather, not particularly mentioned, and were therefore regarded by the Indians as having been excluded.

One of the soldiers' wives, having frequently been told that prisoners taken by the Indians were subjected to tortures worse than death, had from the first expressed a resolution never to be taken, and when a party of savages approached to make her their prisoner she fought with desperation, and though assured of kind treatment and protection, refused to surrender, and was literally cut in pieces, and her mangled remains left on the field.

After the surrender, one of the baggage-wagons, containing twelve children, was assailed by a single savage, and the whole number were massacred. All, without distinction of age or sex, fell at once beneath his murderous tomahawk.

Captain Wells, who had as yet escaped unharmed, saw from a distance the whole of this murderous scene, and being apprised of the stipulation, and on seeing it thus violated, exclaimed aloud so as to be heard by the Pottawatomies around him, whose prisoner he then was: "If this be your game I will kill too!" and turning his horse's head, instantly started for the Pottawatomie camp, which was near what is now the corner of State and Lake, where the squaws and Indian children had been left ere the battle began. He had no sooner started than several Indians followed in his rear and discharged their rifles at him as he galloped the prairie. He laid himself flat on the neck of his horse, and was apparently out of their reach, when the ball of one of his pursuers took effect, killing his horse and wounding him severely. He was again a prisoner—as the savages came up, Winnemeg and Wa-ban-see, two of their number and both his friends, used all their endeavors in order to save hin; they had disengaged him already from his horse, and were supporting him along, when Pee-so-tum, a Pottawatomie Indian, drawing his scalping-knife, stabbed him in the back, and thus inflicted a mortal wound. After struggling for a moment he fell and breathed his last in the arms of his friends, a victim for those he had sought to save—a sacrifice to his own rash, presumptuous, and perhaps indiscreet intentions.

The battle having ended and the prisoners being secured, the latter were conducted to the Pottawatomie camp near the fort. Here the wife of Waw-bee-wee-nah, an Illinois chief, perceiving the exhausted condition of Mrs. Helm, took a kettle and dipping up some water from the stream which flowed sluggishly by them, threw into it some maple sugar, and stirring it up with her hand gave her to drink. "It was," says Mrs. Helm, "the most delicious draught I had ever taken, and her kindness of manner, amid so much atrocity, touched my heart." Her attention, however, was soon directed to other objects. The fort, after the troops had marched out, became a scene of plunder. The cattle were shot down as they ran at large, and lay dead or were dying around her. It called up afresh a remark of Ensign Ronan's made before: "Such," said he, "is to be our own fate--to be shot down like brutes."

The wounded prisoners, we have already remarked, were not included in the stipulation made on the battle-field, as the Indians understood it. On reaching, therefore, the Pottawatomie camp, a scene followed which beggars description.

A wounded soldier lying on the ground was violently assaulted by an old squaw, infuriated by the loss of friends, or excited by the murderous scenes around her, who, seizing a pitchfork, attacked with demoniac ferocity and deliberately murdered, in cold blood, the wretched victim now helpless and exposed to the burning rays of the sun, his wounds already aggravated by its heat, and he writhing in torture. During the succeeding night five other wounded prisoners were tomahawked.

Those wounded remained in the wigwams of their captors. The work of plunder being now completed, the fort next day was set on fire. A fair and equal distribution of all the finery belonging to the garrison had apparently been made, and shawls and ribbons and feathers were scattered about the camp in great profusion. The family of the principal Indian trader having been moved across the river, Black Partridge and Wa-bau-see, with three other friendly Indians, stood sentinels at his door. Everything was now tranquil. Even savage ferocity appeared to be gorged. Soon, however, a party of Indians from the Wabash arrived, the most implacable of all the Pottawatomies.

Runners had been sent to all their villages, and information transmitted thither that the fort was to be evacuated, that its spoils were to be divided among the savages, and its garrison to be massacred; they had therefore hurried on with their utmost speed to participate in the exhilarating and awful scene. On arriving at the Aux Plains they were met by a party returning from Chicago bearing a wounded chief along. Informed by these friends that a battle had been fought and a victory won, that its spoils had been divided among the conquerors, and the prisoners scalped and slain (and they not present), their rage was unbounded. They therefore accelerated their march, and on reaching Chicago blackened their faces in token of their intentions, and entered the parlor of the Indian trader before referred to where the family were assembled with their faithful protectors around, and seated themselves without ceremony in silence upon the floor.

Black Partridge, perceiving in their looks what was passing in their minds, and not daring to remonstrate, observed in an under tone to Wa-bau-see, "We have endeavored to save our friends, but all is in vain—nothing will save them now." At this moment another party of Indians arrived, and a friendly whoop was heard from the opposite shore. Black Partridge sprung upon his feet, and advancing to the river's bank, met their chief as he landed.

"Who," said Black Partridge, "are you?" "A man," replied the chief; "who are you?" "A man like yourself." "But tell me," said Black Partridge, "who are you for?" "I am," said he, "the Sau-ga-nash." "Then make all speed to the house," replied the former; "your friends are in danger, and you only can save them."

Billy Caldwell, the newly arrived chief (for it was he), thereupon hurried immediately thither, entered the parlor with a calm deliberate step, and without the least agitation in his manner, took off his accoutrements, and placing his rifle behind the door, saluted the hostile savages. "How now, my friends?" said he, "a good day to you. I was told there were enemies here; but I am glad to find none but friends. Why have you blackened your faces? Are you mourning for the friends you have lost in the battle (purposely mistaking the token of their evil intentions), or are you fasting? If so, ask our friend here and he will give you to eat. He is the Indians' friend, and never refused them what they had need of."

Taken thus by surprise, the savages were ashamed to acknowledge their bloody purpose; and in a subdued and modest tone said they had come to beg of their friend some white cotton, in which to wrap their dead before interring them. This was given them, with other presents, and they quietly departed.

Captain and Mrs. Heald were sent across the lake to St.

Joseph's after the battle; the former was twice, and the latter seven times wounded in the engagement. The horse rode by Mrs. Heald was a fine spirited animal, and the Indians were anxious to obtain it uninjured. Their shots were therefore principally aimed at the rider. Her captor being about to tear off her bonnet, in order to scalp her, young Chaudonnaire, an Indian of the St. Joseph's tribe, knowing her personally, came to her rescue, and offered a mule he had just taken for her ransom; to this he added a promise of ten bottles of whiskey. The latter was a strong temptation. Her captor, perceiving that she was badly wounded, observed that she might die, and asked him if he would give him the whiskey at all events; he promised to do so, and the bargain was concluded.

Mrs. Heald was afterward put into a boat in company with others, including her children, and a buffalo robe thrown over them. She was then enjoined to be silent, as she valued her life. In this situation she remained, without uttering a sound that could betray her to the savages, who came frequently to the boat in search of prisoners. Captain Heald was captured by an Indian from the Kankakee, who, having a strong personal regard for him, and seeing the wounded and enfeebled condition of his wife, released him without ransom, in order that he might accompany Mrs. Heald to St. Joseph's. To the latter place Mr. and Mrs. Heald were conveyed by Chaudonnaire and his party. The Indian who had so nobly released his prisoner, on returning to his tribe, found them dissatisfied; and their displeasure became so manifest that he resolved to make a journey to St. Joseph's, to reclaim his prisoner. News, however, of his intention preceding him, Mr. and Mrs. Heald, by the aid and influence of To-pa-na-bee and Kee-po-tah, were put into a bark canoe, and paddled by a chief of the Pottawatomies and his wife to Mackinaw, three hundred miles distant, along the eastern coast of Lake Michigan, and delivered to the British commander. They were kindly re-

CHICAGO IN 1830.

ceived, and sent afterward as prisoners to Detroit, where they were finally exchanged.

Lieutenant Helm was wounded in the action, and taken prisoner; he was afterward removed by some friendly Indians to the Au Sable, and from thence to St. Louis, and liberated from captivity through the intervention of Mr. Thomas Forsythe, an Indian trader.

Mrs. Helm was wounded slightly in the ankle, had her horse shot from under her, and after passing through several agonizing scenes, was taken to Detroit.

The soldiers, with their wives and children, were dispersed among the Pottawatomies on the Illinois, the Wabash, and Rock rivers, and some were taken to Milwaukie. In the following spring they were principally collected at Detroit, and ransomed. A part of them, however, remained in captivity another year, and during that period experienced more kindness than they or their friends had anticipated.

CHAPTER III.

The indolent, debauched barbarians were among the most serious obstructions to the progress of the infant town, as their bloody and vengeful ancestors had hindered the early settlement. Men were unwilling to hazard their scalps in unequal contests with these wild savages unless there was some prize to be gained worthy the dangerous venture; and when they had become tamed they were still animals, corrupt and corrrupting. The condition of the muddy banks of Chicago river and the outlaying prairie was not particularly inviting to persons of intelligence, who had been accustomed to the comparative civilization and improvements of the East. But one by one these obstacles disappeared.

The inferior race, made so by ages of ignorance and superstition, must inevitably go down before the superior, exalted by centuries of education and Christian influences. Once, indeed, Teuton and Saxon and Celt were low down in the scale of humanity, scarcely equalling the North American Indian in his best estate; and long periods of revolution and elevation preceded the present high position they occupy in the New World. And now, placed on the borders of civilization, exposed to the low and debasing influences of barbarism, they are liable to descend to a depth of degradation scarcely conceivable. In 1827 an agent of the Government reported Chicago as having no dwellings above kennels and pens, and described the squatters as "a miserable race of men, hardly equal to the Indians." It was therefore a policy of wisdom in the United States, and even of humanity, to remove the savages to a distance from the whites, between whom a mutual degradation was exerted. We submit Parton's description, which graphically tells the story and justifies the action of the authorities, while it enables us to realize some of the gigantic difficulties under which our infant city labored:—

"On a day in September, 1833, seven thousand of them gathered at the village to meet Commissioners of the United States for the purpose of selling their lands in Illinois and Wisconsin. In a large tent on the bank of the river the chiefs signed a treaty which ceded to the United States the best twenty million acres of the Northwest, and agreed to remove twenty days' journey west of the Mississippi. A year later four thousand of the dusky nuisances assembled in Chicago to receive their first annual annuity. The goods to be distributed were heaped upon the prairie, and the Indians were made to sit down around the pile in circles, the squaws sitting demurely in the outer ring. Those who were selected to distribute the merchandise took armfuls from the heap, and tossed the articles to favorites seated on the ground. Those who were overlooked soon grew impatient, rose to their feet, pressed forward, and at last rushed upon the pile, each struggling to seize something from it. So severe was the scramble, that those who had secured an armful could not get away, and the greater number of empty-handed could not get near the heap. Then those on the outside began to hurl heavy articles at the crowd, to clear the way for themselves, and the scramble ended in a fight, in which several of the Indians were killed and a large number wounded. Night closed in on a wild debauch, and when the next morning arrived few of the Indians were the better off for the thirty thousand dollars' worth of goods which had been given them. Similar scenes, with similar bloody results, were enacted in the fall of 1835; but that was the last Indian payment Chicago ever saw.

"In September, 1835, a long train of forty wagons, each drawn by four oxen, conveyed away, across the prairies, the children and effects of the Pottawatomies, the men and able-bodied women walking alongside. In twenty days they crossed the Mississippi, and for twenty days longer continued their westward march, and Chicago was troubled with them no more. Walking in the imposing streets of the Chicago of to-day, how difficult it is to realize that thirty-two years have not elapsed since the red men were dispossessed of the very site on which the city stands, and were 'toted' off in forty days to a point now reached in fifteen hours."

Were there space to insert here, after the above interesting exit of "poor Lo," Judge Ruger's poem nailed to the walls of the Old Block House which was threatened with demolition, we should perceive how fondly the early settlers clung to the relic whose reminiscences were full of painful interest. The Fort was abandoned in consequence of the unsettled state of affairs in the country; but as men would congregate here, it was rebuilt in 1816, and finally demolished in 1856. Our people scarcely have time or space to devote to what is not strictly practicable for present uses, and hence the relics of other days soon fade and perish from neglect or actual violence.

> She boldly faced the daring foe,
> She did her duty well.
> She kept the white men's foes at bay—
> The savage hounds of hell!

CHAPTER IV.

Another difficulty which oppressed the early settlers was the mud, which at times seemed bottomless. Where the city lately prospered in all its glory and grandeur, with clean streets, deep basements, and dry cellars, and buildings rising in towered majesty, the water stood a portion of the year, or teams struggled, helplessly "slewed" in the deep black ooze of the streets and prairies. Often a wagon would sink so far that little but the tongue appeared to indicate where the remainder lay. Or a board was set up with a rude inscription, evidently facetious, "No bottom here." The water was surface water, and little better than if dipped out of a pool by the road-side. The river's mouth was choked by a bar of sand which destroyed the harbor, and communication with the better portions of the country was extremely precarious. More than two centuries after the Pilgrims landed at Plymouth all this vast prairie region was, as it were, a wilderness occupied by wild beasts and still wilder men, and the metropolis of the Northwest yet lay floundering like an infant in swaddling clothes—so slowly does the Creator evolve His plans, and leave something ever fresh and rich for human enterprise to discover and possess. The bold fathers of New England wrung from nature's bosom scanty nourishment; and her cities grew slowly—far more slowly than the western Hercules. When the East had become established in wealth, and overflowed with brains and energy, here was the natural outlet and place for expansion and investment. God had made the flattest spot on the continent the fit location for that "city set on a hill which cannot be hid." For, singularly enough, the rain that falls in this spot finds its way by natural courses partly to the Atlantic by the St. Lawrence and partly to the Gulf of Mexico by the Mississippi. There is Lake Michigan connecting us with the northern seas, and the Illinois river bearing our sewerage to the southern ocean. A few men were gifted with that far-sightedness that enabled them to see how the young child must grow. They had even seen his star in the East, and they came with their gifts of courage, talent, hope, and industry to lay them at his feet, and swear allegiance to the destiny of the promising Infant.

II. THE YOUTH.

CHAPTER V.

Men early undervalued land without trees, and often chose the openings, or groves, the sheltered banks of streams, or hilly locations, in preference to these naked prairies. All lived to regret their choice who saw the development of these portions of the soil which contain the largest and best accumulations of fertility, and offer the easiest opportunities for cultivation. At certain seasons they seem barren and gloomy in their nakedness, but at most periods there is something beautiful in their boundlessness, like the ocean's expanse; and their undulating bosom, like the sea in a storm, is covered with a green spray, or lit up with the golden glory of abundant harvests.

It was doubtless a blessing to our country that the Pilgrims did not, like the early Spaniards, light upon these rich parts of the country, or discover the mineral resources of the Pacific coast. They grew a nobler race in consequence of their tough encounters with savage men, and the rugged shores and hills of New England. We had a basis of moral and mental stability, and political prosperity, when the gates were flung wide and the world invited to pour their masses forth upon these virgin treasures.

The Illinois river flows into the Mississippi, and is connected with Lake Michigan by a canal at La Salle, ninety-six miles from Chicago. This great work was begun in 1836, and completed in 1848, and many thousands were already awaiting its benefits in the young city, where the transshipment of the produce of the Southern counties must furnish employment and create business.

This enterprise gave Chicago its first strong push upward. In later days the ditch has been so deepened that the amber-colored waters of our lake flow through the Chicago river and cleanse out its filth, so long an offence whose rankness smelled to heaven Until last spring, or the early summer of 1871, at all seasons, except when the ice shut down the foul odors, there rose from the bayou or lagoon lying stagnant along its twelve or fifteen miles, unless stirred by the pumps, the vilest stench, which not only disgusted the senses, but attacked the health of our citizens. To strangers it was a perpetual source of raillery. A story is told of a citizen who made a visit in June to the country, and was so overpowered by the fresh air that he fainted, and was revived only upon the application to his nose of a decayed fish. As he rallied, and speech returned, he asked, "Where am I? It smells so much like home." The canal has, therefore, proved a double advantage, never to be overestimated, in bringing us into contact and relations with the wealthy heart of our State, and bearing away from us the sewerage of a populous city.

CHAPTER VI.

In 1832 there was a tax of one hundred and fifty dollars levied on the eight hundred people then dwelling on the banks of Chicago river, and the first public building consumed one-twelfth of the levy—"a pound for stray cattle." The population multiplied from 1833, though in 1837 there were but 4,470 persons here. Government began to dredge out the river, and Nature helped with a freshet that swept away the bar, and made a harbor accessible to the largest vessels. The Infant then became the Youth, and people were wild with excited hopes of sudden riches.

Ford's History of Illinois says: In the spring and summer of 1836 the great land and town-lot speculation of those times had fairly reached and spread over Illinois. It commenced in the State first at Chicago, and was the means of building up that place, in a year or two, from a village of a few houses to be a city of several thousand inhabitants. The story of the sudden fortunes made there excited at first wonder and amazement, next a gambling spirit of adventure, and lastly an all-absorbing desire for sudden and splendid wealth. Chicago had been for some time only a great town market. The plots of towns for a hundred miles around were carried there to be disposed of at auction. The eastern people had caught the mania. Every vessel coming west was loaded with them, their money, and means, bound for Chicago, the great fairyland of fortunes. But, as enough did not come to satisfy the insatiable greediness of Chicago sharpers and speculators, they frequently consigned their wares to eastern markets. Thus, a vessel would be freighted with land and town lots, bound for New York and Boston markets, at less cost than a barrel of flour. In fact, lands and town-lots were the staple of the country, and were the only article of export.

Outside the little town floundering in the mud, there were sturdy farmers wresting from the black and fertile soil their hidden treasures. These men had Chicago for their chief market, and contributed to raise it from the revulsions which cast it down in 1836 and 1837. It is interesting to notice how the country made the city, and this reacted upon the country, so that the whole Northwest is vitally concerned in the prosperity of her metropolis. We turn again to Parton's description of this period, and of the progress in business now steadily observable.

" A little beef had already been salted and sent across the lake; but in 1839 the business began to assume promising proportions, 3,000 cattle having been driven in from the prairies, barrelled, and exported. In 1838 a venturesome trader shipped thirty-nine two-bushel bags of wheat. Next year, nearly 4,000 bushels were exported; the next, 10,000; the next, 40,000. In 1842 the amount rose, all at once, from 40,000 to nearly 600,000, and announced to parties interested that the "hard times" were com ing to an end in Chicago. But the soft times were not. That mountain of grain was brought into this quagmire of a town from far back in the prairies,—twenty, fifty, one hundred, and even one hundred and fifty miles! The season for carrying grain to market is also the season of rain, and many a farmer in those times has seen his load hopelessly "slewed" within what is now Chicago. The streets used often to be utterly choked and impassable from the concourse of wagons, which ground the roads into long vats of blacking. And yet, before there was a railroad begun or a canal finished, Chicago exported two and a quarter millions of bushels of grain in a year, and sent back on the most of the wagons that brought it, part of a load of merchandise."

In 1849 the first locomotive halted ten miles below the city, and heralded the coming of the tide that rolled across the prairies, as the Nile freshets enrich its banks. The immigrants were usually of the better class, and made communities which have no superiors in the civilized world.

CHAPTER VII.

During these years, between 1833 and 1850, men came here who have made the city great by their labors. Many of these noble spirits lived only long enough to see the name of Chicago respected and honored, and escaped the sorrow of witnessing her proud career so cruelly arrested. Others still survive the conflagration, who have lost much of their accumulations; perhaps all is consumed; they are left to an old age of disappointment and want. Of the prominent citizens whose brain and energy gave the city its present pre-eminence, some are yet in the prime of vigorous manhood, and will rally to rebuild and restore that which was at once their pride and joy. They are crippled in resources, but undaunted in spirit. We can give a view of the youth of this region and city with increased vividness by sketching briefly the career of some of these public-spirited men, who became early identified with the fortunes of Chicago. One, of whom all persons familiar with our affairs would be quick to speak and glad to hear, is a gentleman who has experienced very severe losses both here and in the States devastated at the same time.

Wm. B. Ogden, the Railway King of the West, still towers among us, a strong refuge and help in our time of need. From a faithful notice in Biographical Sketches, we glean these items:—

"He arrived at Chicago in June, 1835, having then recently united with friends in the purchase of real estate in this city. He and they foresaw that Chicago was to be a good town, and they purchased largely, including Wolcott's addition, and nearly the half of Kinzie's addition, and the block of land upon which the freight-houses of the Galena and Chicago Union Railroad now stand.

Mr. Ogden was very successful in his operations in 1835–6; but he became embarrassed in 1837-8, by assuming liabilities for friends, several of whom he endeavored to aid, with but partial success. He struggled on with these embarrassments for several years. Finally, in 1842–3, Mr. Ogden escaped from the last of them; and since then his career of pecuniary success has been unclouded. They were gloomy days for Chicago when the old internal improvement system went by the board, and the canal drew its slow length along, and operations upon it were finally suspend ed, leaving the State comparatively nothing to show for the millions squandered in "internal improvements."

His operations in real estate have been immense. He has sold real estate for himself and others to an amount exceeding ten millions of dollars, requiring many thousand deeds and contracts which have been signed by him. The fact that the sales of his house have, for some years past, equalled nearly one million of dollars per annum, will give some idea of the extent of his business. He has literally made the rough places smooth and the crooked ways straight, in Chicago. More than one hundred miles of streets, and hundreds of bridges at street corners, besides several other bridges, including two over the Chicago river, have been made by him, at the private expense of himself and clients, and at a cost of probably hundreds of thousands of dollars.

Mr. Ogden's mind is of a very practical character. The first floating swing bridge over the Chicago river was built by him for the city, on Clark street; (before he ever saw one elsewhere), and answered well its designed purpose. He was early engaged in introducing into extensive use in the West McCormick's reaping and mowing machines, and building up the first large factory for their manufacture—that now owned by the McCormicks. In this manufactory, during Mr. Ogden's connection with it, and at his suggestion, was built the first reaper sent to England, and which at the great Exhibition of 1851, in London, did so much for the credit of American manufactures there.

He was a contractor upon the Illinois and Michigan canal, and his efforts to prevent its suspension, and to resuscitate and complete it, were untiring.

Mr. Ogden is a man of great public spirit, and in enterprise unsurpassed. To recapitulate the public undertakings which have commanded his attention and received his countenance and support, would be to catalogue most of those in this section of the Northwest.

Mr. Ogden has never married. In 1837 he built a delightful residence in the centre of a beautiful lot, thickly covered with fine native growth forest trees, and surrounded by four streets, in that part of the city called North Chicago; and there, when not absent from home, he indulges in that hospitality which is at the same time so cheering to his friends and so agreeable to himself.

What the presence of a man, born like him to command, and organize action, might have done for our stricken city we now know not. As soon as the dreadful tidings reached him, as will be seen from his letter inserted below, he flew to the rescue. Thirty-five years ago Mayor Ogden forecast the future, as men of judgment may do, but this vision did not rush red on his sight Nevertheless, he rallies in youthful zeal, his eye not dimmed nor his natural force abated, to gather up the fragments, and reconstruct out of these shattered remains a city that shall be worthy of the lavish gifts of nature, and the splendid endowments of capital. His words, spoken to the citizens amidst the ruins, and in their exchanges, have all been hopeful, conciliatory, and wise. May Heaven grant him years to see a rehabilitation of our dismantled town, so that, like the patriarch, his last days may be his best. His letter must strike every mind and touch the heart with the sense of the pathos of human life; for, like many others, he had doubtless come to feel that "nothing can stop Chicago now." A change of wind for a few hours on Monday would have fairly blotted us out, and scattered our three hundred thousand people to the four winds of Heaven. In his sublime faith, he says: The Northwest, which made Chicago, and forces her on more and more rapidly, is not, except in her sympathies for our great loss, affected by the Chicago fire, and her borders were never being extended so fast, so broadly, or so far, by railways, by settlements, improvements, and added people and wealth, as now; and Chicago's future and "manifest destiny" as a great metropolitan Western city was never so assured.

"CHICAGO, October 11, 1871.

"I left New York on Monday morning last, and reached this utterly indescribable scene of destruction and ruin on Tuesday evening after dark.

"On the cars I kept hearing of more and more dreadful things until I reached here. The truth cannot well be exceeded by report or imagination. How it could be that neither buildings, men, nor anything could encounter or withstand the torrent of fire without utter destruction is explained by the fact that the fire was accompanied by the fiercest tornado of wind ever known to blow here, and it acted like a perfect blow-pipe, driving the brilliant blaze hundreds of feet with so perfect a combustion that it consumed the smoke, and its heat was so great that fireproof buildings sank before it almost as readily as wood—nothing but earth could withstand it; consequently my brother Mahlon's house is the only unburned dwelling on the North Side, from the river to Lincoln Park, within half or three-quarters of a mile of the lake shore; and the only other unburnt buildings were two down at the end of our north pier. On the South Side, east of the South Branch and north of Harrison street, but two buildings are left.

"The fire advanced almost as fast as you could escape before it, and in a very few hours about one hundred thousand people had to leave their houses and flee for their lives, carrying but little, often nothing, with them.

"When I reached the depot on my arrival here it was quite dark. The burning district had no lamp. Thousands of smouldering fires were all that could be seen, and they added to the mournful gloom of all around you, and do so yet. I saw no one that I knew at the depot, and had as yet no definite knowledge of the extent and details of the ruin. I hired a hack and started for my own house, directing the hackman—who was a stranger—as wel.

as I could. Often, however, I was lost among the unrecognizable ruins, and could not tell where I was. Not a living thing was to be seen. At length, however, more by the burnt trees than anything else, I threaded my way over the fallen *débris*, and past the pale blue flames of the winter's stock of anthracite coal burning in almost every cellar, until I came to the ruined trees and broken basement walls—all that remained of my more than thirty years' pleasant home. All was blackened, solitary, smouldering ruins around, gloomy beyond description, and telling a tale of woe that words cannot.

"I proceeded to learn the fate of Mahlon's and Caroline's beautiful places. Near the ruined water-works on Chicago avenue I saw a lantern; stopped the carriage, got out and made my way, over fallen walls that blocked the street in many places, to it, and there met the engineer of the water-works, whom I knew, and from whom I first learned that Mahlon's house, through the efforts of General Strong, Charley, and others, was the only one unburned in all that region, and I gladly made my way to it. Found Mahlon, General Strong, and Charley there, all the rest of the family having fled to Riverside.

"The wind at the time of the fire was from the southwest, and Mahlon's house being some six hundred to eight hundred feet distant from others across the park in that direction, the flames could not reach it so directly, and the air mingled with them more, and made it possible to live and breathe there while the fiery torrent which so filled the air passed. Everything but the two buildings mentioned is swept from our dock and canal property, and the new piers are considerably injured.

"Aid and sympathy come to us from all quarters with a will that touches our heart to the core, and serves us wonderfully in our hour of need; but the great loss and ruin remain.

"Worse than all here, so far as it goes, is the utter destruction of Peshtego village, with all its houses, factories, mills, stores, machine-shops, horses, cattle, and, sad to say, seventy-five to one hundred or more people. They buried yesterday two hundred and fifty of the farming people around our mills, burned by the tornado on their farms or on their way to the village for safety, and seventy-five more from the village; and it is said others are drowned in the pond.

"My large mills and buildings at the mouth of the river escaped entirely. One of my large barge-vessels was burned, and two others and the steamer that towed them are missing as yet, since the storm, with a million of lumber on them.

"I have been two days, by snatches of time, writing this, with much difficulty.

"W. B. Ogden."

CHAPTER VIII.

One of the sadder features of the conflagration was the loss of property by men who have grown gray in the service of their fellow-men, and whose competence seemed assured. Especially painful was this aspect of the case when men were sorely wounded, whose fortunes have been sacredly held as a legacy from God for the promulgation of truth and the amelioration of human sorrow. Hon. Samuel Hoard belonged to this privileged class whose delight is in promoting the welfare of mankind and the glory of God. He wears the hoary head which is a crown of glory, and has felt the truth of that scriptural saying that riches take to themselves wings and fly away. What he has given he has as an everlasting treasure laid up with Him who loveth a cheerful giver. His life is inseparably bound up with young Chicago, and we take pleasure in reproducing a brief view of his history here, and his general character and influence, for it should be known that the men who did most for the rising West were generally men of integrity and Christian virtue. We have been cursed with many bad men, and blessed with many whose names shine on the scroll of the wise and good.

Becoming infected with the Western fever, he migrated to Illinois, and commenced life in Cook County, upon a prairie farm. In that early day the farmer paid great prices for oxen and seed, and obtained small prices for beef and grain, so that the prospects of sudden wealth vanished, or were dashed with disappointment. One of Mr. Hoard's neighbors spent two days in marketing a load of potatoes, and then, not finding a purchaser who would offer more than ten cents a bushel, he drove to the wharf, dumped his load into the stream, and vowed that he would never bring another potato to that market. *Tempora mutantur!* In 1840 he was appointed to take the State census for the County of Cook. Chicago was then ambitious to be considered a large town. But neither he nor Sheriff Sherman, who took the United States census, could find five thousand persons in that infant city. In 1842 he was elected State Senator, and served in the sessions of 1842–3. Being soon after appointed clerk of the Circuit Court, he removed to the city, and engaged in public affairs and the real estate business until 1845, when he formed a partnership with J. T. Edwards in a jewelry house, where he continued until the first year of the war. The love of country burned in his bosom, and he threw his whole soul into the work of saving the nation from dismemberment and overthrow. He was an indefatigable member of the Union Defence Committee, and gave one year's gratuitous service, as secretary, to the patriotic labors in which they were absorbed. He was appointed by President Lincoln postmaster of Chicago, and retained his position, filling it with eminent success, until Mr. Johnson's general proscription cut him off, with so many others, from the public service. His last official position

SAMUEL HOARD

has been in connection with the Board of Health, where he has rendered the public invaluable benefits in warding off the scourge of cholera, the attack of which was universally dreaded. He has passed through an eventful experience, and in his old age has ample means, abundant honors, and hosts of friends. In personal appearance large and well-formed, with a broad and high forehead, and a dignified yet graceful carriage, Mr. Hoard would be a noticeable gentleman in any company, and command instant respect. In society he is affable and courteous to all classes, and diffuses an agreeable atmosphere and influence wherever he mingles. He exhibits the effect of his association with men of talent and varied culture.

Through his countenance and address shines also his kind and unselfish nature. He is a man who possesses a warm, generous soul, that throbs in sympathy with human experiences, and opens his ear and his hand to every call for attention and succor. Eternity only will reveal the instances of personal kindness, the timely gifts, the encouraging words, the helpful visits, the cordial greeting, which have made him beloved and honored.

It would scarcely be possible to do justice to the Youth of this proud municipality without introducing "Long John," who shipped his trunk by the brig Manhattan from Detroit, and set out on foot to reach the new town then clustering on this spot. He was a New Hampshire boy, and his legs were long, and he soon made his way along the beach from Michigan City—this being the only road at that primitive epoch—and arrived upon the scene of his exploits and triumphs October 25, 1836. The railroad had progressed from Schenectady as far as Utica at that date, and Illinois was farther from the Yankees than Rome or Athens is from us, and almost as mythical a region as these places are now to many.

"John Wentworth is one of the very few men now living who attended the meetings called in the winter of 1836–7, to consider the expediency of applying to the Legislature, in session at Vandalia, for a city charter.

He was secretary of the first political meeting ever called in the First Ward to make nominations preliminary to the first municipal election, and at which meeting Hon. Francis C. Sherman was one of the nominees for alderman. In August, 1837, he was secretary of a convention held at Brush Hill (now of Du Page County), to nominate officers for the then county of Cook, and at which Walter Kimball was nominated for Judge of Probate. In 1838 he was appointed school inspector; and he held the same office, under the new name of Member of the Board of Education, when he was last elected to Congress. He has met among the scholars, whilst making his official visits, the grandchildren of those he met as scholars in his first year of service He was the first corporation printer in Chicago, elected in 1837, and he held the position for about three-fourths of the period of the twenty-five years that he was sole editor, publisher, and proprietor of the *Chicago Democrat*. He commenced making public speeches at our first municipal election, when Hon. W. B. Ogden was elected Mayor." Often Mayor of Chicago, he always gave satisfaction and proved himself an energetic executive officer. To have seen him at the head of police and firemen during the Great Fire would have been a source of joy to the good citizens, and gallant little Phil. Sheridan would have earned no laurels, for Mr. Wentworth would have had no need of military, and would have fired his own powder in arresting the flames. As it was, all things were ready, except our leaders, for the conflagration, and it took its own resistless course, and won its awful victory.

CHAPTER IX.

Among our most influential men, and, alas! heavy losers by the fire, is Governor Bross, whose outline we borrow from a full portrait in the "Western Monthly":

In October, 1846, Mr. Bross started out West, and visited Chicago, St. Louis, and other Western cities. Chicago, though then an apparently unimportant town—not a commercial emporium, but literally a "Garden City"—was recognized by his cultivated eye as the future focus of the great Northwest. He decided to make it his home. He returned to the East, closed his school, and moved to Chicago, arriving here on the 12th of May, 1848, as the active partner in the bookselling firm of Griggs, Bross & Co.

In the fall of 1849, Mr. Bross commenced the publication of the "Prairie Herald," and two years afterwards the "Democratic Press."

The paper was "started" with a definite object—not as a mere shift. The proprietors had carefully canvassed the situation, and come to the conclusion that Chicago and the West were about to enter on a rapid and tremendous growth. They saw that this was inevitable; but they also recognized that the extent of that growth would largely depend upon the impression which Chicago should make abroad. Mr. Bross at once bent himself to a study of the resources of this region, and then set about with equal diligence to let the world know their character and extent. He felt that all that was necessary was to exhibit the facts; that the inference would be irresistible; that the brain and muscle, the energy, enterprise, and capital needed to develop this fruitful scene would roll in like the tide of ocean, if the world was posted in regard to what was being done here and what could be done.

That year was really an epoch in the history of Chicago; it marked the beginning of her real prosperity. In 1852 the city was opened up to direct relationship with the East by the two great iron arteries known as the Michigan Southern and the Michigan Central Railroads. The roads now leading westward from Chicago were also all projected, and some of them begun; the Galena road being pushed as far as Elgin, and the Rock Island road to Joliet, while workmen were busy on the track of the Illinois Central. Our city was emerging from the lethargy which had weighed her down since the panic of 1837, and was asserting her claim to be the great railroad and commercial focus of the Northwest.

Mr. Bross loved to write of Chicago in the then present; but he also delighted to sketch its inevitable future as it appeared to him. Many even among those who believed that Chicago would be a great city, regarded him as a visionary; but the most skeptical have since confessed that he saw and thought accurately, judging of the future from the causes then operating around him, and not fondly guessing or lazily dreaming out visions of grandeur. Our subsequent history has realized almost all that he dared to predict. In his pamphlet of 1854 we find such words as these: "We are now in direct railroad connection with all the Atlantic cities from Portland to Baltimore. Five, and at most eight years, will extend the circle to New Orleans. By that time also, we shall shake hands with the rich copper and iron mines of Lake Superior, both by canal and railroad, and long ere another seventeen years have passed away, we shall have a great national railroad from Chicago to Puget's Sound, with a branch to San Francisco." On another page of the same pamphlet, after speaking of the advantages of the situation, glancing at the light death-rates, and alluding comprehensively to the position of Chicago at the head of the great chain of lakes, as guaranteeing to her a focal point from and to which should flow for all time the articles consumed by, and productions raised in, that immense region of country lying to the westward, he points confidently to the "free navigation of the St. Lawrence, by which means vessels loaded at our docks will be able to make their way to the ocean, and thence direct to the docks of Liverpool." Looking around on the great coal-fields of Illinois, the lead mines of Galena, and the grand copper mines of Lake Superior, he wrote, that they all "point to Chicago as the ultimate seat of extensive manufactures." In the light of our present knowledge we might almost be tempted to think that these expressions were mere antedated history. Our railroad system now connects Chicago with every part of the Continent. Long before the seventeen years have passed over his head, he has lived to see the great Pacific Railroad completed, and ship navigation around Niagara Falls almost a fixed fact. We are already manufacturing Lake Superior iron in our city, and our vessels carry its copper to the East; while our grain and pork trade have long since mounted far up into the millions.

It is difficult to conceive how the burning of his fondly-cherished city must have crushed the heart of one who had done so much to raise it to its late eminence. Harder still to realize his feelings as he saw his own home and property melting and smoking before his eyes, and he powerless to save them! Let us listen to him as he tells the story of his experience in the night of gloom and on the following day:—

"About 2 o'clock on Monday morning, my family and I were aroused by Mrs. Samuel Bowles, the wife of the proprietor of the *Springfield Republican*, who happened to be a guest in our house. We had all gone to bed very tired the night before, and had slept so soundly that we were unaware of the conflagration till it had assumed terrible force. My family were all very much alarmed at the glare which illuminated the sky and lake. I saw at once that a fearful disaster was impending over Chicago, and immediately left the house to determine the locality and extent of the fire. I found that it was then a good deal south of my house and west of the Michigan Southern and Rock Island Railroad depots. I went home considerably reassured in half an hour, and finding my family packing things up, told them that I did not anticipate danger, and requested them to leave off packing. But I said: "The result of this night's work will be awful. At least 10,000 people will want breakfast in the morning; you prepare breakfast for 100." This they proceeded to do, but soon became again alarmed and recommenced packing. Soon after 2½ o'clock I started for *The Tribune* office, to see if it was in any danger. By this time the fire had crossed the river, and that portion of the city south of Harrison street and between Third avenue and the river, seemed in a blaze of fire, as well as on the west side. I reached *The Tribune* office, and seeing no cause for any apprehension as to its safety, I did not remain there more than twenty minutes. On leaving the office I proceeded to the Nevada Hotel (which is my property), at Washington and Franklin streets. I remained there for an hour watching the progress of the flames and contemplating the ruinous destruction of property going on around. The fire had passed east of the hotel, and I hoped that the building was safe; but it soon began to extend in a westerly direction, and the hotel was quickly enveloped in flames. I became seriously alarmed and ran round North street to Randolph street, so as to head off the flames and get back to my house, which was on Michigan avenue, on the shore of the Lake. My house was a part of almost the last block burned.

At this time the fire was the most grandly-magnificent scene that one can conceive. The Court-House, Post-Office, Farwell Hall, the Tremont House, Sherman House, and all the splendid buildings on La Salle and Wells streets, were burning with a sublimity of effect which astounded me. All the adjectives in the language would fail to convey the intensity of its wonders. Crowds of men, women, and children were hurrying away, running first in one direction, then in another, shouting and crying in their terror, and trying to save anything they could lay their hands on, no matter how trivial in value; while every now and then explosions, which seemed almost to shake the solid earth, reverberated through the air and added to the terrors of the poor people. I crossed Lake street bridge to the west, ran north to Kinzie street bridge, and crossed over east to the North Side, hoping to head off the fire. It had, however, already swept north of me, and was travelling faster than I could go, and I soon came to the conclusion that it would be impossible for me to get east in that direction. I accordingly re-crossed Kinzie street bridge and went west as far as Desplaines street, where I fortunately met a gentleman in a buggy, who very kindly drove me over Twelfth street bridge to my house on Michigan avenue. It was by this time getting on toward 5 o'clock, and the day was beginning to break. On my arrival home I found my horses already harnessed and my riding-horse saddled for me. My family and some friends were all busily engaged in packing up and in distributing sandwiches and coffee to all who wanted them or could spare a minute to partake of them.

"I immediately jumped on my horse and rode as fast as I could go to *The Tribune* office. I found everything safe; the men were all there, and we fondly hoped that all danger was passed as far as we were concerned, and for this reason: the blocks in front of *The Tribune* building on Dearborn street, and north on Madison street, had both been burned; the only damage accruing to us being confined to a cracking of some of the plate-glass windows from the heat. But a somewhat curious incident soon set us all in a state of excitement. The fire had unknown to us crawled under the sidewalk from the wooden pavement, and had caught the wood-work of the barber's shop which comprises a portion of our basement. As soon as we ascertained the extent of the mischief we no longer apprehended any special danger, believing, as we did, that the building was fire-proof. My associates, Mr. Medill and Mr. White, were present; and, with the help of some of our employés, we went to work with water and one of Babcock's Fire Extinguishers. The fire was soon put out, and we once more returned to business. The forms had been sent down stairs, and I ordered our foreman, Mr. Keiler, to get all the pressmen together, in order to issue the papers as soon as a paragraph showing how far the fire had then extended could be prepared and inserted. Many kind friends gathered round the office and warmly expressed their gratification at the preservation of our building. Believing all things safe, I again mounted my horse and rode south on State street to see what progress the fire was making, and if it was moving eastward on Dearborn street. To my great surprise and horror, I found that its current had taken an easterly direction, nearly as far as State street, and that it was also advancing in a northerly direction with terrible swiftness and power. I at once saw the danger so imminently threatening us, and with some friends endeavored to obtain some powder for the purpose of blowing up some buildings south of the Palmer House. Failing in finding any powder, I proposed to tear them down. I proceeded to Church's hardware store, and succeeded in procuring about a dozen heavy axes, and handing them to my friends, requested them to mount the buildings with me, and literally 'chop them down.' All but two or three seemed utterly paralyzed at this unexpected change in the course of the fire; and even these, seeing the others stand back, were unwilling to make the effort alone. At this moment I saw that some wooden buildings and a new brick house west of the Palmer House had already caught fire. I saw at a glance that *The Tribune* building was doomed, and I rode back to the office and told them that nothing more could be done to save the building, McVicker's Theatre, or anything else in that vicinity. In this hopeless frame of mind I rode home to look after my residence and family, intently watching the ominous eastward movement of the flames. I at once set to work with my family and friends to move as much of my furniture as possible across the narrow park east of Michigan avenue on to the shore of the lake, a distance of about three hundred feet. At the same time I sent my family to the house of some friends in the south part of the city for safety; my daughter, Miss Jessie Bross, was the last to leave us. The work of carrying our furniture across the avenue to the shore was most difficult and even dangerous. For six or eight hours Michigan avenue was jammed with every description of vehicle containing families escaping from the city, or baggage wagons laden with goods and furniture. The sidewalks were crowded with men, women and children, all carrying something. Some of the things saved and carried away were valueless. One woman carried an empty bird-cage; another, an old work-box; another, some dirty empty baskets, old, useless bedding, anything that could be hurriedly snatched up, seemed to have been carried away without judgment or forethought. In the mean time the fire had lapped up the Palmer House, the theatres, and *The Tribune* building; and, contrary to our expectation, for we thought the current of the fire would pass my residence, judging by the direction of the wind, we saw by the advancing clouds of dense black smoke, and the rapidly-approaching flames, that we were in imminent peril. The fire had already worked so far south and east as to attack the stables in the rear of the Terrace Block, between Van Buren and Congress streets. Many friends rushed into the houses in the block and helped to carry out heavy furniture, such as pianos and book-cases. We succeeded in carrying the bulk of it to the shore, where it now lies stored; much of it, however, is seriously damaged. There I and a few others sat by our household gods, calmly awaiting the contemplation of the coming destruction of our property—one of the most splendid blocks in Chicago. The eleven fine houses which compose the block were occupied by Denton Gurney, Peter L. Yoe, Mrs. Humphreys (owned by Mrs. Walker), William Bross, P. F. W. Peck, S. C. Griggs, Tuthill

King, Judge U. T. Dickey, Gen. Cook, John L. Clarke, and the Hon. J. Y. Scammon.

"The next morning I was of course out early, and found the streets thronged with crowds of people moving in all directions. To me the sight of the ruin, though so sad, was wonderful to a degree, and especially being wrought in so short a space of time. It was the destruction of the entire business portion of one of the greatest cities in the world! Every bank and insurance office, law offices, hotels, theatres, railroad depots, most of the churches, and many of the principal residences of the city a charred mass, and property without estimate gone!

"Mr. White, my associate, like myself, had been burned out of house and home. He had removed his family to a place of safety, and I had no idea where he or any one else connected with *The Tribune* office might be found. My first point to make was naturally the site of our late office; but before I reached it I met two former tenants of our building, who told me that there was a job printing office on Randolph street that could probably be bought. I immediately started for Randolph street. While making my way west through the crowds of people, over the Madison street bridge, desolation stared me in the face at every step. And yet I was much struck with the tone and temper of the people. On all sides I saw evidences of true Chicago spirit. On all sides men said to one another: 'Cheer up; we'll be all right again before long;' and many other plucky things. Their pluck and courage were wonderful. Every one was bright, cheerful, pleasant, hopeful, and even inclined to be jolly in spite of the misery and destitution which surrounded them and which they shared. One and all said Chicago must and should be rebuilt at once. On reaching Canal street, on my way to purchase the printing office I had heard of, I was informed that, while Mr. White and I were saving our families and as much of our furniture as we could on Monday afternoon, Mr. Medill, seeing that *The Tribune* office must inevitably be burned, sought for and purchased Edward's job printing office, No. 15 Canal street, had got out a small paper in the morning, and was then busy organizing things. One after another all hands turned up, and by the afternoon we had improvised the back part of the room into our editorial department, while an old wooden box did duty as a business counter in the front window. We were soon busy as bees, writing editorials and paragraphs, and taking in any number of advertisements. By evening several orders for type and fixtures were made out, and things were generally so far advanced that I left for the depot at Twenty-second street with the intention of coming on to New York. Unfortunately, I missed the train and had to wait till Wednesday morning. We shall get along as best we can till the rebuilding of our office is finished. Going down to the ruins, I found a large section thrown out of the north wall on Madison street. The other three walls are standing; but the east and west walls are so seriously injured that they must be pulled down. The south wall is in good condition. More of our office and the Post Office remains standing than any other buildings that I saw. Our building was put up to stand a thousand years, and it would have done so but for that awful furnace of fire, fanned by an intense gale on the windward side, literally melting it up where it stood."

CHAPTER X.

It was once said by Sir Astley Cooper, to his graduating class of medical students: "Now, gentlemen, give me leave to tell you on what your success in life will depend. Firstly, upon a good and constantly increasing knowledge of your profession; secondly, on an industrious discharge of your duties; thirdly, upon the preservation of your moral character. Unless you possess the first—knowledge—you ought not to succeed, and no honest man can wish you success. Without the second—industry—no one will ever succeed. And unless you preserve your moral character, even if it were possible that you could succeed, it would be impossible you could be happy."

The career of Hon. Charles N. Holden furnishes a practical illustration of the great surgeon's wisdom and correctness in this advice, and a healthful example for the young men of our country. His parents, William C. Holden and Sarah Braynard, emigrated, soon after the war of 1812, from New Hampshire to Fort Covington, in Northern New York, where he was born May 13, 1816. His father was an industrious farmer, and his mother an energetic helpmeet, whose life was given to the welfare of her family. The necessities of that early day prevented him from devoting more than a few months yearly to the district school or village academy, but he progressed so well in his education that at the age of twenty he himself wielded the pedagogue's birch. After spending a year as clerk in a store, where he acquired a taste for business, he left home, with forty dollars in his purse, to make a home in Chicago. July 5, 1837, he landed here with ten dollars in his pocket, and found none of his friends, the Woodburys, who preceded him, and no opening for a young man but the open country. With a brave heart in his bosom, and his clean linen in a bundle, he started to find his uncle, a farmer in Will County. Two days of wandering took him thither and introduced him to Western hospitality. He immediately located a claim, hired a breaking team of five yoke of oxen, with his cousin, a lad of ten, as driver, and commenced life on the prairie. That youthful driver is now President of the Common Council of this city, one of the most prosperous, respected, and noble among the prominent citizens of Chicago—Hon. C. C. P. Holden.

From Fort Covington, Mrs. Woodbury, subsequently Charles' mother-in-law, removed with her mother to Chicago. She was the widow of Major Jesse Woodbury, who was the cousin and associate of United States Senator Levi Woodbury, Johnson's and Van Buren's Secretary of the Treasury, and uncle of Mrs. Montgomery Blair. This accession to Chicago proved a magnet to draw the young farmer to the city, where he was clerk in the lumber office of John H. Kinzie, Esq., whose magnanimity he recollects with gratitude. His leisure hours were spent in reading upon various subjects, which made him a careful observer, and a man of wide general intelligence. In the spring of 1837, with three hundred dollars which he had saved, he commenced business in a log store, near Lake street bridge. Three years afterwards he made another venture, the most successful of his life, and was married to Miss Frances Woodbury.

From his father he derived a sturdy constitution, a full muscular frame, and vigorous health. He seems to have but entered upon the prime of his manhood and powers of usefulness. He has probably been the counsellor and friendly adviser of more persons than any other man in his position, on account of the trust he inspires in the coolness and judicial weight of his opinions. His taciturn and abstract manner sometimes leads to the idea that he is cold, distant, and haughty. But nothing is less true. A tender heart beats in his breast, and he weighs men in the scale of manhood, and delights in doing good. He has given his time and means to education with generous enthusiasm. He was chosen President of the Board of Education, and after his retirement, one of the new school buildings was named in honor of him. He had also manifested profound interest in the higher grade of culture provided for in the University and Baptist Theological Seminary founded in this city. Writing thus of Mr. Holden years ago, the author is now compelled to add that the blow which sent the Young Giant reeling, also smote heavily upon him and his family. It remains also to be said that he stands erect in his sterling manhood to renew the conflict; and to his co-laborers in the church his language is, "We must not begin to retrench with the Lord's cause."

CHAPTER XI.

The administration of justice in that early day was often exceedingly rude, on account of the dissipated habits of the magistrates and lawyers, whose great talents were often marred and wasted by the excesses of frontier life. Judge Spring was one of those brilliant men whose passions were in the ascendancy, and brought him to a premature grave with the delirium-tremens. It was peculiarly unfortunate that the Judge had his high-times of spreeing just at the busiest season, when court was in session and matters were most urgent. It was at such a time that his career came to a tragical end, and under the following circumstances: At the opening of the afternoon court the Judge appeared in the door of the room very promptly, for it was his pride to be prompt, and on each side of him was a lawyer—Ballingall and Phillips. They walked to their places, and the Judge crept up into his seat, and showed to the spectators that he was very drunk. The court was opened in due form, and Ballingall arose, and leaning against a post, turned to the Judge and said, "May it please your honor," and then facing the assembly, whom he imagined to be a jury, he added, "and gentlemen of the jury." Thereupon there was a general smile; and Tracy, also very tight, arose and protested that the lawyer was drunk, and appealed to the Judge to stop him. Mr. B., taken aback by this interruption, ventured an argument. Some gentleman of the bar suggested that the Judge and counsel looked sick, and moved an adjournment of the court until the next day at two. This was assented to, and the proceedings came to an end. The Judge was taken home, and his wife sent for Captain Ruger, of the police, to come in and quiet her husband. Knowing what was going on, he quietly dropped in, and found the Judge sitting in his dining-room, with his feet perched upon the table, and his hand on the coat-collar of his son, a lad ten years old. The Judge spoke to the Captain, and said he was very glad he had come in, as he held a prisoner whom he wished to have locked up. "On what charge?" asked the Captain. "Contempt of court." He promised to have the matter at once attended to, but inquired about a case that had interested the Court in the morning, and found the Judge clear and collected in his judgment. Meanwhile Mrs. Spring is drawing near her son, and watching her opportunity to rescue him. The drunken man commenced an abusive assault upon her, in profane and obscene language. The Captain again provoked a discussion upon that morning's case, and diverted his attention, so that the mother seized her boy and drew him away, and thrust him out of the open door, and the little fellow improved his opportunity to put plenty of distance bteween himself and home. The Judge demanded her arrest for rescuing a prisoner. The Captain said he had his eye on her, and would see that she did not leave the house. The Judge then began to speak of his own situation, and to give the most solemn assurances that if he recovered from this attack he would never be guilty of touching another drop of liquor, and would die a sober man. The Captain left the house, and, returning at ten, he found the poor man a corpse. And such was the end of nearly all the prominent men of that early time, whose brains and culture gave assurance of distinction, honor, and usefulness, while animalism drew them into shame, ignominy, and death

In refreshing contrast were the examples of many who lived Christian lives, and did not lose their religion on the Lakes, as they sailed to the Far West. There were godly preachers of the Gospel, whose labors have helped to lay the foundations of the flourishing institutions which Christianity now uses as the machinery of its advancement in the elevation of man's desires, and the purification of his character.

There is something pathetic in the subjoined words, spoken b an old Methodist minister, in 1837, at "Lake Michigan Huddle," then the nucleus of what was but recently the "unrivalled metropolis of Chicago." He was a venerable person of seventy years, with profuse hair as white as snow. His face, however, was without a wrinkle; and, what was very remarkable, his skin was as fair and smooth as that of a young man of five-and-twenty. The building in which he spoke was constructed of rough pine boards, but it was crowded by devout and not irresponsive or silent listeners.

The only thing about the speaker that was at all weak or fal tering was his voice. It was sufficiently distinct, yet it trembled, and if anything, rather added to the effect of the ending sentences which he uttered. In closing a brief description of the dangers that had beset him in the Far West, and of the benignity of the power which had sustained him through every trial, he said:—

"How often—how often—have I swam my horse across midnight rivers, carrying the glad tidings of salvation to settlements in the wilderness, when the fearful cry of the wolves rang in my ears, and the watch-fires of the hostile Indians blazed beneath the giant pines! How often have I wandered through the tall grass of the prairies, day after day, and night after night, with my overcoat for my evening pillow, and the star-gemmed vault of heaven for the curtains of my rest! I was sad, but I was comforted. I was thirsty, but my spirit had refreshment. I was weary, but the arm of the Omnipotent sustained my fainting footsteps, and I laid my head upon the bosom of Peace. I was far from man—in silence—alone; and yet not alone, for my God was with me—the Saviour was by my side. . . .

"This is the last time, dear friends, that my circuit will bring

THE COURT-HOUSE BELL.

CHICAGO IN 1836—KINZIE'S HOUSE.

me before you. In a little while I shall depart hence, and be no more seen."

Here the speaker clasped his hands, looked upward through his tearful eyes, and closed with the verse—

> " Then in a nobler, sweeter song,
> I'll sing thy power to save,
> When this poor lisping, stammering tongue
> Lies silent in the grave ! "

III.—THE YOUNG GIANT!

CHAPTER XII.

THE infant, conceived by Providence in the womb of Time, came to birth amidst the pangs and throes of travail, grew feebly and discouragingly, and even had no special promise of greatness to ordinary eyes, until it sprang into sudden manhood and girded its loins for a great destiny. In 1850 there were less than thirty thousand people here; in 1851 the increase had been six thousand; and from that time the Young Giant advanced with amazing strides, distancing all competitors, and hastening to overtake the oldest and most prosperous cities of the Union. In 1857 there were gathered beside the offensive waters of this stagnant stream 100,000 human souls. In 1871, by census returns carefully made out, and giving the names and local habitations, there had congregated on this level plain 334,000 persons, and Chicago was the fourth city of this country. When was there such a growth in so short a period, and a progress so real and substantial? People who immigrated hither to make money and return to their Eastern homes to enjoy their fortunes, came to regard this city as the most desirable home, for themselves and their children, to be found on the green earth. The East was flooding in upon us to admire, and praise, and covet our situation, privileges, and opportunities. Nature had been improved by art. Chicago no longer lay deeply engulphed in water half the year. Her citizens were not compelled to drink water pumped from the edge of the Lake and half filled with little fish, or particles of earth and filth. The smell of "Bridgeport" was a painful memory only, and the river itself had become sweet and clear. Common schools, academies, colleges, seminaries, universities, societies for the encouragement of art, and science, and history; churches and missions for the extension of religion and morality; galleries, opera-houses, theatres, libraries, and every luxury and appointment of modern times for the cultivation and entertainment of men, had here their best representatives and specimens, or the beginnings that gave noblest promise. The progress of improvements, partially arrested by the war, received new impulse when the cloud rolled over our heads, and the sky again beamed with the radiance of peace.

The following is a summary of the various branches of trade which have ministered to the city's wealth and population. The total exhibits the receipts and shipments of the articles named, for the year 1870, together with the total valuation of receipts.

The estimated value of the receipts of the articles named for the year 1871 is as follows:

ARTICLE.	VALUE.
Flour	$8,000,000
Wheat	18,000,000
Corn	13,000,000
Oats	4,000,000
Pork	2,000,000
Dressed Hogs	6,000,000
Live Hogs	45,000,000
Tobacco	6,000,000
Cattle	22,000,000
Coal	8,000,000
Lumber	16,000,000
Iron Ore	15,000,000
Shingles	2,500,000
Lath	1,000,000

ARTICLE.	VALUE.
Highwines	$6,000,000
Boots and Shoes	8,000,000
Drugs and Chemicals	4,000,000
Hardware	5,000,000
Jewelry	6,000,000
Dry Goods	35,000,000
Groceries	53,000,000

The total trade is estimated at $400,000,000, showing an increase of some nine per cent. on a gold basis over that of the previous year. We had before the fire seventeen large grain elevators, having an aggregate capacity of 11,580,000 bushels, the largest accommodating 1,700,000 bushels.

To carry on this immense traffic, eighteen banks were in operation, with an aggregate capital of nearly $10,000,000, with nearly $17,000,000 of deposits. The total amount of checks passing through the Clearing House during the year 1870 was $810,000,000.

To accommodate this traffic and the vast travel, not less than 100 passenger trains and 120 freight trains arrive and depart daily, while full seventy-five vessels load and unload every day at our wharves.

For the municipal year of 1870–71, the total assessed valuation of the city was $277,000,000, of which $224,000,000 was real and $53,000,000 personal. This, however, represents scarcely more than half of the actual value, which was in excess of $500,000,000. The taxes collected for that year were $3,000,000, besides nearly an equal amount for special improvements, grading, paving, and curbing. The personal property was classed as follows: Individual personal property, $43,647,920; bank personal property, $7,511,600; vessels, $1,183,430. The whole number of persons assessed for taxes on personal property was 14,633.

The area of the city, according to the last arrangement of

boundaries, including parks, public squares, etc., was about 35 square miles, or 22,400 acres. The number of dwellings, according to the last enumeration, was nearly 60,000, of which about 40,000 were wood.

CHAPTER XIII.

WAR came upon our country, bringing terror and agony to the hearts of all good men; but its results, under Abraham Lincoln's wise and honest administration, were so beneficent and sublime, that we cheerfully bear our losses and burdens, and feel that the sacrifices so freely and grandly offered on the altar of patriotism, were a sweet savor to God and an honor to this century of progress. Chicago gave to the army thirty thousand brave men, immense treasures, and a perpetual benefit. In our midst was established, with perfect confidence in the people's loyalty, a camp for rebel prisoners, named, in honor of our great fellow-citizen, the lamented Stephen A. Douglas, Camp Douglas. Unlike the Ark of God in the house of Obed-Edom, which brought a blessing, this establishment came near proving, like the wooden horse with which the Greeks captured Troy, our destruction and the loss of the Union. The story is told by Eddy, in "The Patriotism of Illinois."

CONSPIRACY.

Tidings of a great organization, opposed to the Republic and friendly to the Confederacy, with officers and five hundred thousand enrolled members, were floating about. Their object was to rise together in various States of the Northwest, and co-operate with the Rebel armies from the South. "The first objective point was Camp Douglas, the real strategic importance of which was in the twofold fact that it was the place where eight thousand rebel prisoners were held in durance, and that the abolition city of Chicago would afford admirable foraging ground. The prisoners were to be liberated and joined by Canadian refugees, Missouri bushwhackers, and the five thousand members of the order in Chicago—in all a force of nearly twenty thousand men—which would be a nucleus for the conspirators in other parts of Illinois; these being joined by the prisoners liberated from other camps, and members of the order from other States, would form an army a hundred thousand strong. So fully had everything been foreseen and provided for, that the leaders expected to gather and organize this vast body of men within the space of a fortnight! The United States could bring into the field no force capable of withstanding the progress of such an army. The consequences would be that the whole character of the war would be changed—its theatre would be shifted from the border to the heart of the Free States; and Southern independence, and the beginning at the North of that process of disintegration so confidently counted on by the rebel leaders at the outbreak of hostilities, would have followed. It was a bold scheme, and might have wrought mischief.

"General Orme had been succeeded in command of Camp Douglas by Colonel Sweet, of Wisconsin, a gallant officer, who had been severely wounded in the shoulder at Perryville, and disabled for field duty. The camp, which included about sixty acres of sandy soil, was inclosed by a board fence an inch thick, and fourteen feet high. The garrison ostensibly consisted of two regiments of Veteran Reserves, but could not muster more than seven hundred men fit for the duty of guarding eight thousand prisoners Among these were men of noted daring and ferocity—Morgan's freebooters, Texan rangers, guerillas—reckless, and ready for adventure. Many of the minor offices of the camp were performed by prisoners, who were thus in possession of the resources of the commandant. Letters passing through the camp post-office, enigmatically worded, first roused his suspicions. Subsequently he became convinced that it was designed to take advantage of a great convention to be held in the city, and convene the outside allies, who might at that time come to the city without suspicion, and carry out the plan. Prompt measures were taken, such as convinced the leaders that an attempt would be dangerous, as it was supposed. The Presidential election was approaching, and the commandant prepared to go home to take part in the canvass, when he felt, he knew not why, that he must stay at his post, and did so. The next day showed why he was needed. Another writer makes this statement: 'On the 2d of November, a well-known citizen of St. Louis, openly a secessionist, but secretly a loyal man, acting as a detective for the Government, left that city in pursuit of a criminal. He followed him to Springfield, traced him from there to Chicago, and on the morning of November 4th, about the hour the commandant had the singular impression I have spoken of, arrived in the latter city. He soon learned that the bird had again flown.

"'While passing along the street (I now quote from his report to the Provost-Marshal General of Missouri), and trying to decide what course to pursue—whether to follow this man to New York, or to return to St. Louis—I met an old acquaintance, a member of the order of American Knights, who informed me that Marmaduke was in Chicago. After conversing with him a while I started up the street, and about one block further on met Dr. E. W. Edwards, a practising physician in Chicago (another old acquaintance), who asked me if I knew of Southern soldiers being in town. I told him I did; that Marmaduke was there. He seemed very much astonished, and asked how I knew. I told him. He laughed, and then said that Marmaduke was at his house, under the assumed name of Burling, and mentioned, as a good joke, that he had a British passport, vised by the United States Consul, under that name. I gave Edwards my card to hand to Marmaduke (who was another old acquaintance), and told him I was stopping at the Briggs House.

"'That same evening I again met Dr. Edwards on the street, going to my hotel. He said Marmaduke desired to see me, and I accompanied him to his house. There, in the course of a conversation, Marmaduke told me that he and several rebel officers were in Chicago to co-operate with other parties in relieving the prisoners of Camp Douglas and other prisoners, and in inaugurating a rebellion at the North. He said the movement was under the auspices of the order of American Knights (to which order the society of the Illini belonged), and was to begin operations by an attack on Camp Douglas on election day.'

"The detective did not know the commandant, but he soon made his acquaintance, and told him the story. 'The young man,' he says, 'rested his head upon his hand, and looked as if he had lost his mother,' and well he might! A mine had opened at his feet; with but eight hundred men in the garrison, it was to be sprung upon him. Only seventy hours were left! What would he not give for twice as many? Then he might secure reinforcements. He walked the room for a time in silence; then, turning to the detective, said, 'Do you know where the other leaders are?' 'I do not.' 'Can't you find out from Marmaduke?' 'I think not. He said what he did say voluntarily. If I were to question him he would suspect me.' That was true, and Marmaduke was not of the stuff that betrays a comrade on compulsion. His arrest, therefore, would profit nothing, and might hasten the attack for which the commandant was so poorly prepared. He sat down and wrote a hurried dispatch to his general. Troops! troops! for God's sake, troops! was its burden. Sending it off by a courier —the telegraph told tales—he rose, and again walked the room in silence. After a while, with a heavy heart, the detective said, Good night,' and left him."

From another quarter he obtained a full statement of the scheme, which was gigantic in detail, and contemplated a general uprising through the North, while Hood should move upon Nashville, Buckner upon Louisville, and Price upon St. Louis, and the blow was to be struck in Chicago on the night of the 8th of November.

The commandant took prompt measures, secured the police, and arranged his plans, and at two in the morning made his descent. When daylight came a hundred of the suspected leaders were in custody. The official report of the commandant says: "Have made during the night the following arrests of rebel officers, escaped prisoners of war, and citizens in connection with them:—

"Morgan's Adjutant-General, Colonel G. St. Leger Grenfell, in company with J. T. Shanks (the Texan), an escaped prisoner of war, at Richmond House; Colonel Vincent Marmaduke, brother of General Marmaduke; Brigadier-General Charles Walsh, of the Sons of Liberty; Captain Cantrill, of Morgan's command; Charles Traverse (Butternut). Cantrill and Traverse were arrested in Walsh's house, in which were found two cart-loads of large-size revolvers loaded and capped, 200 stand of loaded muskets and ammunition. Also seized two boxes of guns concealed in a room in the city. Also arrested Buck Morris, Treasurer of the Sons of Liberty, having complete proof of his assisting Shanks to escape, and plotting to release prisoners at this camp.

"Most of these rebel officers were in the city on the same errand in August last, their plan being to raise an insurrection and release the prisoners of war at this camp. There are many strangers and suspicious persons in this city, believed to be guerrillas and rebel soldiers. Their plan was to attack the camp on election night. All prisoners arrested are in camp. Captains Nelson and A. C. Coventry, of the police, rendered very efficient service. B. J. Sweet, Colonel Commanding.

"Camp Douglass, Nov. 7th, 4 A.M."

The city was horrified, and none knew certainly that the storm would not yet burst. Husbands and fathers shuddered at the thought of the city given up to the brutal control of that mob of eight thousand prisoners, and their more brutal allies.

Never were so many citizens armed in Chicago as that day. Patrols rode to and fro, and the city wore the appearance of a military camp. The election progressed peacefully, additional arrests were made, and arms seized; but the life was gone, and the conspiracy collapsed.

The sealed findings of the Court which tried the prisoners arrested for conspiracy, were as follows: "Charles Walsh, Brigadier-General of the Sons of Liberty, guilty, and sentenced to three years' imprisonment with hard labor in the Ohio State Penitentiary; Buckner L. Morris, not guilty; Vincent Marmaduke, not guilty; G. St. Leger Grenfell, guilty of both charges and specifications, and sentenced to the extremest penalty, death; Raphael S. Semmes, guilty, and sentenced to two years' imprisonment. The prisoner Anderson, on the 19th of February, committed suicide by shooting himself while confined in McLean Barracks; and on the 16th of the same month, Traverse, *alias* Daniels, escaped from the custody of a careless guard, during a momentary recess of the Court, in the Court House." Thus another of the city's vicissitudes was safely passed, and the way was open to swift and sure prosperity.

CHAPTER XIV.

Among the first necessities recognized by the Creator in providing a home for His creatures upon the globe, is an abundant supply of pure water, which flows from myriads of fountains, sparkles in running brooks, rushes in rivers, tosses in lakes, and lies in the bosom of the earth everywhere under their feet, ready to bubble up at their stroke. There is danger in some new countries, or sections of primitive regions, that settlers rely on the first basin below the surface for their drinking water, and hence imbibe slow but certain poison from the vegetable matter accumulated through ages. This often accounts for the sickness which attacks persons in becoming acclimated in the West. When the city of Chicago began its rapid growth it was felt that a prime necessity was good water, and the subject received careful attention, which resulted in the use of the Lake water, which is clear and healthful in the highest degree, and cool enough for use as a beverage in the heat of summer. It was at first pumped from wells at the shore; but impurities unavoidably filtered in from the wash of the shore, the fish that swarmed in millions, and the sewerage of the river. Then the gigantic plan was conceived and executed of drawing the water from the bosom of the Lake through a tunnel, connecting with a well two miles out from shore, and directly east of the old works, by which arrangement boundless supplies of the crystal fluid would be accessible. Other cities bring the water of rivers and lakes for many miles through pipes into reservoirs, from whence distribution is made to the population; but this plan superseded any such necessity, and gave us an element of health and power which must forever contribute to the advancement of this city. Her Young Giant can never drink up the contents of Lake Michigan, however vast his wants become in the great future.

It must be a source of interest to the public to follow the progress of this new enterprise, and see the mode by which so many million gallons of this fluid are furnished daily to our people for the innumerable purposes of life. And while the reader wonders at the boldness and energy, skill and success of the projectors and contractors, he will also perceive how futile were all the efforts of man to provide against such a catastrophe as that which prostrated us into the dust, and left us dependent—helpless in the hour of direst extremity, when fires were raging and human mouths were thirsting. The works were commenced in 1852. In 1863 the daily average consumption of water was 6,500,000 gallons, and it had immensely increased in 1871, when a new and more powerful engine was in process of erection within the buildings where the fire wrought such mischievous effects on the morning of October 9th. A description of these works is given us by Engineer Cregier, who has been in charge from the beginning, or since the old Hydraulic works at the foot of Lake street were abandoned. They are situated on the North Side, and bounded by Chicago avenue, Pine street, Pearson street, and the Lake. They have a frontage on Pine street of 218 feet, and extend from the Lake west a distance of 571 feet. They are connected with reservoirs, throughout all divisions of the city, by immense iron pipes laid below the frost and under the river, and through those the engines propel streams of water day and night; and under the pressure of the column in the water tower, it rises to the upper stories and becomes one of the conveniences of city life, the loss of which was keenly felt during the week after the fire.

"The style of architecture is castellated Gothic. The dimensions of the engine-room are one hundred and forty-two feet long, sixty feet wide, and thirty-six feet in the clear from the main floor to ceiling. A projection of twenty-four by fifty-six feet forms the centre of the main front. This portion is divided into two stories. The upper part is devoted to drawing-rooms and sleeping apartments for the engineers. The lower part is divided by the main entrance, the floor of which is tiled. On the south side of the vestibule is a large room designed for commissioner's or reception room. On the north side are offices and other conveniences for engineers. All the walls are two feet thick. The walls of the interior of the main building are rough cast, blocked off representing cut-stone work. The ceiling is divided into square panels, formed by projecting moulded purlins, supported by large Gothic brackets resting on heavy corbels built in the wall. The roof of the main building is constructed of massive timbers, covered with slate and pierced with the necessary ventilators, etc.

Midway between floor and ceiling, and extending around the entire interior space of the building, there is a handsome and substantial gallery or balcony, protected by fancy Gothic iron railing, the whole resting upon brackets of like style built into the walls. From this point a pleasing view of the operations of the engines is obtained. This gallery is reached by two flights of spiral stairways constructed entirely of iron. Below the main floor of the principal building there is a space extending over the whole area, and nine feet high in the clear. Here are located the pumps, delivery mains, stop-valves, etc., of the several engines, also store-rooms and other conveniences. From the floor of this large room the pump-wells connected with the Lake Tunnel descend. The south well, intended for additional engines, was sunk to place in October last. The form and constructions of this curb, as well as the mode of sinking it to its place, is similar to that adopted for the north well; it is, however, larger. The outside diameter is forty-four and one-third feet at the bottom, forty-three and one-third feet at the top, and twenty-two feet from the top of the cast-iron shoe to the top of the coping. The outside has a batter of six inches. The vertical bond consisted of forty-eight one-and-a-half-inch bolts.

The boiler-rooms are placed nineteen feet apart, and are located in the rear of the main building. They are forty-six and a half feet long, thirty-six feet wide, and twenty-five feet from the floor to the ceiling. The floor is of stone, and the roof is wholly of iron and slate, thus rendering them fire-proof."

If so, Mr. Cregier, why did they succumb so readily when they were most needed? The answer might be returned, that this extraordinary conflagration melted iron into shapeless masses, and consumed stones into dust, and mocked at fire-proofs. The facts, probably, are these. The ventilators in the roof were left open, and the raging shower of sparks, cinders, and flames poured down these inlets into the engine-room, where vast timbers were built into stagings for the accommodation of mechanics in placing the new engine, drove out the workmen and watchmen, and consigned everything to speedy destruction. There should have been no wood in the construction of the interior; the windows should have had iron shutters, though they be unsightly; and the ventilators and all openings in the roof should have been covered with wire screens, impervious to fire, and the men in charge should have been early reinforced with ample resources for their entire safety amidst all possible contingencies and exigencies. How clear all this becomes after the event!

THE WATER-TOWER

"Is the most imposing feature among the whole mass of buildings comprising the works, and is without doubt the most substantial and elaborate structure of the kind on this continent. Its centre is 106 feet west of the main buildings, upon ground purchased for the purpose in 1865; 168 piles, capped with 12-inch oak timbers, the spaces filled with concrete, constitute the foundation up to the surface of the water; from thence to a point six feet below grade, solid, massive dimension-stones laid in cement intervene. At this point the gate, pit, and arched ways on each corner for mains (large pipes of iron), are formed. The base of the tower is 22 feet square. The exterior of the shaft is octagonal, and rises 154 feet from the ground to the top of the stone-work, which terminates in a battlemented cornice. The whole is surmounted by an iron cupola (not yet finished), pierced with numerous windows, from whence may be obtained a magnificent view of the lake, the city, and surrounding country. The exterior of the tower is divided into five sections. The first section is 40 feet square, exclusive of battlements, turrets, etc., and surrounds the base of the shaft, forming a continuous vestibule nine feet wide on the four sides, with a grand entrance on each side. The floor and roof of this portion is of massive stone. The roof forms a balcony. The walls are plastered and blocked off like those of the engine-room. The ceiling is groined and corniced, and the sides are ornamented with tablet drinking fountains, etc. The other sections of the exterior recede from each other in graceful proportion, each having turreted cornice, battlements, etc.

The bottom of the exterior is hexagonal; here the base-piece of stand-pipe (a casting weighing six tons) is placed, having six openings, supplied with 30-inch gates, to which the water mains are connected. From this base, a 36-inch wrought-iron stand-pipe ascends to a height of 138 feet. Around this pipe is an easy and substantial spiral stairway leading to the cupola on the top, and lighted throughout with alternating windows.

The whole structure is thoroughly fire-proof, being constructed wholly of stone, brick, and iron.

THE LAKE TUNNEL.

The work was commenced at the land-shaft on the 17th of March, 1864, the delay since the date of the contract having been caused by waiting for the cast-iron cylinders for the first 30 feet. These cylinders are nine feet internal diameter, 1½ inches thick, and in three sections, each ten feet long. The bottom of the lowest section has a cutting edge. The sections were united by internal flanges, bolts, and rust-joints. The top flange of the cylinder was fitted to receive an air lock, in case that should have proved necessary in the prosecution of the work.

It was intended originally to make the lining of the land-shaft of brick, clear to the top, but the Board feared trouble from the quicksand which extended down about 14 feet from the surface, and particularly as the inlet through which the city was supplied was not only in this quicksand, but very near the shaft. Owing to the want of suitable pumps, there was unexpected delay in sinking the cylinders, but as soon as the clay had been penetrated a few feet all serious difficulty ended, and the remainder of the shaft was sunk to its proper depth, through clay of various degrees of tenacity, from very soft near the top to indurated near the bottom. The shaft was walled up eight feet in diameter, with masonry 12 inches thick, to the bottom of the cast-iron, the inside of which was laid with masonry to the top of the lowest section. At the bottom of the shaft there was a sump six feet deep, below the bottom of the tunnel. This had to be emptied generally twice a day during the whole progress of the work, as the quantity of water discharged from a spring there continued very uniform.

From the bottom of the shaft a drift, at first only intended to be temporary, was made about 50 feet long westward, with a chamber at the end, with fixtures for mounting a transit. The regular tunnel work eastward was commenced May 26th, 1864.

Here much pains were taken to introduce a curved surface in the masonry, between the shaft and upper side of the tunnel, and it was satisfactorily accomplished. The entrance to the tunnel was made six feet in diameter, and tapered down to five feet in a distance of twenty feet. The masonry on this portion was made of three shells of brick-work, each four inches thick, with cement joints half an inch thick between. The rest of the tunnel proper was lined with two shells of brick-work. It was intended at first to fill the cavities around the outside of the brick-work with well-tamped earth, but it was soon found impossible to get this done in a satisfactory manner. For this reason, solid masonry was almost immediately substituted for the tamped earth. The upper arch was built on a ribbed centre of boiler iron, which diminished the open space inside of the tunnel only 4¼ inches, and thus

CHICAGO WATER-WORKS. FROM THE NORTH-WEST

CROSBY'S OPERA HOUSE.

allowed the cars which conveyed away the earth to go up to the face of the excavation, usually kept from ten to tweety feet ahead of the masonry. The iron centre was 30 inches long, in the direction of the tunnel. About two feet in length of masonry was usually made at a time, and, as a rule, it was found safe to strike the centre within fifteen minutes after the arch was keyed. At first it was supposed necessary to excavate nearly a foot above the top of the brick-work, in order to give the masons room to build the upper arch; but very soon it was found that they could build it perfectly well, generally, without making the excavation any larger than the space required for the brick-work. This was done by driving the last four or five top courses of brick into well-tempered cement mortar first thrown into the cavity. The driving of the bricks effectually filled up the spaces which could not otherwise have been reached by hand. The ends of the masonry were left "toothing," and thus furnished a guide in driving the bricks on the upper arch. The lower arch was built by templets or patterns, as ordinary sewers are, and usually kept some six feet in advance of the upper arch, to allow of greater convenience in loading the cars with earth, which the miners had to keep at some distance behind them, and which the shovellers could not throw into the cars very well, when they stood under the brick-work.

The excavation was generally through stiff, blue clay, but with the irregularities of character peculiar to the drift. It very seldom required bracing when not left to support itself more than thirty-six hours. Sometimes sand-pockets were met, and when those were over the upper arch they would empty themselves partly, leaving cavities to be filled with masonry, but these were seldom of much importance. Sometimes small bodies of quicksand were encountered, but they occurred only in pockets, and not in strata, and therefore gave no serious trouble. Sometimes the clay would be soft enough for a miner to run his arm into it, but with the exception of requiring a little more "trimming" for the masonry, this gave no trouble. Sometimes boulders weighing several hundred pounds were met, and interfered a little with the regular progress of the work, but seldom more than a little.

The greatest and most dangerous difficulty met with was one that was not anticipated at first, and that was inflammable and explosive gas. Early in the progress of the work several accidents occurred from this cause, but fortunately without fatal results to the workmen, though there were several narrow escapes. Very soon the miners learned to detect the proximity of cavities containing this gas from the sound produced by striking over them with their picks. When a cavity was thus detected, it was bored into with a small auger, and the gas ignited as soon as it began to escape. In this way explosions were prevented which otherwise took place when large bodies of gas were suddenly allowed to mix with the air. The explosions that did occur were slight in character, but left a body of flame in the upper part of the tunnel. At such times the miners fell with their faces to the ground, and thus escaped without any greater injury than singed beards and eyelashes and blistered faces, except in the first severe case, when a miner was badly burnt. At this time the gas kept the miners out of the tunnel three days.

CHAMBERS.

With trifling exceptions, this work was prosecuted day and night by means of two sets of miners and one of masons, working eight hours each in every twenty-four, for six days in the week, till the 16th of October, when a point about 750 feet from the centre of the shaft was reached. Here it was determined to make two temporary chambers, one on each side of the tunnel, with which they were to be connected by small and short openings. It took about one week to construct these chambers and connections, all of which were supported by timbers and planks. In the tunnel, and at the connection between the chambers, a turn-table was placed. This arrangement permitted not only the passage of cars by each other, but also making up of trains, which soon became an absolute necessity for the economical and rapid execution of the work. By means of such chambers it was practicable to carry on the work a mile or more out under the lake as fast as could be done near the bottom of the land shaft; in fact, the progress upwards of a mile out was really greater than it was near the shore, owing to the greater skill and experience acquired on the way. A gap of about six feet in the masonry of the tunnel was left at the connection between these, the first chambers to be built in after the completion of the rest of the work. After two or three weeks several cracks, entirely around the tunnel, were discovered in the brick work within a distance of about twelve feet on each side of the turn-table. There were various conjectures as to the cause of these cracks, for up to this time repeated careful observations had shown no indications whatever of any movement in the masonry after the keying of the upper arch. Occasionally, in soft ground, the sides of the lower arch had been pressed in an inch or two before the upper arch was built, but no transverse crack was ever discovered except those near the chambers. The conclusion was that they were probably caused by the yielding of the earth in a pit of the turn-table; yet no settlement in the masonry was observed.

The second set of chambers was made one thousand feet beyond the first, and the character of the work, as well as the mode of carrying it on, continued the same, except that the use of mules was substituted for men in the transportation of earth and materials for the masonry. Stout abutments were built on each side of this turn-table to prevent the cracking of the brick work, observed on each side of the first, but just the same number and character of cracks occurred, notwithstanding.

It then became evident that these cracks were owing to a tendency in clay to move, or "creep," as it is sometimes expressed, towards any cavity made in it. The gap left at the connection between the chambers being only temporarily supported with wood, could not wholly prevent this creeping movement. It was therefore determined afterwards to continue the brick work over and around the next turn-table, and to brick around the connections between the chambers, groining carefully in their intersections with the tunnel. After this method of constructing the chamber connections was carried out, all trouble from cracks ceased. In this manner, placing the sets of chambers about a thousand feet apart, the work was continued to about a mile and a half from the land shaft.

The character of the work continued throughout very much the same.

The greatest progress made during any one week was ninety-three feet. Only once was a boulder so large as to require blasting met with. There was a little nervousness as to the effect of a blast under the Lake, but it caused no serious disturbance either of the ground or the masonry.

VENTILATION.

The ventilation of the tunnel was effected by means of tin pipes, through which the foul air was drawn out and fresh air consequently drawn in through the main opening. At first a six-inch pipe was used, and this was connected with the furnace of the hoisting engine. Later it became necessary to provide an engine and fan expressly for the purpose, and to put in larger pipes. Eight-inch ones were introduced. It was difficult to keep the joints of the pipes, which were only of ordinary tin, very tight, especially near the chambers, where the mules struck them with their heads in turning. Still they answered a very good purpose, and the air, a mile and a half out, was about as good as it was much nearer to the land shaft.

Ordinarily there was so much smoke from the miner's lamps and vapor from the heat of the workmen, as to make it impossible to see distinctly enough to run the lines and levels required to keep the tunnel in the right direction. On Sunday nights, however, and on other holidays, the air became so clear as to cause sperm candles to burn with a beautiful silver brightness, visible sometimes two thousand feet.

ALIGNMENT.

To determine the position of the lake shaft and the line of the tunnel, much pains were taken to establish an accurate base on the shore for the purpose of triangulation. Owing to the buildings in the way, this was no easy task. For the alignment of the tunnel, an astromonical transit of four-inch aperture, by Pike, of New York, was mounted on a tower built for the purpose, 166 feet westward of the land-shaft, and sometimes used in the chamber below already described. To aid in placing the lake-shaft beyond all doubt in the line of the tunnel, a six-inch tube was sunk 280 feet eastward of the land shaft, after the masonry had been carried beyond that point. By plumbing up through this tube, a "range" of great accuracy for such a purpose was obtained. The astromonical transit could only be used on the tower above, or in the chamber below. As soon as the work had been carried so far that the sperm candles used in the alignment could not be seen at "the face" of the work, the centre line was produced from point to point by means of a goniometer with two telescopes, which, when in perfect adjustment, could be made to "reverse" on the same point, which was thus proved to be in a straight line with the instrument and the "back-sight."

Mr. Kroeschell, an educated and experienced mining engineer, was principal inspector of mining, and directed the "brimming" shift, which worked the eight hours immediately before the masons commenced. He set the "patterns" by which the masonry was built, producing for this purpose the lines and levels given by the engineer in charge, by means of plummets, ranges with sperm candles, and spirit levels. His shift consisted usually of four miners and four other men, who at first pushed the loaded cars to and from the shaft, but afterwards to and from the nearest chambers, from which they were hauled by mules to the shaft and back again, either empty or loaded with brick, cement, or sand.

Only two of the miners usually were regularly trained men, the others being but picked laborers, who soon learned to use mining tools in the clay.

The general custom was for two miners to work together for ten or fifteen minutes at a time with more than common vigor, and then rest.

The pushers loaded the excavated earth into the cars, brought as near the face as possible on a movable truck.

This shift, besides frequently carrying the face of the excavation five feet ahead, did all the trimming necessary to form the interior of the excavation as nearly as possible to the exact outside shape of the masonry. The next or mason's shift usually consisted of three masons, one mortar mixer, and four to six helpers, according to the distance between the chamber and the work. The water for mixing the cement mortar was all brought from the top of the land shaft in tank-cars, made especially for the purpose. The average length of masonry laid by this shift was twelve feet a day for the entire distance, but for the first 2,000 feet the greatest progress scarcely equalled this rate. Afterwards it sometimes reached 15⅓ feet a day; but this latter rate could only be attained by putting on a couple of miners during the shift; but this course enabled the contractors to advance the whole work two feet more a day than they could have done without it.

PLAN AND CONSTRUCTION OF THE CRIB.

Preparations for commencing operations at the outer end of the tunnel were early made, but owing to disappointments of the contractors in getting the necessary timber for the crib, and other delays, the foundations of the outer, and only one it was found necessary to build, were not laid till May, 1864. This was done on the north side of the river, about 800 feet west of the Light-House.

The dimensions of the crib, as required by the specifications, are fifty-eight feet horizontal measurement on each of the five sides, and forty feet high. The inner portion, or well, has sides parallel with the outer ones, and twenty-two feet long each, leaving the distance between the inner and outer faces of the crib, or thickness of the breakwater, twenty-five feet. This breakwater was built on a flooring of twelve-inch white pine timber laid close together. The outer and inner vertical faces and the middle wall between them were all of solid twelve-inch white pine timber, except the upper ten feet of the outside, which was of white oak, to withstand better the action of ice. Across the angles of the outer and middle walls were placed brace walls about ten feet long, of solid twelve-inch timber. The middle wall on each side of the crib was continued straight through to the outside wall.

Connecting the outer and inner walls, and passing through the middle wall, were cross-ties of twelve-inch timber, placed horizontally about nine feet, and vertically one foot apart. The ends of the timbers, where they passed through the outer and inner walls, were dovetailed, and notched half and half into the timbers of the middle wall.

All of the timbers used were carefully inspected and well jointed, which was mostly done by hewing, though nearly all of it was first sawed. It was found impossible, however, to get sawed timber of perfectly uniform dimensions. The floor timbers were laid on ground timbers placed directly under the outer, middle and inner walls of the crib. Round one and a half inch bolts, thirty-six inches long, with large washers at the bottom, were placed vertically, four feet apart, to hold the ground and floor timbers firmly to the first two courses of wall timbers above the flooring. All of the wall timbers were fastened to each other by one and a quarter square inch bolts thirty-four inches long, pointed and driven somewhat slanting into one and a quarter inch auger-holes about five feet apart. The slant was given in opposite directions to the bolts nearest each other, to avoid the possibility of their being drawn out by the buoyancy of the timber, an accident which once occurred to a somewhat similar structure in the West.

Three rectangular openings, each four feet wide and five feet high, were made through the breakwater at different depths below the surface of the Lake, so that the water could be drawn from near the bottom, middle or top, as future experience might show to be best. These openings, and wells four feet square from them to the top of the breakwater, were timbered around in the same careful manner as the rest of the crib. Each well was provided on its inner face with slides for a temporary gate to cut off the water whenever thought necessary.

The floor and walls of the crib were all carefully calked. The interior of the breakwater was divided into seven water-tight compartments, made so by the calking already mentioned, and "matched sheathing" between the walls. The object of these water-tight compartments was to make it easy to build solid masonry in the whole of the breakwater at any time within the course of a few years, if it should be thought best.

The whole of the outside surfaces of the outer and inner walls were sheeted with two-inch pine plank carefully jointed, placed vertically, and spiked on. Instead of pine, three-inch white oak was used for the upper portion of the outside, to resist the ice. The upper ten feet of each outside corner was protected by angle irons, extending each way two feet, and firmly fastened by two-inch round bolts. From the bottom to the top of the crib, and into which the ends of the angle irons were let, there were ten pieces of white oak, 5 x 14 inches, fastened every two feet to the middle wall with two-inch round bolts.

Similar pieces, 3 x 12 inches, thirty-nine feet long, reaching from the top of the crib to the flooring, were fastened by the same bolts to the inside of the middle wall. It will thus be seen that apparently excessive care was taken to make the crib strong, but subsequent experience showed that this care was none too great.

The crib, when built, was in a horizontal position. In order to launch it, it was raised by screws, and inclined at an angle of one in twelve towards the water. Seven ways were placed under it, and extended out sixty-four feet into the river on trestle work. The river portion of the ways gave a great deal of trouble, on account of the uneven and stony character of the bottom, and accidents caused by passing vessels. Everything being ready, the launch took place on the 24th day of July, 1865, when the crib glided without accident or delay gracefully into the water in the presence of a large number of spectators. Immediately after the launch, the contractors towed the crib out to its position in the Lake. As soon as the bar was passed, three small gates near the bottom of the crib were opened, and the draft of water, which at first was but a little over eight feet, increased soon after reaching the anchoring ground to twenty-one feet. A mooring screw, opposite the intended position of each angle of the crib, had been placed under the direction of Mr. Clarke. To each mooring screw a one and a half inch chain cable was attached, and the loose end of the chain fastened to a buoy. Unfortunately, lake propellers had destroyed three of these buoys, and it was thought most expedient to substitute for the sunken chains ordinary anchors and hemp cables. As soon as the crib was brought near its position, the work of filling with loose rubble was commenced. Very soon the crib got "out of trim," and one corner of it rested on one of the low bars, peculiar to the Lake at this distance from the shore. After some time had been lost in vain efforts to get the crib righted, and into its exact position, the Board became alarmed for its safety, in case a severe storm should arise, and directed that no expense be spared that might seem necessary to the engineers to secure it with the utmost despatch.

A wrecking pump was at once employed. By means of this, sufficient water was pumped into or out of the crib, as occasion required, to right it. The partitions between the compartments failed, and it was a matter of rejoicing that they did, for otherwise the removal of the wrecking pump from one compartment to another could not have been made in time.

Three powerful tugs were hired, which, by the aid of sufficient tackle, finally towed the crib to its exact position. Immediately the contractors resumed the operation of filling the crib with stone, but very soon after a violent storm set in, and drove the vessels loaded with stone into the harbor. This storm continued for three days, and threatened, before it abated, to do serious, if not fatal, injury to the crib. In order to hold it in its position as firmly as possible, the wrecking pump was kept at work to fill it with water, the stone thrown in previously not being sufficient to hold it down. During the height of the storm, every wave caused a perceptible rocking of the crib. The angle joints of the inner and middle walls began to separate, and for a time caused intense anxiety. When the storm was over, two of the inner angle joints had parted an inch on top, and the entire crib had worked against wind and waves thirteen feet, and the northwest angle was three and a half feet lower than the southeast.

The great difficulty there would have been in restoring the crib to its exact position, and the fear there might be another storm in the meantime, prevented any attempt of the kind from being made. The very slight deflection this rendered necessary in the line of the tunnel, was of no practical importance whatever, though regretted, and the variations of the sides of the crib from perpendicular, though a constant eyesore, did not affect its stability.

The filling of the crib with stones was proceeded with as fast as the contractors could, and since it was completed, about the middle of August, no variation whatever in the position of this structure has ever been perceived. A slight tremor is sometimes felt during severe storms, and when large fields of ice are passing. The rubbing of the field-ice against the crib is occasionally accompanied with a fearful noise. At such times the crib appears to a spectator on it to be an immense plough moving through the ice. On several occasions the broken masses lodged on the south side of the crib, forming banks several hundred feet long, and reaching from the bottom of the Lake to ten or fifteen feet above the surface.

The breakwater portion of the crib being filled with stone, the contractors erected over it a temporary wooden covering, with a light-house on top, and rooms above and below for the accommodation of their own men, as well as the inspectors employed by the Board. It may be said, in passing, that the air was so pure at this dwelling-place as to cause complaints, at first, from the cook of the voracious appetites of the men. The reputation of the crib for healthfulness is still maintained, the present keeper being now quite vigorous and hearty, although apparently a feeble consumptive when he went there to live, about eighteen months ago.

CYLINDER AND LAKE SHAFT.

The cast-iron cylinder for the lake shaft was made in Pittsburg, by Messrs. James Marshall & Co., who also made the one for the land shaft. It consists of seven sections, each nine feet in length, nine feet internal diameter, two and a quarter inches thick, and in all other respects like the one for the land shaft, except that the lowest section was turned on the outside, to make it penetrate the clay more easily, and the upper end was provided with two gateways, for the introduction or exclusion of the lake water. The gateways are each fifty-four inches high by thirty-two inches wide, and placed with their tops below the lowest known level of the lake. Each gateway was provided with a sliding gate on the outside of the cylinder, raised by a screw worked at the top of the cylinder. Provisional arrangements were made at each gate-opening for forming chambers on each side, in case it should ever be necessary to repair either gate, by simply sliding in temporary gates. The sliding faces for those temporary gates, as well as of the permanent ones, were made of "composition." Inclined ways were placed inside of the crib during its construction, to aid in lowering the cylinder to its place, but the storm already mentioned destroyed them. The lowest and next cylinder-sections were put together on an incline. They were held in place, when required, by chains on the outside, secured to the lower end of the bottom section, and a brake over the upper side of the cylinder. They were lowered gradually on the incline by means of screws attached to the upper flange. These screws had to be removed, of course, for every new section put on. Care was taken to have sections enough together before removing the chains from the bottom of the cylinder, to reach above the water. This required five, or forty-five feet altogether, to be sure. A false bottom of wood was put into the cylinder at its lowest section, to keep out as much water as practicable. This gave the cylinder great buoyancy when sunk to a depth of thirty feet, and made it very easy to handle with blocks and falls placed overhead. On being lowered the cylinder sunk by its own weight two or three feet into the clay, when the false bottom stopped it. A hole was then bored through the false bottom, and the cylinder went down several feet further by its own weight. After the sixth or gate-section was put on, and the false bottom removed, and excavation made within, the cylinder continued to sink by its own weight. After the top section was put on a moderate force only was necessary to push the cylinder down twenty-three feet below the bottom of the Lake. Below this point the work of sinking the shaft was substantially a repetition of that at the shore end of the tunnel, except that no water was met with, and no pump ever put in or required. The little leakage that occurred was easily removed in buckets.

An extension eastward, about fifty feet, was made, in anticipation of the possible extension of the tunnel, at some future day, still further out into the Lake. This was provided with the necessary sump and bottom on which to place another iron cylinder. The extension was of great service during the construction of the work as a turn out of the cars, and afforded, by means of a six-inch tube, sunk perpendicular from above the surface of the Lake to its outer end, an excellent opportunity to start the line of the tunnel below with great accuracy towards the deflecting point in the middle.

TUNNELLING FROM LAKE SHAFT.

THE work of tunnelling was carried on from this end in very much the same manner, and about as rapidly as it was on the first 2,000 feet from the land shaft. The average progress made was 9½ feet a day till a point 2,290 feet from the lake shaft was reached, when operations in this direction ceased. When the work from the land shaft was within 100 feet of the same point, it was thought best to stop the masonry there and run a small timbered drift through to the east face to be certain as to how the lines were going to meet. The two faces were brought together on the 30th of November, 1866, when it was found that the masonry at the east face was only about 7½ inches out of the line from the west end. The horizontal measurements were only three inches longer than was estimated by triangulation. This result, considering the great difficulty of getting a clear atmosphere in the tunnel, was deemed very good, and much better than was generally expected.

The last of the masonry in the regular tunnel, when the two faces were brought together, was completed on the 6th of December, and a stone commemorative of the event placed there by the Mayor of the city, in the presence of the City Council and Board of Public Works, both of which bodies, together with a number of citizens, passed from the shore through the tunnel to the crib, and then by a tug to the city on that day.

The ventilation of the east end of the tunnel was effected by means of six-inch tin pipes, connecting with the furnace of the hoisting engine. The pipes extended to the end of the masonry. Occasionally, ashes from the furnace would stop the ventilation, which would soon be discovered at the face.

GATE-CHAMBER, CONNECTIONS, AND COMPLETION.

In December the work of filling up the chambers was commenced, and also that of connecting the tunnel with the pumping wells. Much had been done previously towards constructing a gate-chamber between the land shaft and the pumping wells. This was made nineteen and a third feet exterior, and sixteen feet interior diameter, and divided into five compartments, separated by walls twenty inches thick. The outer walls were first built on a boiler-iron shoe, or curb, and then sunk by excavating within. An old abandoned inlet gave a great deal of trouble by letting in water; and the boiler-iron shoe, which was adopted for the sake of economy, proved more expensive in the end than a cast-iron one would have been. The foundations were on a bed of concrete twenty-four inches thick, on which the footings of the exterior and division walls, all of brick, were built. Through the bottom of each division wall there were left rectangular gate openings, three feet wide and five feet high. The tops of these openings are 23½ feet below low-water in the Lake. In each opening a cast-iron gate frame was built. The gates themselves are tapering. The frames were fitted with wedging grooves or ways projecting beyond the walls, just sufficient to free the gates when raised or lowered. It was hoped that the wedging grooves would allow the gates to be screwed down perfectly tight; but in practice they have given more trouble than was anticipated, and it is believed now that in making future structures of the kind for the city, if any should be required, it would be safer and better to put a gate on each side of the wall, so that the pressure of the water could always be used to keep the gate tight. The gates are operated by means of rods, stayed at intervals, and by screws with hand-wheels at the top of the walls.

The connection between the land shaft and the gate chamber was of precisely the same size and form as the main tunnel. The connections with the old and new pumping wells, and a partial one with the provisional or south pumping well, as also about 180 feet of a provisional connection with the Lake shore, are all four and a half feet interior diameter, and were tunnelled through soft clay without any difficulty, except a little trouble in working under and through the piling beneath the old pumping well. The connection with the Lake shore, or rather the old inlet basin, is to be used, in case it should ever be necessary to suspend the supply through the main tunnel, either to examine, cleanse, or repair it.

A temporary connection between the land shaft and the mouth of the old inlet was made by means of a timbered drift through the clay, and a brick well four feet interior diameter and thirty feet deep, provided with a curb built above the water on an iron shoe, held together by iron rods, and sunk by means of the same dredging apparatus that was used for sinking the curb of the new pumping well. Two wooden gates were left in the top of the curb, just below the surface of the water. A small area, enclosing the well and the inlet, were coffer-dammed around as far as necessary to cut them off from the flow of the Lake whenever desired.

The work of filling the chambers of the main tunnel, and the cleansing of that structure having been completed, water was first let into it on the 8th of March, 1867, when only the horizontal portion was filled, this precaution being taken to avoid too sudden a pressure on the masonry. By the morning of the 11th, the shafts were filled to the level of the Lake. For the purpose of ascertaining if any defective workmanship existed where cavities on the outside of the masonry had been filled in, the water was pumped out of the tunnel sufficiently to permit the engineer and three representatives of the city press to go upwards of half way through the tunnel. Not a brick was observed to be out of its place or to have started. The party not being able to push their boat any further without great discomfort, returned, but were upset and left in total darkness about 600 feet from the Lake shaft, to which they walked. Had this accident occurred a mile out, it would have proved very serious, if not fatal, to most of the party, as the water was too cold to be endured long.

After the examination the tunnel was again filled, and on the 24th, about 4 P.M., the mouth of the old inlet was cut off from the Lake. Immediately the pumps, which were not stopped at all, drew down the surface of the water at the mouth of the inlet upwards of a foot. For a moment it seemed to some of the bystanders as if the tunnel would not perform its intended office, but the next instant the water began to bubble up beautiful and clear at the top of the well, and continued to do so till the temporary connection was no longer needed; when this most pleasing and unexpected feature of the works ceased to delight the public.

THE NEW PACI

. IN CHICAGO.

The formal celebration of the completion of the tunnel and introduction of pure Lake water, by appropriate public ceremonies, took place March 25, 1867. From that time to this there has been no cessation in the supply except three times, when stoppages of a few hours by anchor ice occurred. The experience thus far gained in this respect is believed to be sufficient to show how to prevent the recurrence of such accidents.

Careful observations have frequently been made to ascertain the head required to deliver given quantities of water through the tunnel, and it is found to exceed in capacity the original estimates. No indications whatever of internal injury to the structure have yet been observed.

The original estimate of the probable cost of the work was $307,552. The actual cost, including all preliminary and other expenses of whatever nature chargeable to the Lake tunnel, up to April, 1867, was $457,844.95.

Thus was completed this important series of works which deliver to us the pure crystal contents of our mighty inland seas. What man's forethought could devise was here planned and prepared to guard against the possibility of failure to supply the city in its largest need. But the fire, which tried every man's work of what sort it was, could be stayed by nothing human. How needful that every man keep in his own place of humble dependence on the Almighty. "Blessed is every one who trusteth in Him!"

CHAPTER XV.

THE river makes the harbor, and the harbor determines the location and greatness of the city. This cut into the shore, or bayou, first extends west about one mile, and then forks north and south, dividing the place naturally into three sides, north, south, and west. The first lies between the Lake, the north branch, and the main river. The West comprises all west of the two forks or branches, and the South is bounded by the Lake, the river, and the South Branch. The great ships and propellers sail grandly into the harbor, and float away inland, yet remain ever in close proximity to the business and the people. This gives peculiar facilities for commerce, but impedes intercourse between the three sections, because the river is crossed by swinging bridges which whirl upon a table in the centre of the stream. When the bridges are " open " the people must wait till they shut. Tugboats draw fleets of vessels through the sluggish water while the impatient multitudes gather, and long lines of teams stand waiting, each eager to be on the move. We have sometimes considered this enforced delay a blessing, inasmuch as it gave the hurrying Chicagoans a breathing spell and moment to collect their thoughts. It is to be feared that the majority do not indulge pious thoughts, or confine themselves to words of gentleness, while they fume and fret over the impediment to their onward pursuit after money or pleasure. To obviate this difficulty and facilitate the necessary interchange between the three great divisions, tunnels have been constructed underneath the river and South Branch, both for vehicles and foot-passengers. They were finished none too soon, for in our present distress they have paid for themselves. Business was mainly in the South Division till that was ravaged by destruction, and people were compelled to transfer the chief portion of it to the West Side, where also the large majority of dwellings stood and still remain. In the time of the conflagration, when bridges fell into the river smouldering masses, these became necessary for the escape of those who were driven before the fire. They were the scenes of some peculiar experiences on that dreadful night. When the gas works let off the gas to prevent explosion after midnight, the Washington street tunnel was full of vehicles, and the footway crowded with fugitives bearing away their families and possessions to a place of refuge. This being illuminated by gas was light and safe, even with a dense throng pressing excitedly through it. But in an instant the flickering flames were extinguished, and the place was dark as midnight without a star. One shriek of anguish rang out along the arches, and then the voices of men were heard quieting the people. "Be still," "move slow," "there is no danger," "do not push or crowd," were some of the directions given and carried out, so that the whole procession felt their way composedly •o the farther end, and not a person was trodden down or injured. Reports went over the land by telegraph that hundreds of people were smothered in the tunnels and fearful deeds were perpetrated in their darkened passages; all of which proved false, and these excavations were coverts from the storm of flame, and by their means many escaped who otherwise must have become victims of the fiery demon.

Those who sought exit from the furnace by the La Salle street tunnel were less fortunate, according to the succeeding description, for which, however, we do not vouch, although every word is possibly true:

"One of the most dramatic and impressive scenes of the fire not yet recorded, was the flight through the new La Salle street tunnel, under the river, during Sunday night. It was about 2 o'clock when this strange hegira began, and in ten minutes it became a furious rout. The bridges on both sides were on fire, and the flames were writhing over the decks of the brigs in the river, and winding their fierce arms of flame around the masts and through the rigging like a monstrous luminous devil-fish. The awful canopy of fire drew down and closed over Water street as the shrieking multitude rushed for the tunnel, the only avenue of escape. There was no light in any house, save the illumination which lighted up only to destroy. But into the darkened cave rushed pell-mell, from all directions, the frenzied crowd – bankers, thieves, draymen, wives, children—in every stage of undress, as they had leaped from burning lodgings, a howling, imploring, cursing, praying, waiting mob, making their desperate dive under the river. It was as dark in the tunnel as it is in the centre of the earth, perhaps darker. Hundreds of the fugitives were laden with furniture, household goods, utensils, loaves of bread, and pieces of meat, and their rush through the almost suffocating tunnel was fearful in the extreme. They knocked each other down, and the strong trod on the helpless. Nothing was heard at the mouth of the cavernous prison but a muffled howl of rage and anguish. Several came forth with broken limbs and terrible bruises, as they scattered and resumed their flight under the blazing sky to the North."

The tunnels having become an established institution, will be multiplied as the necessities of the future require, till the river shall become no barrier to the intercourse of the inhabitants in every part of the city. They exist as a monument of the unconquerable energy of the people under whose patronage they have been constructed.

CHAPTER XVI.

Our city has enjoyed an unenviable reputation abroad for wickedness. Doubtless the sins of our people are a cause of reproach, weakness, and shame. "Righteousness exalteth a nation, but sin is a reproach to any people." We have had all the vices of large populations, both open and secret. The divorce business is especially a crying evil of the time, and our foreign element have been peculiarly given to disregard of the marriage tie. It was said of us that the newsboys cried at the approach of railway trains: "Chicago, fifteen minutes for divorces!" The annals of crime have been full and red. Only the week preceding the fire a mysterious murder occurred which sent a thrill throughout the community. Drinking was carried to excess, even Sunday dram-selling being tolerated by the executive to an alarming and shameful extent. Covetousness also prevailed under the forms of prodigality and avarice. The Sabbath was terribly profaned by our foreign population, and the demoralization ran along all orders of society. Sinful or doubtful amusements received the devotion of multitudes. At an expense of eighty thousand dollars Crosby's Opera House had been refitted, and the winter was expected to be one of unusual gayety and excitement. Money was to have flowed like water from the rich and the poor alike. We had our low resorts in great numbers. The harlot plied her trade with success and profit. Blocks of buildings were occupied by young men who had their orgies and debauches, where young women were welcome visitors. The secret immoralities of a great city are innumerable and shocking. None but the Omniscient can spy them out. Occasional revelations are like flashes of lightning upon a stormy sea, disclosing the rush of black billows and the seething of bottomless eddies of corruption. Alas for our city! Pompeii could scarcely excel the madness of its passion, though law gives no sanction to iniquity, as it did in that vile nest of heathen immorality.

While we thus glance at the darker aspect of life here, in order to be just and true to facts, we turn gladly and boldly to another side of the picture, and hold up a people whose liberality, generosity, piety, and morals will compare in their fruit—their actual outworkings—with those of any other people under the sun. It must be remembered that in the new West everything has had to be done, as it were, at once—every necessity to be provided for within a generation. We have not had two and a half centuries to grow all these institutions and make the improvements needful to our comfort. True, we have had the benefits of other men's capital and culture, and used them well. We claim nothing more, and demand the recognition of this from our fellow-citizens in older and well-regulated communities. The "almighty dollar" has not absorbed attention and made us forgetful of the higher interests, nor have we failed to recognize the immutable principles of justice and honesty in our political or commercial relations. Some of the best church edifices in the country were and are still standing in our city, and these were and are carefully attended and liberally supported. In mission work among the poor and neglected we have not fallen behind our brethren elsewhere. The names of our workers and their labors have become famous not only in America, but abroad; and the good report has had no small share of influence in bringing to our city a better class of people, and inspiring confidence. D. L. Moody's enterprise as a missionary and a leader in the Young Men's Christian Association had so widely affected the public mind, that contributions to rebuild his burned edifice come pouring in upon him from all quarters. The various churches have been awake and earnest in their fields, to gather the harvest for God's kingdom. The North Star Mission found friends familiar with its holy fame, who generously came forward to restore it upon a good foundation of usefulness. On a late Sunday in October, the Sabbath preceding the catastrophe, in the Second Baptist Church audience-room were collected a vast number of children. First came the infant class of the Home School to the front, and took their places; then the middle classes followed, and lastly the Bible Classes filed into the centre of the house. Upon the one side marched in 600 German youth and infants; upon the other, Danish, Swedish, and others, from one mission. A company clean and bright came from Bridgeport, and another from the Union Stock Yards. It was a gallant array, of whose conduct and appearance the earnest, self-sacrificing workers were justly proud. These were allowed to sing, and, after listening to the speeches, to depart. Carriages and cars had been provided, and great pains taken to make all comfortable and happy, that they might join in a welcome to the pastor, who had just returned from a long absence. This is a specimen of the manner in which Christians in Chicago have labored and sacrificed to build up the youth into a maturity of knowledge, religion, and virtue. The inner life of the churches is sweet and vigorous, and their beneficence has begun to bear golden fruit. They have given their energies, talents, and money to found and endow institutions of learning. The Methodists have a University, Female College, and Seminary in one of our beautiful suburbs, Evanston; the Congregationalists have a noble Seminary for the education of Ministers; the Presbyterians, also, have the same; the Baptists have a University and Seminary in the city, at Cottage Grove, already educating hundreds; the Catholics had several institutions, and all Christians had their organs of the press, their organizations and associations for disseminating their views and evangelizing the world. There is a pleasant fraternity of feeling manifesting itself in a variety of forms, and especially through the Young Men's Christian Association. There are many living, devoted men of God, laborious, prayerful servants of Christ, benevolent, helpful followers of Him who went about doing good. If the devil is active, his opponents are thoroughly awake and ready to give him battle on every side. Since the disaster which destroyed so many sanctuaries and crippled the benevolent, one of the first thoughts has been to re-establish these institutions of religion and save the seats of learning. This fact speaks volumes for the character of our people, showing their appreciation of the value of Christianity and their profound interest in its progress. Many of them, though burned out or injured, sought out the Lord's treasury and divided their little remnant of money for the care of their church servants and services.

Besides all this, they have manifested great kindness and honor in the hour of mutual adversity, and are seeking to do the thing that is right between man and man. There is much reason to take pride in such a people, who have gathered here from all quarters, and scarcely learned to know and appreciate one another. It was but simple justice which led an eminent writer to say: "It is my impression that human nature there is subject to influences as favorable to its health and progress as in any city of the world, and that a family going to reside in Chicago from one of our older cities will be likely to find itself in a better place than that from which it came."

A gentleman who spent a Sabbath here and spoke in the evening at Farwell Hall, in giving an account of what occurred, said that he thought, as he saw the liquor saloons open and thronged, that Chicago was the worst place he was ever in. But before he reached the Hall several young men met him, and invited him to go to church, and addressed him and others with great courtesy and earnestness. He said he concluded that if the devil was well served, certainly the Lord's people were the most devoted workers he had ever met.

CHAPTER XVII.

OUR Republic has become an asylum for strangers from all nations. Ancient Rome drew to itself, by conquest, representatives of many countries, and trade attracted others, so that it became a Babel. Chicago has been the star in the West by whose beams multitudes have been guided to the Valley of the Mississippi, from almost every nation under the whole heaven. It has offered a home to many, and a market to others. The country makes the city, and the city develops the country. Thus they act and react perpetually upon each other in respect to all the various interests and concerns of life. It was a hazardous thing in the eyes of some persons to encourage railroads, lest they should divert and scatter trade from Chicago all along their lines. It took but a slight experience tc demonstrate that if ever the city attained greatness, steam must have the glory of it. Railroad therefore followed railroad, till now, from having forty miles in 1850, this metropolis has already grown to be a chief railway centre of the world. More than 8,000 miles of rail centre here, and fifteen trunk lines radiate every way, each from three hundred to one thousand miles in length, and still they come. These marvellous facilities make us the focal point of the great West, and bring to our doors all peoples, languages, and colors. The grain trade, as we have shown, is very great, and our advantages for handling it are unsurpassed. All persons have heard of the elevators, and we subjoin an account of one lately built.

"The building is 312 feet long, 84 feet wide, and 130 feet high; machinery is driven by a 400 horse-power engine. It is divided into 150 bins 65 feet deep, with a storage capacity of 1,250,000 bushels. The yard will hold 300 or 400 cars.

"Two switch engines, when in full operation, are required to put in and take out cars.

"Two tracks receive each ten cars, unloaded at once, in six to eight minutes, each car having its elevator, conveying the grain to its large hopper scale in the top of the building. There weighed, it is spouted to the bin appropriated to that kind and quality. To carry grain to the several bins renders the elevation necessary. Allowing fifteen minutes to unload each set of ten cars, four hundred are unloaded in ten hours, about 140,000 bushels.

"Shipping facilities equal receiving, there being six elevators for that work, each handling 300 bushels per hour, or 180,000 bushels in ten hours. The grain is run out of the bins to another set of elevators, which throw it into large hoppers at the top of the building, in which it is weighed, and sent down in spouts into the hold of the vessel.

"The same company have another elevator on the opposite side of the slip—for a slip at right angles to the South Branch is cut to lay vessels alongside the warehouse—and ten other large elevators and five smaller, afford the same facilities. Any one of thirteen of them, too, will unload a canal-boat of 5,000 or of 6,000 bushels, in an hour and a half to two hours; an aggregate from 65 canal boats alone of 357,000 bushels in ten hours."

Modern invention economizes the results of industry and the productions of the earth, as well as human muscle and time. Many are not aware of the process by which corn can be stored and preserved, with an immense saving from waste and deterioration. The subjoined brief picture of the dryer and its operations may interest a large class of readers:

"A tower seventy-five feet high, built of brick and iron, fire-proof, receives the grain at the bottom, where it is elevated to the top, and passes slowly down over perforated iron plates, the motion of the falling grain being constant and uniform, regulated by slides or valves at the bottom.

"The grain in motion forms a solid column seven feet wide and three inches deep. There are two columns of grain, and a furnace at the bottom supplies hot air, which is evenly distributed by suction-fans, so as to pass constantly and equally through the grain the entire height of the kiln. Temperature is regulated by thermometers set in the walls at several points, avoiding all danger of over-heating. Impurities or foreign substances are passed off in vapor or steam. Then it is thoroughly cooled before being passed to the bins in the elevator by the same process, except cold air instead of hot is used, which contributes further to dry as well as cool."

It is a marvel to many how drainage has been secured upon so flat a plain, the highest point of the city being but twenty-five feet above the surface of the river. This important and essential end has been achieved by raising the whole land some fourteen feet. High stone walls are built, the interior is filled with earth, and the pavement laid upon that. The sidewalks are built from each wall to the yards or fronts of the lots, and the houses are raised up to grade. This gigantic operation is still going forward, and miles of wooden streets offer their noiseless surfaces to the wheel of the vehicle. This elevation of grade has made the ways of our city rather uneven, and suggested forcibly the ups and downs of Chicago. A man in New York, arrested for drunkenness, pleaded not guity, and said that being just from Chicago, where the sidewalks were so uneven, his gait was mistaken for that of an intoxicated man. Gradually men are bringing themselves up to level, mending their ways, and making pedestrianism less dangerous and more agreeable. We lost in the fire one hundred and twenty-two miles of sidewalk, which gives some idea of the extent of territory it traversed, and the amount of labor required to remove the traces of its progress.

Among modern precautions against fire, the fire-alarm telegraph occupies a conspicuous place, and has been for some years in full operation in Chicago. Wires are stretched over house-tops throughout the city, and boxes placed at frequent intervals for the use of these wires by citizens, who wish to call the attention of the Fire Department to any outbreak of fire in their vicinity. The turning of the handle in the box is felt at the rooms in the Court-House, and the number of the district indicated to the operator, who sends it to the engine houses, where horses are standing harnessed day and night, ready to speed the steam-engine to the point of attack. There is also a watchman in the cupola of the Court-House, who sends word to the operator of any fire he may see, and rings the great bell (now, alas! forever silent), to warn the firemen and people of the location of the fire. Suppose the conflagration is in district one hundred and twenty-three; he strikes the bell once, then rests a moment, strikes it twice, then rests again, strikes it three times, and then, after a longer interval, repeats this process, till the city is made fully aware of the situation of the danger, nearly every house having a printed list of the fire districts. In this connection, the following statement of one of the operators on duty the night and morning of the Great Fire, is full of interest: "I arrived at the office 12.30, A. M. While I was on duty, Stations 19, 13, and 10 were turned in, and struck by me in rapid succession. About this time some man came into the office and notified me that the fire had crossed to the south side of the river. At the same time the watchman in the tower told me that the wooden ventilators on the west wing were on fire. I then asked the man on duty in the Central Station (Policeman Vesey) to send me a fire-extinguisher, which he did. With the aid of the extinguisher and the assistance of the two watchmen in the tower, I managed to keep down the small fires which were constantly appearing on the wooden tower and ventilators, until about half-past 1 o'clock A.M., when a ball of tar, or a piece of tarred paper, came through the windows under the balcony of the dome, and fell on the stairs, just where some plastering had been pulled off. I started up the stairs to put it out, but before I could reach it, the lathing and some dry material under the roof had ignited. I then called loudly for Mr. Deneson, the watchman, to come down from the tower, which he did, making a narrow escape with his life. Knowing by the appearance of things that the building was doomed, I returned to the office and struck my electric repeater, striking upwards of seventy blows on the outside bells, thinking that, perhaps, the noise would awaken some of the many sleepers with whom I knew many of the blocks were filled. Previous to this, I caused the Court-House bell to be rung by hand. As the office was by this time full of smoke, and the heat was becoming intense, I was obliged to switch off my repeaters and leave the office, which I did, with one or two others, by way of the west wing, stopping to close the fire-doors between the two buildings. Once out of the building, I procured a fire-hat, and worked until 3 o'clock on Monday afternoon, at the south end of the fire, when I went home to get some sleep."

Brave fellow! he had earned his coveted repose, and we do well to honor the men who keep watch and ward over our dwellings and lives, while we sleep and while we wake. His ingenious mechanism, and all the appurtenances of his department, must be renewed, and even upon a grander scale, in the Chicago of the future.

Recognizing the value of universal education, our city has provided, partly through State liberality, a splendid system of common-school instruction free to every child, of every nationality, religion, and condition among us. The officers and teachers of these schools are persons, many of them, of the highest intelligence, culture, and skill, and generally we are admirably served. A vast throng of children gather in the buildings devoted to this purpose, which are, almost all of them, noble, commanding, commodious edifices, capable of providing room for all the youth who choose these facilities. In addition, there are numerous private schools and academies, both for primary and higher education, which find ample patronage from a people who prize the power of knowledge and despise ignorance as weakness.

It has been the honorable aim of our city to place the highest objects of ambition among men upon a footing worthy their preeminence. Religion, morality, knowledge, culture, and social enjoyment have their temples and seats, paraphernalia and apparatus, in as advanced a state of perfection as in any community under the sun. And all this, be it ever remembered, has been the growth of a heterogeneous people, upon a new soil, within the period of a generation. Their enterprise and its results constitute a fitting symbol and monument of the age.

Want of space must prevent an elaborate account of those

splendid blocks which had sprung up on every hand, built both by home and foreign capital, many of them rivalling in beauty the finest models of architecture in the Old World; those grand hotels, both old and new, which had a national reputation and a promise of eclipsing the world; those beautiful homes, where taste and wealth combined their resources to provide elegance and comfort; those public buildings, stored with the trophies of genius and the results of scientific research; those sanctuaries, proclaiming the purpose of the people to give God the best; together with a myriad tokens of prosperity, so many of which are now level with the ground, or stand in unsightliness and ruin to mock the pride of man. At the height of a proud and princely position the Young Giant stood erect, beckoning the world to his arms, when the fatal decree went forth, and his might, touched by the flaming breath of Omnipotence, shrivelled and shrunk, and he lay prone like a tree, storm-bent and fire-scathed.

IV.—THROUGH FIRE.

CHAPTER XVIII.

The churches were just dismissing their devout worshippers after evening service, when the fire-bells rang their loud alarum. The evening before, a fire had raged of unparalleled violence, and the embers still glared in the darkness, and people were easily roused to intense alarm. Many hastened from the House of God to the scene of the fire, fearing that the high wind might imperil even larger districts of the city. None dared to dread any such devastation as that which followed.

It was a period of peculiar drought in the whole western country, and the dryness of the atmosphere was so remarkable that an intelligent physician, observing that his plants became desiccated in a few hours after the most profuse watering from the hydrant, trembled all day Sunday lest a spark of fire should drop near his dwelling. There was a strange lack of moisture in the air, which condition did not change until Monday afternoon. On Saturday evening, October 7, about 11 o'clock, a fire caught in a planing-mill, west of the river and within a block of it, in the neighborhood of a wooden district full of frame-houses, lumber, and coal-yards, and every kind of combustible material. Some contend that it originated in a beer saloon, and thence was communicated to the planing-mill.

In the almost inflammable state of the atmosphere, and under the propulsion of a strong wind, the tinder-boxes on every side ignited, and ruin rioted for hours over a space of twenty acres, and destroyed a million dollars' worth of property. Grand and awful as this conflagration seemed to the thronging thousands, who crowded every approach and standpoint where a view could be obtained, it paled and faded away in comparison with that of the following night; but, as the event proved, this first fire saved the remainder of the West Division of the city, for when the raging element came leaping and roaring onward it found nothing to burn, and then paused and was stayed, while it rushed across the river, and satiated itself upon the noblest and best portion of the town, east and north.

Of this eventful period so many writers have wrought out descriptions which are unapproachable in graphic delineation and powerful word-painting, that simple justice to our readers demands that we collate from these all that is necessary to present the whole mournful subject in its many-sided aspects. Like a great battle, with its multitudinous features unobservable by any combatant or spectator, this conflagration presented so many phases that each was absorbed in what he saw, while matters of unspeakable interest were occurring on every side beyond his ken. Let, then, many testimonies combine to set forth to the gaze of mankind what has perhaps never been equalled, and certainly never surpassed in the checkered experience of humanity. We bring together around this terrific scene the sketches of the press published in Chicago and elsewhere, and individual experiences.

THAT KICKING COW.

The reporters gave the world to understand that a woman named Scully had gone to milk her cow or tend a sick calf in her stable—a crazy wooden shanty filled with loose hay—bearing a candle or lamp in her hand. Stories varied as to these details, but all agreed that the light had been overturned, and that the building had on the instant burst into flames. So rapid was the progress of the fire that in less than ten minutes two blocks between Jefferson and Clinton streets were all ablaze.

VIEW FROM THE COURT-HOUSE, LOOKING SOUTH.

VIEW FROM THE COURT-HOUSE, LOOKING SOUTH-EAST.

DRAKE AND FARWELL BLOCK, WABASH AVENUE.

UNITY (MR. COLLYER'S) AND NEW ENGLAND CHURCHES

Upon this report the London *Punch* becomes funny, and kindly too:—

"We suppose that the most costly pail of milk ever heard of in the world was the pail which burned Chicago. The gallant Americans are the last people to cry over spilt milk or burned cities. Chicago will quickly be *Rediviva.* She has very likely accepted the omen that she will soon be flowing again with milk—and honey—has elected in her cheery way to call herself the Cow City. Therefore, Bull, evince the affection of a relative; show that you have what *Benedick* calls "an Amiable Low" (needless to say that we do not allude to any keeper of the public purse), and that you come of the stock of the Golden Bull. With which sweet, choice, and dainty conceits to lighten the way, let the pensive public be off to the Mansion House with their help for the homeless by Lake Michigan. The Americans remembered us in the time of Ireland's hunger and of the cotton famine, and must now allow us to remember them. And let's be quick about it, or the city will be rebuilt before the money gets there. 'Right away—this very now,' as they say."

We thank Mr. *Punch* for his generous confidence and witty appeal, and assure him that this is our purpose, to revive in more than former splendor and power, that our city may be able to help the poor, and empty its cornucopia into the lap of the world. The story of this origin of the disaster may be true, in spite of affidavits to the contrary, or may have but a spark of truth in its fabric; at all events, the fire commenced at the barn, and grew into

THE GREAT CONFLAGRATION.

Before we summon our eye-witnesses, we are willing to allow the inveterate joker of the *Hartford Post* to have his bit of fun at their expense, since he is a newspaper man and cannot be expected to "set down aught in malice" against his brethren.

"The reporters and correspondents did try to 'do the subject justice' in writing up the Chicago fire. We can imagine them looking on the roaring sea of flames and the crazed multitudes seeking refuge from them, and making up their minds deliberately that in the matter of describing the fury of the fire and the wild tumult of the crowd, nothing was left to exaggeration; they must climb up by dizzy successions of polysyllabic adjectives as nearly as possible to the heights of the great occasion, and feel then that words were unequal to it; that they had not and could not exaggerate it. Of course it piqued their ambitious pens. It occurred at length to one of them that it was an exceedingly proper time for bloodshed, that in all this chaos there was a lack—to the reporter a painful lack—of devilishness. It was a horrible picture, but it lacked murder to make it complete. What so good time as this for hangings and lynchings, and other such bloody carryings on. It was such a happy thought, that the first reporter interpolated forthwith into his account the shooting down of an incendiary. It took. The reading public licked its intellectual chops and said: 'Ah, now it begins to be congruous and coherent-like. This is something like it.' And the reporter thereupon, after the manner of the menagerie man tossing raw beef to the tigers, jerked into his account the sweet little sentence: 'Seven men have just been shot down in the act of kindling incendiary fires.'

"'Only seven,' growled the public. 'There must be more than that; the fire was a very large one.'

"The reporter was equal to the occasion. 'Forty-seven men have already been shot,' he telegraphed; 'no arrests are made. Incendiaries are shot down wherever taken.' He had kindled to it. The raging public wanted blood. He could furnish it. Then it occurred to him to heighten the interest by giving names—it wanted local and personal color. So with a dash of the pen he strung Barney Aaron, the pugilist, to a lamp-post, and shot another notoriety named Tracy, with a file of muskets. He was doing well. The fire was subsiding, but there never was such an opportunity for murderers, never a man so handy at inventing them. But the fire was the biggest thing the world ever saw, and these were only ordinary murders. He had not worked brutality enough into the picture. And so, to finish and crown all, he strung up a boy by the heels, head downward, and described, with horrible minuteness, how the crowd amused itself by stoning him to death. And then that reporter retired from business. Next day General Sheridan, who was in command, in reply to some sort of a telegram, possibly asking him if it was not feasible to quench the flames with the human gore this sanguinary reporter had set running, said it was very quiet there, and no disturbance of any account. But a blood-thirsty public was not to be so deceived. 'Ah!' they said, 'Sheridan is so used to blood! This is nothing to him. To a man who has swam his horse through it in the Shenandoah a mere streetful of blood is nothing. Ah, ah! Oh, yes! "very quiet"—that's good; but of course, as a matter of fact, they have shot incendiaries, and hung thieves to lamp-posts and stoned them to death, and there is no doubt that Barney Aaron and Tracy were killed, for the telegraph has distinctly said so.'"

And yet, ten days after the event, it turns out that the boy was not inverted and hung and stoned to death, and that the soldiers did not shoot anybody, and that nothing of the sort happened. And Barney Aaron, who was hung to a lamp-post, sits on the steps of a New York gambling-house, and asseverates that he was not killed.

That reporter rose to the occasion. He writes with a harrow.

Had this hard joker, who rightly takes off sensational writing, been a spectator and sufferer on that woful night, doubtless he would have felt that a pen dipped in Tartarean flames would have been needed to adequately depict the scenes that transpired.

"None but an eye-witness can form an idea of the fury and

power of the fire among the buildings and warehouses on the South Side, with the wind blowing a hurricane. At times it seemed but the work of a moment for the fire to enter the south ends of buildings, fronting on Randolph, Lake, and Water streets, and reappear at the north doors and windows, belching forth in fierce flames which often reached the opposite buildings, and then the flames, issuing forth from the buildings on both sides of the street, would unite, and present a solid mass of fire, completely filling the street from side to side, and shooting upward a hundred feet into the air. Thus was street after street filled with flame. Huge walls would topple and fall into the sea of fire, without apparently giving a sound, as the roar of the fierce element was so great that all minor sounds were swallowed up, and the fall of walls was only perceptible to the eyes. Many of the buildings situated along South Water street buried their red-hot rear walls in the water of the river, into which they plunged with a hiss. The heat was so intense at times from some of the burning buildings that they could not be approached within 150 feet, which accounts for the manner in which the fire worked back and often against the wind. The fire, after reaching the business portion of Randolph and South Water streets, leaped the river to the North Side in an incredibly short space of time, and thence among the wooden buildings on that side, reached the lake shore after destroying block after block of happy dwellings. A scene of such utter powerlessness in the face of an enemy was never presented as that of this people trying to combat the flames.

"Now was to be seen the most remarkable sight ever beheld in this or any country. There were from 50,000 to 75,000 men, women, and children fleeing, by every available street and alley, to the southward and westward, attempting to save their clothing and their lives. Every available vehicle was brought into requisition for use, for which enormous prices were paid. Thousands of persons inextricably commingled with horses and vehicles, poor people of all colors and shades, and of every nationality—from Europe, China, and Africa—mad with excitement, struggled with each other to get away. Many were trampled under foot. Men and women were loaded with bundles, to whose skirts children were clinging, half-dressed and barefooted, all seeking a place of safety. Hours afterwards, these people might have been seen in vacant lots, or on the streets far out in the suburbs, stretched in the dust. These are the homeless and destitute, who now call on the rich world for food and clothing. One of the most pitiful sights was that of a middle-aged woman on State street, loaded with bundles, struggling through a crowd, singing the Mother Goose melody,

'Chickery, Chickery, Crany Crow,
I went to the well to wash my toe!'

"There were hundreds of others likewise distracted, and many, made desperate by whiskey and beer, which, from excess of thirst and in the absence of water, they drank in great quantities, spread themselves in every direction, a terror to all they met."

Instead, therefore, of considering these descriptions which follow as exaggerations, we do well to remember that all concur in declaring that language fails to do justice to the roar and rush of the elemental forces, combining to demolish the proudest monument of American enterprise, the glory and boast of our country, and the wonder of the world. All things concurred to make this the climax of triumph for the fire-fiend.

Sunday evening seemed to have been designed purposely for a repetition of the horrors of Moscow, or the "calamitous and piteous spectacle" of old London. A strong wind, rising at times to a hurricane, blew across the city. Every roof was baked dry as tinder by fourteen rainless weeks. The power to disseminate and the readiness to receive were there, and but one spark was needed to blot out a city and blacken the prairie with houseless heads.

CHAPTER XIX.

Every thinking man inquires for a philosophy of the fire, and the world wishes to guard itself from a recurrence of the calamity that has fallen like a thunderblast on the Great West. Mr. Charles Barnard thus writes of

THE REMOTE CAUSE OF THE WESTERN FIRES.

Chicago has burned down, and whole square miles of western land are burned up. That misguided cow and unhappy lamp have been berated enough. If the barn had been damp with recent rains perhaps the fire had gone no farther. Certain is it that if the roof-tops had not been baked dry by a summer's drought Chicago would not have mourned her lost children and ruined homes.

Had not those Wisconsin fields been as ashes in the dry wind, had plentiful rains drenched the Michigan woods, the country would have been happier to-day. Everything there was as dry as tinder, say all the papers.

Now whose fault was it? People with more piety than wisdom may say, in a horrified way: "What a question! Do you arraign the acts of Providence?" No. There has been blame somewhere. We are not inclined to shift it upon heaven. Men, not Providence, brought this calamity upon us. It is we who have created these dry summers. Had there been no drought there had been no such wide ruin.

The time was when such long-continued dry seasons were not known. Men can and do change the character of climates. We can cause the rain to fall, or drive away the clouds. Men have altered the temperature and moved the dew-point. The farmers of the Northern States are, in a measure, responsible for the series of dry summers that have prevailed for the last ten years.

Meteorology is beginning to take a high position. We have mapped the winds, and can signal the coming storm to the sailor and farmer. The laws of the weather are no longer a matter of guess-work. Cause and effect are as sure in the clouds as on the ground. Observing the effect, we can trace the cause. Given this series of dry summers, science points to the cause—our denuded forests.

In our foolish American haste we have wastefully cut down the trees, dried up the springs, raised the temperature, so that precipitation of moisture is reduced, and have driven the rain away in useless clouds or invisible vapor over the Atlantic. Chicago is burned down, and we are solemnly saying, "How heavy is the hand of heaven upon us!" We have prayed for rain one day of the week, and driven it away with an axe on six.

The mischief is done, and the best thing we can now do is to examine the matter with a view to future prevention. How shall we bring back the rain? How restore our forests? Simply by planting our woods anew.

This is not a new or untried idea. Artificial woods are no longer a novelty in Europe. There this whole matter is well understood. In parts of the Continent foresters are appointed by government. It is their duty to inspect all standing forests. Schools of arboriculture are established. The habits of the trees are considered, the soil examined, and tree-planting carried on over hundreds of square miles. For every tree cut down one or more new ones must be set. Nurseries, producing millions of young trees, do thriving business in supplying this material. Under the advice of the foresters the new forests extend year by year. On the rocky hills of Scotland the oak, maple, and chestnut are planted; the willow is set out by the million on the marsh-like "polders" of Holland; about Utrecht, and on the sandy plains of Zelderland, near Arnheim, the traveller passes artificial pine-forests by the hour.

In view of these western fires it is high time we prepared to imitate our transatlantic friends. At once the great cost of such an undertaking comes up. Now we think it can be shown that the thing will pay to do. If there is money in it, it will get itself done fast enough.

The land used for such forest is generally fit for nothing else. We have millions of acres that are barren wastes—an eyesore and a tax on the owners. By examining the most flourishing trees growing in similar soil in the neighborhood, we can decide what to plant. By sowing the seed or buying young trees a year old, we can soon start a forest that in twenty years will bring a cash return that will cover the cost of planting, interest, and taxes, and leave a margin of profit besides.

To come down to details, let me present an estimate prepared for a gentleman who had a hundred acres of nearly valueless land in Eastern Massachusetts. It was a continual tax-bill, and brought no return whatever. The land was valued at fifty dollars an acre. The interest for twenty years would be $6,000; the taxes, $5,000. If he did nothing to the land he would be $6,000 out of pocket at the end of that time. There was a fence round the whole lot that it was estimated would cost twenty dollars a year to maintain. Each acre would hold five hundred trees, or fifty thousand in all. The trees could be bought for $1,500. The planting would cost about $600. The trees, at the present price of posts and sleepers, would be worth at least seventy-five cents each. To sum up:—

Interest	$6,000
Taxes	5,000
Fencing	400
Oversight, at $50 per year	1,000
Fifty thousand trees	1,500
Planting	600
	$14,500
Fifty thousand trees at 75 cts	37,500
Five per cent. loss	7,360
	$30,200
Cost	14,500
	$15,700

The care would be slight, as there is no culture of any kind. Certainly this would be a nice little piece of property to leave to the children, or set them up in life with. Were the trees cut down, the place could be replanted. With better kinds of trees and more time, a greater price could be obtained. The trees to be used were maples and chestnuts. The Scotch are noted for minding the "mickle" that brings the "muckle," and the Zelderlanders are the closest-fisted people in Europe. That they plant trees in countless thousands proves they have an eye on the above cheerful pennies.

CHAPTER XX.

Whatever the indirect cause of the fire, it is plain that the immediate aggravating conditions were such as rarely occur. Long-continued positive drought, peculiar dryness of the atmosphere, a heavy wind that increased to a tornado, vast masses of pine wood and coal, weary firemen, and finally utter loss of water to feed the engines, account for what followed, and prepare us to accept the glowing paragraphs and solemn lines which tell the tale of general and individual woe.

THE POST'S VERSION.

At 9.32 an alarm was sounded, summoning the brigade to the corner of Jefferson and DeKoven streets. Ere the first engine was on the ground, the flame had enveloped half a dozen outbuildings, and was pouring its columns upon the city to the southward and eastward with the resistless grandeur and celerity of a barbaric invasion.

The firemen, convinced of the impossibility of saving anything in the district now attacked, confined their efforts to checking the northward march of the fire. Heroic as these efforts were, they were in vain. The flames ran along the wooden sidewalks, and whole tenements would burst into flames as simultaneously as if a regiment of incendiaries were at work. The narrow streets were crowded with appalled spectators, half-dressed women with aprons thrown over their heads running distractedly hither and thither, and men tearing furniture to pieces in the furious haste with which they flung it out of doors or dragged it through the crowd. The element had the best of the battle so far. Engine No. 14, driven back foot by foot, was penned in a narrow alley; in another moment a gush of flame came from the rear, and the firemen could only cover their eyes from the blinding heat and stagger desperately to safety through the burning belt that fringed them round, abandoning the engine. Still they fought on gallantly. The advance of the fire was strongly defined in two great columns running north, one between Jefferson and Clinton streets, the other between Clinton and Canal streets. The latter led the way, and as one o'clock struck, had seized the buildings on Van Buren street, while the other was spreading more slowly along West Harrison.

One o'clock had just struck, and a sudden puff of the variable wind blew down a curved wing of the great golden-red cloud above our heads. It fell like the sheer of a sabre, and in a second a red glare shot up on the South Side, as if the blow had fallen on a helmet and sent up a glitter of sparks and a spurt of blood. The fire had overleaped the narrow river and lodged itself in the very heart of the South Division. The angry bell tolled out, and in a moment the bridges were choked with a roaring, struggling crowd, through which the engines cleft a difficult way toward the new peril. The wind had piled up a pyramid of rustling flame and smoke into the mid-air. Lower currents at times varied and drove tides of fire athwart the great roaring stream. When these met, eddies that made the eye dizzy were formed, which sucked up blazing brands and embers into their momentary whirl, and then flung them earthward. In such a fiery mælstrom had a shower of sparks and large fragments of detached roofing been hurled into the neighborhood of the old Armory. The skirmish ing was over, and man and fire were now grappling in earnest where the prize was millions of money and hundreds of lives.

When once the fire had established itself in the South Division the task of following the course or describing its ravages in detail became an utter impossibility. As well might a private soldier endeavor to paint Waterloo, Sedan, or Gravelotte. All that the writer can say is that everybody was mad, and everything was hell. The earth and sky were fire and flames; the atmosphere was smoke. A perfect hurricane was blowing, and drew the fiery billows with a screech through the narrow alleys between the tall buildings as if it were sucking them through a tube; great sheets of flames literally flapped in the air like sails on shipboard. The sidewalks were all ablaze, and the fire ran along them almost as rapidly as a man could walk. The wooden block pavements, filled with an inflammable composition, were burning in parallel lines like a gridiron. Showers of sparks, intermingled with blazing brands, were borne aloft by one eddy of the breeze, and rained down into the street by the next, while each glowed a moment and was gone, or burned sullenly, like the glare of an angry eye. Roofing became detached in great sheets, and drove down the sky like huge blazing arrows. The dust and smoke filled one's eyes and nostrils with bitter and irritating clouds. There was fire everywhere, under foot, overhead, around. It ran along tindery roofs, it sent out curling wisps of blue smoke from under eaves, it smashed glass with an angry crackle, and gushed out in a torrent of red and black; it climbed in delicate tracery up the fronts of buildings, licking up with a serpent tongue little bits of woodwork; it burst through roofs with a rattling rush, and hung out towering blood-red signals of victory. The flames were of all colors, pale pink, gold, scarlet, crimson, blood-hued, amber. In one place, on a tower covered with galvanized iron sheets, the whole roof burned of a light green, while the copper nails were of a beautiful sparkling ruby. Over all was the frowning sky, covered with clouds varied by an occasional undazzled star.

The brute creation was crazed. The horses, maddened by heat and noise, and irritated by falling sparks, neighed and screamed with affright and anger, and reared, and kicked, and bit each other, or stood with drooping tails and rigid legs, ears laid back, and eyes wild with amazement, shivering as if with cold. The dogs ran wildly hither and thither, snuffing eagerly at every one, and occasionally sitting down on their haunches to howl dismally. When there was a lull in the fire, far-away dogs could be heard barking, and cocks crowing at the unwonted light. Cats ran along ridge-poles in the bright glare, and came pattering into the street with dropsical tails. Great brown rats with bead-like eyes were ferreted out from under the sidewalks by the flames, and scurried hither and thither along the streets, kicked at, trampled upon, hunted down. Flocks of beautiful pigeons, so plentiful in the city, wheeled into the air aimlessly, circled blindly once or twice, and were drawn into the maw of the fiery hell raging beneath. At one bird-fancier's store on Madison street, near La Salle, the wails of the scorched birds, as the fire caught them, were piteous as those of children.

The firemen labored like heroes. Grimy, dusty, hoarse, soaked with water, time after time they charged up to the blazing foe only to be driven back to another position by its increasing fierceness or to abandon as hopeless their task. Or, while hard at work, suddenly the wind would shift, a puff of smoke would come from a building behind them, followed by belching flames, and then they would see that they were far outflanked. There was nothing for it then but to gather up the hose, pull helmets down on their heads, and with voice and lash to urge the snorting horses through the flame to safety beyond.

The people were mad. Despite the police—indeed the police were powerless—they crowded upon frail coigns of vantage, as fences, and high sidewalks propped on rotten piles, which fell beneath their weight and hurled them, bruised and bleeding, into the dust. They stumbled over broken furniture and fell, and were trampled under foot. Seized with wild and causeless panics they surged together backwards and forwards in the narrow streets, cursing, threatening, imploring, fighting to get free. Liquor flowed like water, for the saloons were broken open and despoiled, and men on all sides were to be seen frenzied with drink. Fourth avenue and Griswold street had emptied their denizens into the throng. Ill-omened and obscene birds of night were they. Villanous, haggard with debauch, and pinched with misery, flitted through the crowd, collarless, ragged, dirty, unkempt, these negroes with stolid faces, and white men who fatten on the wages of shame; glided through the mass like vultures in search of prey. They smashed windows reckless of the severe wounds inflicted on their naked hands, and with bloody fingers rifled impartially till, shelf, and cellar, fighting viciously for the spoils of their forays. Women, hollow-eyed and brazen-faced, with foul drapery tied over their heads, their dresses half torn from their skinny bosoms, and their feet thrust into trodden-down slippers, moved here and there, stealing, scolding shrilly, and laughing with one another at some particularly "splendid" gush of flame, or "beautiful" falling-in of a roof. One woman on Adams street was drawn out of a burning house three times, and rushed back wildly into the blazing ruin each time, insane for the moment. Everywhere dust, smoke, flame, heat, thunder of falling walls, crackle of fire, hissing of water, panting of engines, shouts, braying of trumpets, roar of wind, tumult, confusion, and uproar.

From the roof of a tall stable and warehouse to which the writer clambered the sight was one of unparalleled sublimity and terror. He was above almost the whole fire, for the buildings in the locality were all small wooden structures. The crowds directly under him could not be distinguished because of the curling volumes of crimsoned smoke through which an occasional scarlet lift could be seen. He could feel the heat and smoke, and hear the maddened Babel of sounds, and it required but little imagination to believe one's self looking over the adamantine bulwarks of hell into the bottomless pit. On the left, where two tall buildings were in a blaze, the flame piled up high over our heads, making a lurid background against which were limned in strong relief the people on the roofs between. Fire was a strong painter and dealt in weird effects, using only black and red, and laying them boldly on. We could note the very smallest actions of these figures—a branch-man wiping the sweat from his brow with his cuff and resetting his helmet, a spectator shading his eyes with his hand to peer into the fiery sea. Another gesticulating wildly with clenched fist brought down on the palm of his hand, as he pointed toward some unseen thing. To the right the faces of the crowd in the street could be seen, but not their bodies. All were white and upturned, and every feature was as strongly marked as if it had been part of an alabaster mask. Far away, indeed for miles around, could be seen, ringed by a circle of red light, the sea of housetops broken by spires and tall chimneys, and the black and angry lake on which were a few pale, white sails.

As many as a dozen different fires were raging at once; the flames on Wells, Franklin, and Market streets marched steadily toward the north-east, crossing Madison street, below Wells. But before they had reached this point, the Union Bank and Oriental Building were on fire, the Chamber of Commerce was seamed with thin wreaths of smoke, the low brick block opposite the Sherman House was ablaze, and the roof of the Court House was strewn with embers, each of which sank out of sight to be succeeded by ominous puffs of pale-blue smoke, slowly reddening.

It was this peculiar progress of the flames which lent to the great fire a distinctive and terrible character. The flames advanced like the advance of an army. Single Uhlans skirmished here and there far in front, then small detachments cut off the weaker and outlying forces, then well developed battles took place around the stout buildings, which stood firm like the squares of the Old Guard amid the rout at Waterloo, and finally the main body of fire came up and swept these solitary resisting eddies into the great general tide of ruin. So while the scenes in one street and at one hour might stand for those in the city generally and through the whole night, yet around each of the great buildings, as the Court House and the gigantic hotels, episodes of peculiar and thrilling interest took place.

At the Court House the fire had communicated with the roof and dome several times, only to be extinguished. Finally it caught such a hold that the tower had to be abandoned. The great bell, which had been clanging fitfully all night, now kept up one incessant rattle, the machinery having been set by the keeper as he descended. The buildings on all sides were in flames, and the streets filled with the ruins of fallen walls. The prisoners in the County Jail, almost suffocated with smoke, ran to the doors of their cells and shook the iron bars with the strength of frenzy, uttering dreadful yells and imprecations of despair, as a horrid fear that they were to be burnt alive possessed them. Captain Hickey, seeing that there was no hope of saving the building, ordered the cells to be unlocked, and in a moment the released prisoners, all bareheaded, many barefooted, rushed into the street, yelling like demons. A large truck, loaded with ready-made clothing, was passing the corner of Randolph street at the time, and in a moment the convicts swarmed upon it, emptied it of the contents, and fled to remoter alleys and dark passages to don their plunder and disguise themselves. Not all, however, escaped. Those charged with murder, except Nealy, accused of murdering a man on Canal street, were securely handcuffed and led away between guards, scowling and downcast. Meanwhile the bell still jangled, the flames lit up the faces of the great clock with more than noontide light, the building glowed without and within like a furnace.

Suddenly, when the hands of the clock pointed to 3.10, the dome sank a little, rocked, then fell with a tremendous crash and clang while a pyramid of red fire and black cloud towered up for a moment and then melted into the general blaze.

The Sherman House, with its hundreds of windows, resisted stoutly. The flames were around it and beyond, but it stood up majestic, its white walls rosy and its windows bright with the reflected glare. The roof and woodwork were smoking in places, but for nearly an hour the house held good. Suddenly a spurt of flame came from a window in the third story on the southern face, another and another followed, and in twenty minutes, from every window hung out a red festoon, while great coils of black smoke twisted around the eaves and met above the roof with the flames already bursting through. Then all was over, and people could only watch it burn.

It was broad day now, and the sun was up. At least a small crimson ball hung in a pall of smoke, and people said that was the sun. For the rest, all consciousness of the hour and date was lost. The wind had freshened, and the tumult increased. The fire had pursued its inexorable march in the van of the south-west wind across the south side of the river. Toward the west it had burned more slowly, and it was nearly noon before the distilleries at Madison street bridge yielded. The north side was already attacked in a dozen places. Of the south division, between State street and the river, all the slighter buildings had been wiped out, many of the larger edifices were in ruins, and a few of the stoutest were still ablaze, islands of fire. Streets and blocks were no longer distinguishable. The gap beween the ruins were, it is true, still filled with people, but they were not working to save anything. There was nothing to save, no place whence to escape. The tumult was still loud, but it was changed in its character. It was now the wailing of children seeking their parents, of mothers seeking their families, of men maudlin with liquor and stupefied with

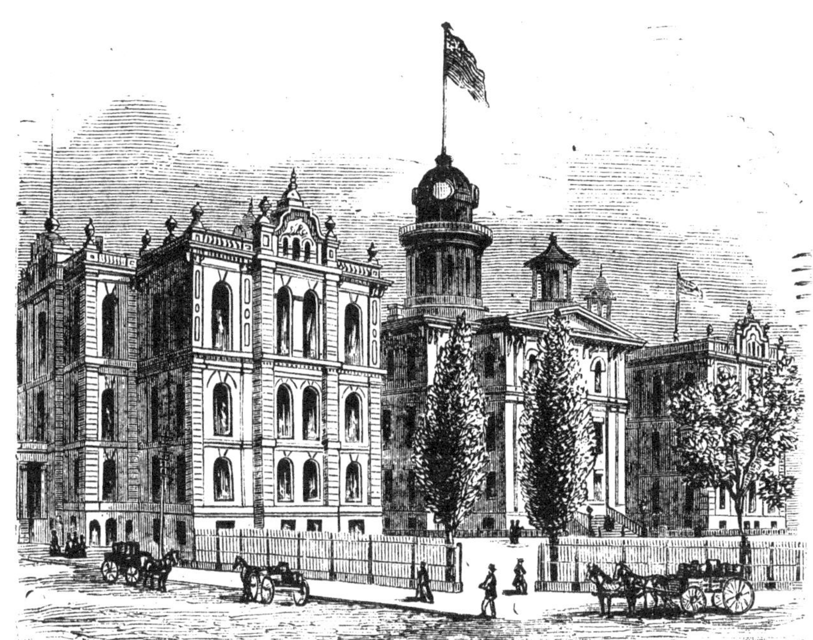

THE COURT-HOUSE.

THE CHAMBER OF COMMERCE.

SHERMAN HOUSE.

CLARK STREET, SOUTH FROM WASHINGTON STREET

grief bewailing their losses. The curious now pressed forward to see, and the dishonest to steal. These coming from the west and extreme south, met the throngs flying from the north, and made human eddies in every street. But the fire was practically over, the battle had rolled away to the northward, leaving behind it its ruins, through which poured the fugitive and the wounded, those who came on errands of curiosity or mercy, and those who prowled about to pillage and destroy.

ON THE EXTREME SOUTH.

That a fire of considerable proportions was raging on the West Side was known at ten o'clock on Sunday night to persons residing on the South Side, but the fact created so little apprehension that people sought their beds, and many never knew of the awful destruction until their usual rising hour in the morning. This, however, was not true of people living north of Twelfth street, for long before daybreak they were fully warned of the destruction which came upon most and threatened all. At two o'clock a reporter of *The Post* ran from his residence to Polk street bridge. The fire at that time had not crossed the river so far south, but to those residing between the river and the lake it seemed, from the flames, that the fire was immediately upon them. No one knew the extent the disaster had attained even at that hour; none would have believed it. From the bridge the West Side seemed all in flames. The crowd cried, Is the river a barrier? Will it stay the stalking fiend? The answer came from the flame itself. It did not cross the bridge, for that had been swung open, it leaped the river at a single leap, and caught in a hot and destructive embrace the lumber yard lying south of Polk street. So sudden was its crossing that numbers of persons standing upon the approach to the bridge narrowly escaped suffocation, and saved themselves only by a hasty retreat through the hot, black smoke that already swept across the street. On the north side were the old Bridewell buildings, which were being used as the headquarters of the First Precinct Police. The buildings were of wood. In a moment they were in flames. In the lock-up were twenty-five prisoners. The keeper opened the door and bid them run for their lives. They leaped from the crackling ruin and ran from death with a fleetness that they never displayed with a policeman pursuing. One prisoner was lying upon the floor stupidly drunk. The keeper could not rouse him. To Sherman street and Clark, to Fourth and Third avenues, to State street and Wabash avenue ran back the cry, "The flames are upon us! God alone can stop them!" That cry of horror woke every one to frenzied exertions, and, for blocks and blocks, the people who inhabited the houses did nothing but throw out furniture from the homes that they felt were certain to be doomed. The gas ceased to burn, but the fierce fire furnished a ghastly light by which every one could work. The streets were crowded by half-clad multitudes.

Frightened horses were hastily harnessed into wagons, and every one who could command a vehicle commenced to move. Hurried on by the howling wind, the flames spread northward and swept away block upon block of the wooden tenements which were crowded into that quarter of the city; but though the general direction of the fire was northward, yet the fierce heat fought in the face of the blast, and though slowly, yet surely, gained in the south. Running down Clark to Taylor, and on Taylor to the river, the writer found himself south of the fire. From Polk street the flame had eaten back until it had found Gurney's tannery, which, with its cords upon cords of dry bark, made a morsel that was soon devoured. On the West Side, the immense brick walls of the Chicago Dock Company's storehouse presented a formidable barrier to the further southward progress of the flames, but along the dock the sheds were burning. The framework seemed of harder wood than the coverings, for while the boards were rapidly consumed the beams were but slowly devoured. The framework fretted with fire looked like a golden grapery. Upon the building a stream from a single engine was pouring, but as well might one oppose the straw of a pigmy to the sword of a giant. Looking down the river, Polk street bridge was seen tumbling into the stream that quenched its burning timbers. Burning rafts floated upon the water. Tugs with steam up essayed to reach the brig Fontinella, which was lying at the dock near the burning tannery. Twice they made the attempt and twice fell back. A third was useless. The flames boarded her, ran up her rigging, cut her loose to float from the dock, and left her a blackened hull. The stone-yard of the Illinois Stone Company prevented the fire running southward on the river side, but the wooden houses on Wells street were quickly in flames. Looking northward, the street was a fiery vista. A lot of Norwegian emigrants were grouped about. They were stupid with fear, and had to be almost forced from the street. Returning as he went the writer reached the corner of Clark and Polk streets, where St. Peter's German Catholic Church is located. To it as to the sanctuaries in the old feudal times the people had crowded for safety. Its portals were piled up with the Lares and Penates of many a burning home. A block across, the flame was seen running up the golden cross that topped St. Louis Church. A moment later the church was in ashes. On the west of Sherman street, running from Taylor to Polk, from Polk to Harrison, and terminating on Van Buren street in the magnificent passenger depot, were the long freight houses of the Michigan Southern Railroad. Those who had the coolness to think thought that these would save the district east of them, a hope that could hardly be entertained in the face of the fact that the massive stone passenger depot was toppling into ruin; and yet these brick depots did save everything between them and the lake. A portion of the massive walls of the Pacific Hotel was seen to tumble, and to the East and North nothing was visible but crackling ruin, nothing heard but the roar of the flames which sounded just like the roar of the sea. It was nearly daylight. The water supply had given out, but no one in the south part of the city dreamed that the water had ceased because a mile and a half away the walls of the Water Works had tumbled upon the engines. People merely supposed that the fire engines had exhausted the supply. Even then the man who would have predicted the burning of the North Side would have been considered a madman. Anxious to see the situation down town, the writer essayed to proceed thither by Clark street. He could not reach Van Buren. State was open as far as Madison. Potter Palmer's buildings were tumbling in. Hissing and hurrying on came the flames. They laughed and crackled and roared with demoniac humor. Darting at huge piles of masonry they kissed them with fatal fervor, and rushing on with hellish appetite they embraced whole blocks of brick and marble, leaving them dust and ashes. Driven back on State street, the writer reached the Palmer House. Porters stationed at the doors refused entrance to any but recognized reporters. The Sherman was gone, the Tremont was in ashes, the Briggs had shared the common ruin, the massive Pacific was a red-hot ruin, the Bigelow in the next block was crackling; the question was, Shall we have a hotel left? And the people in the Palmer had the madness to believe that the Palmer would be saved. In half an hour it too was a shapeless mass of stone and mortar.

It was broad day. The wind had not lulled nor the fire ceased. On and on sped the flames in their hurried and horrible march of death and desolation. Strong men who loved Chicago better than they loved many a friend, bowed their heads and wept at her destruction. Terror was written upon the face of some; despair stared from the countenance of others. Many for the moment believed the last day had come. People prayed, and cursed, and hurried on, and at their backs was the ever-consuming, horrid hell of flame

It is proper to narrate how the flames were stayed in their progress southward. At the corner of Clark and Harrison streets the Jones school was burned. A wooden primary on the same lot escaped destruction. Why it escaped would be curious to know. The flames, as if weary of the awful race they had run, did not cross the street. At the corner of Fourth avenue and Harrison street the Jewish Synagogue burned fiercely, but the Otis block of brick buildings, on the northeast corner of the street, did not burn. At the corner of Third avenue and Harrison, men with chains pulled down a wooden residence which, though it was consumed, did not burn fiercely. At the corner of State and Harrison, O'Neil's brick block was blown up by powder, and prevented the further spread in that direction. At the corner of Harrison and Wabash avenue the Methodist Church stood as if defying the flames, and as though it uttered with the voice of authority, "Thus far shalt thou go, and no farther." The flames did not cross Wabash avenue south of Congress street, one block north of Harrison; and the south side of Congress was saved, the Michigan Avenue Hotel standing upon the corner like the huge battlement of a fortress that had withstood a siege. By noon the fire had ceased in its progress southward, and, except by uncertain rumor (and during all the fire many-tongued rumor spread its baleful tales more rapidly than ran the wild fire), no one south of Harrison street knew the desolation which reigned in the North Division. Nor was it known that the city's situation had excited the active sympathy of its neighbors, and that steam engines had upon the wings of steam flown to our rescue.

The lake front was filled with household goods piled in the utmost confusion. Weary watchers stood guard about their little all; and hundreds of people, homeless and without property of any kind, were lying about exhausted. The last was a grievous annoyance, but the roar of the fire was a positive terror which drove minor considerations from the mind. From the lake front, the destruction of the palatial block of residences known as Terrace Row was watched with intense interest. Its burning, although occurring in the day-time, when the spectacular effect of fire is greatly lost, was one of the remarkable scenes of the great tragedy. If it alone had burned, all the rhetoric at the command of the writers on the press would have been used in its description.

IN THE NORTH DIVISION.

The citizens of the North Division, up to three o'clock on that terrible Monday morning, put their trust in the river and Providence, hoping that their side of the city, at least, would escape. This was not to be. The rolling Hudson itself could hardly have stayed that tempest-driven tide of flame which was hurled irresistibly to the main branch of the Chicago river. Already, at three o'clock, the court-house bell had tolled the funeral requiem of Chicago, the gas-works had exploded, the hotels had succumbed. The air was hot with the breath of fiends, and the fiery brands that crossed the city on the wings of the storm obscured the stars above, and rendered blood-red the flood beneath, while they rained a lava-shower on the roofs of dwellings, factories, and storehouses—a shower that to describe would need the pen of the great novelist who has chronicled the desolation of Pompeii. Ere yet the bridge-railings on the south side of the river had ignited, North Water street was blazing, almost along the entire line. The terror on the North Side now became a panic. The thousands who had crossed the river to see the fire in the West and South Divisions, came pouring back over the bridges and through the tunnel, all hurrying to their homes and friends—all flying from the furious enemy that roared and howled behind them. The noise of the exploding material used in blowing up houses in the track of the flames reminded one of the booming of heavy siege guns, and the commune and the reign of terror were being realized in the very heart of the Garden City of the West.

Wells and State street bridges were caught by the flames, and were soon enveloped by them from one end to the other. La Salle street tunnel drew in the mighty volume of flame from the south, and became a submarine hell. With electric velocity the flames seized upon the frame blocks fronting the river on the north, and leaped from square to square faster than an Arab steed could gallop. The brands formed a kind of infernal skirmish line, feeling the way for the grand attack. The storm howled with the fury of a maniac, the flames raged and roared with the unchained malice of a million fiends. Nothing human could stand before, or check these combined elements of annihilation. They defied man's greatest efforts, and appeared to be kindled and fed by the arch-demon himself.

When the fire had passed Kinzie street the terror was something indescribable. Every available means of conveyance—wagons, buggies, drays, carriages, hacks, and even hearses—were used to convey from danger the terror-stricken people and such household goods as they could bear away. Thousands, hastily summoned from their beds, escaped from their already burning homes in their night-garments. The Nicholson pavement in the streets was on fire in every direction. The flames did not advance in a solid column as on the south side, but broke into sections, starting conflagrations here and there, while the great main fire rushed upon what was left, and made havoc of the whole. The fire spared one corner of Kinzie street, a few houses between Market street and the bridge, one elevator (Newberry's), a few lumber yards, and a coal yard or two. With this exception it swept along the North Branch to the gas-works, taking every stick and stone that lay in its line. If it forgot anything by accident, it would return like an unsated hyena, and lick up the miserable remains. It did not take a regular course on the north side. Some streets were ablaze half a dozen squares ahead of the big fire. It worked with the wind and against it, with a frightful impartiality. It held a direct northward course to Division street bridge, near the gas-works, where there are some large vacant lots, rather damp, and without any combustible surroundings. At this point it took an oblique turn eastward, toward Lincoln Park, leaving the Newberry School on North avenue, and sweeping along to Lincoln avenue to Dr. Dyer's new house, where, on that side, it halted, having burned itself out. It left a couple of frame buildings in front of the park entrance, sparing the fine park itself, hardly a shrub being injured. Not so with the old cemeteries, Protestant and Catholic. The grass on the graves was burned, the wooden crosses were consumed, and the gravestones were splintered into dust. The trees were withered like dry leaves, hardly a skeleton remaining, while furniture piled there for safety by the earlier fugitives only served to make a funeral pyre. The very pest-house, down on the lake shore, was burned to the ground, the miserable patients being obliged to seek in the water the fate from which they fled. The affrighted fugitives in the cemeteries fled madly towards the park, while the air resounded with their cries and lamentations. Meanwhile the conflagration swept eastward to the lake, taking everything that lay before it. By this time daylight was beginning to dawn, and with it the great water works, the pride of the city, were discovered to be charred and unrecognizable ruins.

To describe this fire in its details through the North Division would be utterly impossible. It was like a battle, where all was din, smoke, confusion, and turmoil. Each individual of the vast, fleeing tide can tell a different story of peril and escape. Before that awful front of flame the streets yet unburned were packed and jammed with myriads of human beings of every age, sex, and condition. It reminded one of a disastrous retreat, the baggage blocking up the highways, while the very horses were burned to death beneath the loads of household goods crowded upon their wagons. Hundreds of the affrighted animals ran away, mad with pain and terror, crushing in their flight men, women, and children. The principal lines of retreat for the north side community living west of Clark street and north of Oak street were over Erie and Indiana street, Chicago avenue, and North avenue bridges. They retired to the prairie in the neighborhood of the rolling mills, or else took refuge with their terrified and trembling friends in the West Division. The North Side, taking a line from Canal street north, was completely annihilated. The little portion that escaped belonged more properly to the north-western section.

On Erie street and Chicago avenue the loss of life was fearful. The bridges were choked with fugitives and baggage. The wagons became entangled, and the frightened people either plunged into the river and were drowned, or else fell down never to rise, suffocated by the frightful smoke. The scene was enough to unnerve the stoutest heart.

Through the hellish splendor of mingled gloom and fire the tall church steeples loomed proudly against the fiery firmament. The first spire that went down was that of the Holy Name—Roman Catholic—Church, on State street. The crash was fearful, and was only exceeded by the terrific noise produced by the falling of the North Presbyterian Church, on Cass street, a moment later. It was a sad sight to see the beautiful little church of Robert Collyer succumb to the pitiless enemy; and the hardly less beautiful German Catholic Church of St. Joseph met the same untimely doom. And sad was it to see the fine rows of stately trees, which formed the shade of the North Side streets, go down like grass, withered and blackened. The marble can be replaced and the stone can be laid afresh, but many a long year must pass ere we shall see again the maples and poplars and elms.

Those of the North Side inhabitants who lived in that section lying between Clark street on the west and Lake on the east, and between Chicago avenue on the north and the river on the south, were the last to suffer. They expected that the flames would pass by them, as they had already burned up to the Newberry school before Rush street was engulphed. This hope, like so many others, was doomed to be of short duration. Very soon the cry arose that Rush street bridge was burning, while the large reaping machine factory of C. H. McCormick was discovered to be a blazing ruin. Presently the old Lake House, built in 1837, and situated on Michigan, near the corner of Rush street, shot up a column of flame, which proclaimed that the fiend had seized upon it.

This was the signal for a general stampede. The roughs that infested the lower streets, near the river, broke into the saloons and drank what liquor they could find. Many of these ruffians were draymen and wharf-rats, and their conduct was ruffianly in the extreme. Hell seemed to have vomited these wretches forth as fitting denizens of the fiery air around them. The robbers broke into and sacked many houses, the inhabitants thereof being only too glad to get away at any price. Retreat to the north was cut off, for already the flames had fired the water works and were burning the pier at the foot of Superior street. The destruction of Rush street bridge precluded a southward flight, and, besides, the South Side was one ocean of fire. Everything was burned on a line with Rush street, and that was already beginning to go. Language cannot portray the scenes that ensued. Everything was placed on some kind of vehicle, horses were let loose from their stables, children were flung into carts with their half crazy mothers, the lower orders were raging drunk, while the respectable people were wholly demoralized. For a time it looked as if the final day had come for all these thousands, for the fire was rushing down upon them like an avenging spirit. On most faces was depicted terror; on the fewer calm indifference or detestable brutality. Women cried out for aid to save their little ones. Their entreaties were disregarded, or else were made the theme for ribald jokes by the inebriated ruffians from the purlieus of North Water and Kinzie streets. Happy were those women and children who had husbands and father to protect them. Where were all these affrighted beings tending to? The cry of "To the sands! To the sands!" was heard on every side, and to the sands everybody fled as by common intuition.

The "Sands" have long been notorious in .the annals of the city. They used to be infested with the vilest of vile rookeries until long John Wentworth, when he was Mayor of Chicago, became a justifiable incendiary and burned them all out. Since then they have been almost deserted. They are that portion ot the lake shore lying between St. Clair street and Lake Michigan, and between the North Pier and the Water Works. A more desolate place could hardly be imagined. The sand there has been drifted into small mountains, which half conceal knots of miserable shanties, wherein the Arabs of the North Side used to dwell. In most parts these houses reached nearly to the water's edge. In a few places there was an extent of some hundred yards in width. The place might have been comparatively safe from the fire, only that at the foot of Erie street was the large wooden bath house, dry as tinder, and along the southern section, toward the pier, stretched an immense varnish factory, an oil refinery, and a long range of sheds in which pitch and tar were stored in large barrels. All this made the situation anything but pleasant, and very far from secure. All the space unoccupied by houses and lumber was, on that eventful morning, crowded with trunks, bedsteads, mattrasses, pianos, chairs, tables, bundles of clothing, feather-beds, people, horses, wagons, and almost everything that goes to make up a large city; besides there were numerous barrels of whiskey which had been rolled down from the hell shops further up by the dissolute wretches.

Day was just breaking when the conflagration had reached the edge of the sands. The gale continued to drive with fury, and the sand and smoke combined to pelt the very eyes out of the wretched thousands crowded on that desolate place. Soon the smoke became so dense that the sands were dark as at midnight. The strongest constitution could not look that wind in the teeth and remain alive. The people fled down to the very water, while the flames burst through the dense smoke and leaped after them. The fiery brands fell amid the furniture and bed clothing, soon setting the entire shore in a blaze. Hundreds of horses broke from their owners and ran into the lake; the wagons, which were run into the water for safety, took fire where they stood, and burned to the water's edge. Scores of horses perished in the waves, which, even against the wind, leaped upon the shore like mad things of life.

At nine o'clock on Monday morning, sixteen hours after the breaking out of the conflagration, the varnish factory and the rest took fire, raising a wall of flame between the people and the west. All now gave themselves up for lost. The brands came down by thousands, causing the water to hiss where they fell. The clothes of women caught fire from this fatal shower, and one old woman, named McAvoy, was burned to death before she could be rescued.

The smoke grew more dense every moment, and the sense of suffocation was dreadful. Women screamed in utter despair, while the poor children were stricken mute with terror. A number of people were smothered at the bath house. Thousands threw themselves on their faces in the hot sand, while hundreds rushed into the lake up to their necks. The final day could not have brought more terror with its dawn. The great fear was that the north pier itself would go, in which event hundreds, if not thousands, of people must have perished. Fortunately, between the varnish factory and the foot of the pier there lay a broad expanse of sand, and the people on the pier used their hats and a few buckets to extinguish the brands that continued to fall upon the structure. At eleven o'clock that morning the factory was burned out, the pier was saved, and the people began to hope. There was no food and no prospect of any. Five large steamers—Goodrich's—were standing out near the crib in the lake, and a score of steamers were lying to, under bare poles, watching the tableau on shore. Not a sail ventured to approach the sands. The afternoon wore away and the evening shadows were coming to lend a deeper gloom to the smoke-wreaths when a fleet of tug-boats, sent down by the Mayor, came to the relief of the unfortunates. Most of them were taken off and landed, up through the heated river, at Kinzie street bridge, while the others slept that night on the shore, guarding the few household articles that remained to them. The wreck of home comforts lay along that sorrow-laden beach, and some human beings lay there dead. When the sun went down that Monday night, the 10th of October, 1871, he set upon a waste of ruined homes, the lost treasures of grief-wrung hearts, all that remained of world-renowed Chicago.

CHAPTER XXII.

MEN are always anxious to search out the origin of things that interest and concern them. They spend their energies in the investigation of the origin of the human species, and some are even willing to trace their ancestry back to the monkey, or to lower animals. The old Scripture remark is verified once more—"How great a matter a little fire kindleth!" and we are reminded of the indefinite influence of trifles upon human destiny. To a very humble and mean source must we trace the fire that consumed the great city; and we confess that if God had any retributive design, He employed an instrument well calculated to humble our pride. The reporters are doubtless disposed to throw an aii of tragedy around what is commonplace, or to set forth by ludicrous description the comedy of the

CRADLE OF THE FIRE.

The *Times* said :—Flames were discovered in a small stable in the rear of a house on the corner of De Koven and Jefferson streets Living at the place indicated was an old Irishwoman, who had for many years been a pensioner on the county. It was her weekly custom to apply to the county agent for relief, which in all cases was freely granted her. Her very appearance indicated great poverty. She was apparently about seventy years of age, and was bent almost double with the weight of many years of toil, and trouble, and privation. Her dress corresponded with her demands, being ragged and dirty in the extreme.

One day an old man entered the county agent's office and asked that a load of wood be sent to his house, on the West side. On being questioned, he acknowledged the ownership of considerable property, but said he was no better off than Mrs. So and so, referring to the old woman. This remark led to further inquiries, when the agent learned to his astonishment that his supposed pauper owned the ground and the house in which she lived, and was besides the proprietor of a famous milch cow, which furnished enough of the lacteal fluid to supply innumerable neighbors. As a matter of course the agent at once cut off her supplies, and when he took her to task for having deceived him, the old hag swore she would be revenged on a city that would deny her a bit of wood or a pound of bacon. How well she kept her word is not known, but there are those who insist the woman set the barn on fire, and thus inaugurated the most terrible calamity in the history of nations. In justice, however, to the old lady, her own story is given.

On the morning of the fire she was found sitting on the front steps of her own house. Her attenuated form was bent forward, her head resting on her hands. She was rocking to and fro, moaning and groaning, and crying aloud after the manner of her countrywomen when in great trouble. At first she refused to speak one word about the fire, but only screamed at the top of her voice, "My poor cow; my poor cow. She is gone, and I have nothing left in the world." Finally she was induced to talk, and this is what she said: It had been her regular nightly habit to visit the stable and see if her cow was all right. On Sunday night, about half-past nine o'clock, she took a lamp in her hands, and went out to have a look at her pet. Then she took a notion the cow must have some salt, and she set down the lamp and went in the house for some. In a moment the cow had accidentally kicked over the lamp, an explosion followed, and in an instant the structure was enveloped in flames.

The house on the corner, owned by the old hag who had caused all the desolation, was untouched. It stood there yesterday, and it stands there to-day, a sad monument of the past. It rears its lowly front on the borders of an almost destroyed city, and is the only survivor of hundreds of neighbors like itself, lowly in appearance, but the all of many a working man. Alas! how miserable a monument it is, and how sickening the thought that it alone should escape the sea of fire!

The *New York Tribune's* correspondent thus immortalizes the humble scene: I have here before me six miles, more or less, of the finest conflagration ever seen. I have smoking ruins and ruins which have broken themselves of smoking; churches as romantic in their dilapidation as Melrose by moonlight; mountains of brick and mortar, and forests of springing chimneys; but I turned from them all this morning to hunt for the spot where the fire started. It is the greatest and most brilliant apparition of the nineteenth century—more reckless than Fisk, more remorseless than Bismarck. Some details of its early life might not be without edification. There may be lessons in its cradle and its grave. These were the thoughts that justified me in going to De Koven street, though the real reason was that I was curious to see the first footprint of the monster who had trampled a grea city out of existence in a day.

Nothing could be more ignoble and commonplace than this quarter of Chicago. I reached it by crossing over the long draw-bridge at Twelfth street, which was swinging gracefully on its pivot as I came. The streets were all filled with wagons loaded down with furniture, which exposed to the gaze of the loungers the broken life of the family. The air of the quarter was wholly foreign, and not quite reputable. Even the little church of St. Wenzel added to the Bohemian air of the district. German volunteers were guarding the relief stores from hungry Czechs, who would make irregular forays on the provisions. Both sides thought their dignity required they should speak English instead of their native tongue. "Keep your fingers von dem pretzels off, or you'll git a het on you." "Yes! I bet you got a heap o' style, don't it." These colloquies sometimes give us moments of conjecture as to the final doom of our language. I found De Koven street at last, a mean little street of shabby wooden houses, with dirty door-yards and unpainted fences falling to decay. It had no look of Chicago about it. Take it up bodily and drop it out on the prairie, and its name might be Lickskillet Station as well as anything else. The street was unpaved and littered with old boxes and mildewed papers, and a dozen absurd geese wandered about with rustic familiarity. Slatternly women lounged at the gates, and bare-legged children kept up an evidently traditional warfare of skirmishing with the geese. On the south side of the street not a house was touched. On the north only one remained. All the rest were simply ashes. There were no piles of ruin here. The wooden hovels left no landmarks except here and there a stunted chimney too squat to fall. The grade had been raised

FIELD, LEITER & CO.'S BUILDING, STATE STREET.

BOOKSELLERS' ROW, STATE STREET.

TRIBUNE BUILDING.

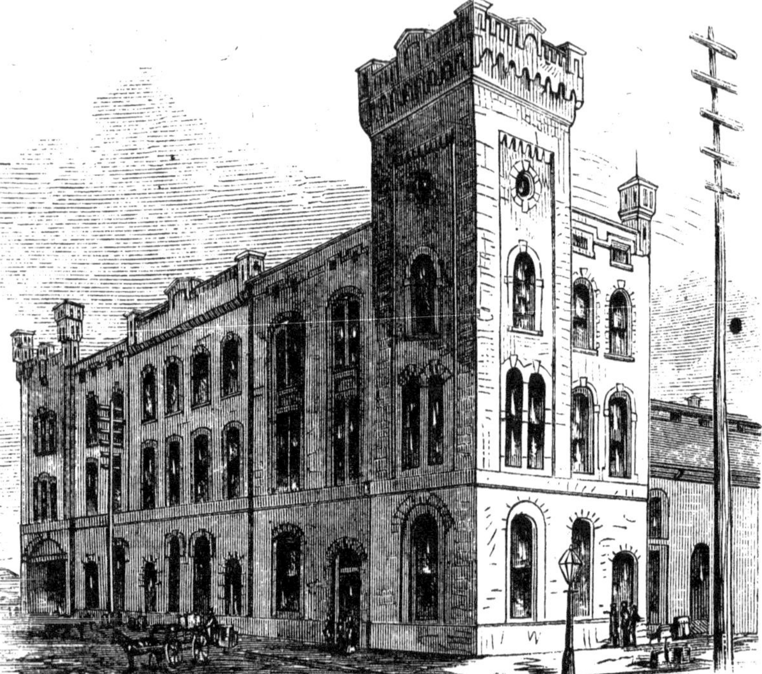
ILLINOIS AND MICHIGAN CENTRAL R.R. DEPOT.

in places and left untouched in others, so that now, as in the North Division, the roads seemed like viaducts, and scorched and blackened trees seemed growing out of sodded cellars. But of all the miserable plain stretching out before me to the burning coal-heaps in the northern distance, I was only interested in the narrow block between De Koven and Taylor streets, now quite flat and cool, with small gutter-boys marching through the lots, some kicking with bare feet in the light ashes for suspected and sporadic coals, and others prudently mounted on stilts, which sunk from time to time in the spongy soil and caused the young acrobats to descend ignominiously and pull them out. This was the Mecca of my pilgrimage, for here the fire began. One squalid little hovel alone remained intact in all that vast expanse. A warped and weather-beaten shanty of two rooms, perched on thin piles, with tin plates nailed half way down them like dirty pantalets. There was no shabbier hut in Chicago nor in Tipperary. But it stood there safe, while a city had perished before it and around it. It was preserved by its own destructive significance. It was made sacred by the curse that rested on it—a curse more deadly than that which darkened the lintels of the house of Thyestes. For out of that house, last Sunday night, came a woman with a lamp to the barn behind the house, to milk the cow with the crumpled temper, that kicked the lamp, that spilled the kerosene, that fired the straw, that burned Chicago. And there to this hour stands that craven little house, holding on tightly to its miserable existence.

I stood on the sidewalk opposite, as in duty bound, calling up the appropriate emotions. A strange, wrinkled face on a dwarfish body came up and said, "That's a dhreadful sight." I assented, and he continued in a melancholy croon: "Forty year I've lived here—and there wasn't a brick house but wan, and that was the Lakeside House, and it's gone now; an' av ye'll belave me, Soor, I niver see a fire loike that." I believed him thoroughly, and he went away. My emotions not being satisfactory from a front view of the shanty, I went around to the rear, and there found the man of the house sitting with two of his friends. His wife, Our Lady of the Lamp—freighted with heavier disaster than that which Psyche carried to the bed-side of Eros—sat at the window, knitting. I approached the man of the house and gave him good-day. He glanced up with sleepy, furtive eyes. I asked him what he knew about the origin of the fire. He glanced at his friends and said, civilly, he knew very little; he was waked up about 9 o'clock by the alarm, and fought from that time to save his house; at every sentence he turned to his friends and said, "I can prove it by them," to which they nodded assent. He seemed fearful that all Chicago was coming down upon him for prompt and integral payment of that $200,000,000 his cow had kicked over. His neighbors say this story is an invention dating from the second day of the fire.

CHAPTER XXII.

A City Sovereign in the golden West,
  But yesterday magnificent in pride,
To-day the wail of anguish from her breast
  Wakes echoes to each mighty ocean's tide.

A wail of anguish, rung out by the flames
  That licked her splendors level to the dust,
Aad blazoned hers the chief of ill-starred names
  That history holds in melancholy trust.

Her matchless miracle of sudden rise,
  That mocked at fable and enchantment's art,
Is peerless now no more in our sad eyes,
  That see her glories like a dream depart.

Her palaces were poems wrought in stone—
  Her marts, like Egypt's, for the world poured grain.
Her prairies girt her with a golden zone:
  Her fame seemed that of Carthage come again.

But Roman legions at Chicago's breast
  Hurled no red bolts that hapless Carthage rent;
In peace the hot cup to her lips was prest,
  And shrieking to her funeral pyre she went.

O day of horror! day of ruthless woe,
  That stripped the West's young queen of all her pride;
Her stately domes and lofty towers laid low,
  And 'whelmed her homes in terror's crimson tide.

Checked are the currents of her boundless trade,
  Her giant granaries smoke with smoldering wheat;
Her daughters, in her silks no more arrayed,
  Half clad and homeless, shiver on the street.

If of her magic growth her heart beat proud,
  And in her stones and stocks she took delight,—
If rivals lightly called her fast and loud,
  None grudge her tears of pity in her plight.

Proud, but beneficent, and fast to spend
  The easy gold her skill was swift to make;
Of arts and toil at royal rate the friend,
  And wisdom's lover for its own sweet sake.

Ah, luckless queen—her strength and beauty scarred
  She lies to-day on ashes for her bed;
And all the land in her despoil is marred,
  And all its joy in her despair is dead.

The East and West their eager hands stretch forth,
  To pour their wine and oil at her scorched feet,
In love and largess blend the South and North—
  A people's pain and pity swift to meet.

Her sons her crumbled greatness will rebuild,
  When the blanched terror flies their kindling lips,
And the glad glow of pride again shall gild
  Their Queen's fair face, now prone in foul eclipse.

W. C. Richards.

In accordance with our purpose to allow the reader to see this terrible panorama in all its length and breadth, under the best lights available, we place at this point—

THE TIMES' REPORT.

Hardly had the first alarm sounded than it was followed by another from the same box, and this in turn by a third, or general alarm, which summoned to that vicinity every available steam-engine in the city.

But all the engines in the country were powerless to have prevented the disaster which already seemed inevitable. The wind was blowing a perfect gale from the south-southwest. With terrible effect the flames leaped around in mad delight, and seized upon everything combustible. Shed after shed went down, and dwelling-houses followed in rapid succession. With a fierceness perfectly indescribable the fiery fiend reached out its red-hot tongue and licked up the dry material. Block after block gave way, and family after family were driven from their homes. The fire department were powerless to prevent the spreading of the calamity. The red demon of destruction was let loose, and in all his fierceness increased by a long restraint, it seized upon every destructible object and blotted it from the face of the earth.

At first it was one structure on fire; then another and another were swallowed up in a whirlpool of flames.

The wind continued its roaring, driving fierceness, and house after house was burned. To the left the fire spread forth its heat like the leaves of a fan until all of the eastern side of Jefferson street was enveloped in the furnace. To the right it had been driven with great fierceness, and Clinton street and Canal street, and Beach street, and then the railroads which run along the western shore of the south branch were in its grasp. Now was the fire at its fiercest. Upward of twenty blocks were burn ing. Upward of 1,500 buildings, including outhouses, were on fire. Upward of 500 families were fleeing from the seeming wrath to come. The streets were almost impassable. Carriages and wagons, and drays, and carts, and all sorts of vehicles were brought into requisition, and were speedily loaded with household goods. Empty wagons were filled with freight, and where there were no beasts of burden to draw the load human hands sprang to the rescue and dragged the property toward the north. Then the fire reached over the street, and while that terrible southwestern wind howled in mad delight, it forced its way into the planing-mills, and the chair-factories, and all the other shops which skirted the creek in that portion of West Chicago. Then it got into the lumber-yards, and into the railroad shops, and the round houses were soon wrapped in its dead embrace. The bricks themselves seemed only additional fuel. The rolling-stock in the railroad yards seemed but a bit of kindling which helped along a fire already fiercely intense.

But, worst of all, the elevators were next in danger. For a few moments it seemed as though one or two of the largest ones would resist the flames and pass through the fire ordeal unscathed. But this thought was not of long duration, for an instant later and the immense piles were in flames from top to bottom.

Like the advance of a great army the fire

MOVED FORWARD IN SEVERAL COLUMNS,

and like a well-whipped, but unconquered foe, the fire department slowly retreated. But they stubbornly contested every foot of ground, however, and would not surrender, although often almost entirely surrounded by the dread enemy. Then they would cut their way out and retreat for a short distance, only to turn again and hurl their charges of thousands of gallons of water full into the face of the enemy. But no power on earth could stem the torrent. Never did firemen fight more fiercely to conquer, and never before did their heroic efforts seem so utterly in vain. Polk street was reached, and here a desperate stand was made. One steamer, the Frank Sherman, stood at the plug on the corner of Polk and Clinton streets until the heat had scorched hair from the impatient horses' limbs, and the brave engineer and the plucky stoker had almost lost all their whiskers. Then the word was given to retreat and run. As they went the pipemen faced the foe and shouted to the driver to stop at the first plug and let them at it again. Hope street proved a sad misnomer for the firemen, and the poor folks who lived thereon, like those entering Dante's hell, were forced to leave all hope behind.

And now to add to the terrible reality of the dread scene it was discovered that a building was on fire away to the rear. Between Gurley and Harrison streets a barn was all ablaze, and before a steamer could reach the spot other barns innumerable were fiercely burning. It was the onslaught of a cavalry corps on the retreating army's rear, and all seemed hopeless. There was one thing noticeable, however, and worthy of special mention. The fierce wind had veered around toward the west somewhat, and now the fire was skipping some houses on the western outskirts of the block bounded by Jefferson and Clinton streets. To be sure there were not many of these escapes, but the fact was apparent, and it cheered the soul of every one. Every one seemed to think it would surely stop at the river, so far as the eastern wing of the advancing flame was concerned, and now that the western wing seemed willing to be lenient, it only depended on its front when a permanent check would be placed upon it. It was only about three blocks to Van Buren street, and here commenced the burnt district of the night before. No one supposed it would be able to go farther in that direction. There was nothing for it to feed upon. The four blocks of fire which had raged with such fierceness on Saturday night had left no supplies for the invaders, and its further march would either have to stop or continue over a barren desert. This latter could not be, and more and more hopeful grew the immense concourse of citizens.

Across Harrison street and Tyler street and along Van Buren street the monster ran, carrying destruction in its fiery course. At the approach to Van Buren street bridge stood the steamer, Fred Gund, a first-class Amoskeag engine, with a complement of officers and men in skill and daring second to none in the land. The steamer was completely surrounded by fire, and for their very lives the boys were forced to fly. They left their engine, but they have the proud consciousness of knowing she went down in a sea of fire with steam up and while fiercely fighting the advancing foe.

Here and there, and almost everywhere, lay thousands of feet of hose stretched to its utmost tension with watery ammunition, which the powerful engines were constantly throwing on the blaze. The fire had now reached what was supposed its limits.

TO THE NORTH,

illuminated by the great light of thousands of burning buildings, lay stretched out those four or five immense blocks of blackened ruins. It was not possible for the fire to continue further in that direction. It seemed hardly possible for it to reach across the river at this point. The width of the stream precluded such a thought. The wind was blowing the sparks and large firebrands toward the north and east, but, while all feared for them, no one supposed for an instant the sequel. The newspaper reporters, who had been from the first alarm fighting with fire and with human beings in the endeavor to obtain authentic information as to losses and insurance, and, failing in that, were only dealing in general results, hastened to their respective offices to "write up" the grandest blaze they had ever seen. Only one man was left to watch the final result and take to the office, as was then supposed, the going down of the fire. Blackened with smoke, with hair and clothing scorched, tired and thirsty, the weary reporters for *The Times* sought their carriages and were driven ever so fast to the office on Dearborn street, South Side. Hardly had they started, however, than away to the north and east, fully five blocks distant, a small flame broke forth and lighted up the already brilliant heavens. The sight sent an awful shudder to the soul of every man, woman, and child who saw it. For a moment every one was spell-bound and speechless. Just where it was, was as yet unknown; but it seemed to be in the neighborhood of the South Side gas works, and there was no one in all that vast concourse of people but who knew the great danger which was already threatening the other side of the river. Every moment witnessed an increase in the blaze, and presently the outlines of the immense reservoir told the story of its immediate vicinity. Fire-Marshal Williams at once sent every available engine to the South Side, and prepared to follow with the remainder immediately. But the flames mounted higher, and the fire grew fiercer, and spread itself out in all directions, until it was impossible to stay its further progress.

In the South Division as early as twelve o'clock the air was hot with the fierce breath of the conflagration. The gale blew savagely, and upon its wings were borne pelting cinders, black driving smoke, blazing bits of timber, and glowing coals. These swept in a torrid rain over the river, drifting upon housetops and drying the wooden buildings along the southern terminus of Market, Franklin, Adams, Monroe, and Madison streets, still closer to the combustion point for which they were already too well prepared.

The housetops were covered with anxious workers, and cistern streams, tubs, and buckets were in constant use to subdue the flying bits of fire that were constantly clinging to shingles and cornices.

Passing eastward over the Madison street bridge, at this hour, was an undertaking accompanied with the risk of suffocation, while once across, the hot wind tore so fiercely along the thoroughfare in question, as to wrench off signs and topple over sheds. The streets were now swarming in this portion of the city with the wretched people who had been driven from their homes by the fire in the West Division. A large portion of these were directing their way toward the North Side, and one of the most pitiable sequences of the continued conflagration was that hundreds of poor families were forced, on several occasions, from the places where they had vainly hoped to find rest, after having been burnt out before.

The writer, near the corner of Madison and Wells streets, aided a Swede in extinguishing a blazing pile of bed clothing which had ignited, as he was rushing along with his burden, from a brand of burning wood that might have been whirled through the air a mile or more. Several similar incidents were noted, and, in the frightful rapidity with which the clothes of the hurrying pedestrians and the more exposed portions of the smaller buildings took fire, a terrible premonition was afforded of what would be the fate of this portion of the city if the conflagration should but once obtain a hold within its precincts.

Van Buren street was soon crossed; the gale continued to increase; the air was flecked with burning cinders as high as the eye could reach; immense firebrands were carried for a distance of more than a mile, dropping them all over the eastern portion of the South Side, and then were the first misgivings felt that the destruction would not stop at the river—apprehensions destined but too soon to be fully realized.

The first foothold obtained by the destroying angel in the South Division was in the tar works adjacent to the gas works, just south of Adams street, and nearly opposite the armory. Almost instantaneously the structure was one livid sheet of flame, emitting a dense volume of thick black smoke that curtained this portion of the city as with the pall of doom. Faster than a man could walk the flames leaped from house to house until Fifth avenue (Wells street) was reached. A steamer or two were sent around, but their previous experiences were only repeated, and no perceptible check was given to the onward progress of the flames. From the gas works to the point it had now reached, nearly the entire space was filled with small wooden structures, and their demolition was the work of but a few minutes.

Apparently but a few minutes subsequent to the ignition of the gas works the wooden buildings south of the armory were found to be on fire, forming the apex of another widening track of desolation, and very soon joining with the other, the two uniting like twin demons of destruction, the armory helping to glut their fiendish cravings. Its massive walls soon yielded, and were tumbled into a shapeless mass.

It might be of interest here to note the peculiarities of the wind currents and their effects, which were such as could only have been produced by such a conflagration as is being described. During all this time, as during the entire continuance of the fire, the wind was blowing a gale from a southwesterly direction; and above the tops of the buildings its course from midnight until four or five o'clock varied but little, not veering more than one or two points of the compass. To the observer on the street, however, traversing the main thoroughfares and the alleys, the wind would seem to come from every direction. This is easily explained. New centres of intense heat were being continually formed; and the sudden rarefication of the air in the different localities, and its consequent displacement, caused continually artificial currents, which swept around the corners and through the alleys in every direction, often with the fury of a tornado. This will account partly for the rapid widening of the tracks of devastation from their apex to the Lake, as well as the phenomenon of fire—to use a nautical phrase—"eating into the wind."

The grand Pacific Hotel, upon which the roof had but just been placed, and which, like the still-born child, was created only for the grave, was among the first of the better class of structures assaulted by the fire. Angered at its imposing front, and scorning the implied durability of its superb dimensions, the flames stormed relentlessly in, above and around it, until, assured that it was at their absolute mercy, they left it tottering to the earth, and crawled luridly along the street in search of further prey. It was now that the waves of fire began to take upon themselves the mightiest of proportions.

How it was that while even a hundred buildings might be blazing, others, far in advance of the track of the storm, could not be protected, has not been understood by those who were not despairingly following the course of destruction. It was partly on account of the artificial currents already mentioned, and because the huge tongues of flame actually stretched themselves out upon the pinions of the wind for acres. Sheets of fire would reach over entire blocks, wrapping in every building inclosed by the four streets bounding them, and scarcely allowing the dwellers in the houses time to dash away unscorched. Hardly twenty minutes had elapsed from the burning of the Pacific Hotel before the fire had cut its hot swathe through every one of the magnificent buildings intervening upon La Salle street, and had fallen mercilessly upon the Chamber of Commerce. The few heroic workers of the police and fire department who had not already dropped out of the ranks of fighters from sheer exhaustion, sought to once more check the progress of devastation by the aid of powder. A number of kegs were thrown into the basement of the grand business palace of the Merchants' Insurance Company. A slow match was applied, and as the crowd drew back the explosion ensued. A broad, black chasm was opened in the face of the street; but with as little attention to the space intervening as though it had only been across an ordinary alley, the arms of flame swung over the gap, and tore lustily at the rows of banking houses and insurance structures beyond.

The Court-House was now faced with a swaying front of fire on the south and west sides. But as the building was in the centre of an open square, and solidly constructed, it was taken as a matter of course that it would be able to survive, if nothing else should be left standing around it.

"Talk about the Court-House," said a leading banker, among the spectators, whose own establishment had already been melted to the very foundations, "it will show to be about the only sound building on the South Side to morrow." And yet, in another five minutes, a great burning timber, wrenched from the tumbling ruins of a La Salle street edifice, had been hurled in wild fury at the wooden dome of the Court-House. As if a thousand slaves of the fire-king had hidden within the fatal structure awaiting this signal, the flames seemed to leap to simultaneous life in every part of the building, and soon the hot, smirched walls alone remained. The course of the fire was now directed almost due east for a few minutes, and Hooley's Opera House, the *Republican* office, and the whole of Washington street to Dearborn, was consumed. Crosby's Opera House came next in order. Renovations to the extent of $80,000 had just been instituted in this edifice, and the place was to have been re-dedicated that same night by the Thomas Orchestra. The combustible nature of the building caused it to burn with astonishing rapidity, and soon its walls surged in, carrying with them, among other treasures, the contents of three mammoth piano houses and a number of art treasures, including paintings by some of the leading masters of the Old and New Worlds. The St. James Hotel was next fired, and here, at the corner of State and Madison streets, the two savage currents of fire that had parted company near the Chamber of Commerce joined hideous issue once more. The course of one of these currents has been indicated. The other had swept down Franklin.

Wells, and La Salle streets to the main banks of the river, swallowing elevators, banks, trade palaces, the Briggs, Sherman, Tremont, and other large hotels, Wood's Museum, the beautiful structures of Lake and Randolph streets, and the entire surface comprised between Market, South Water, Washington, and State streets. Many lives were known to have been lost up to this time. But in the infernal furnace into which Chicago had been turned, it was impossible to conjecture or dare to imagine how many. The heat, more intense than anything that had ever been recorded in the annals of broad-spread conflagrations in the past, had fairly crumbled to hot dust and ashes the heaviest of building stone. What chance was there then of ever finding the remains of lost humanity by those who were already inquiring with mad anxiety for the missing ones?

But all thoughts of others soon began to vanish in fears for the safety of the living.

The stoutest of masonry and thickest of iron had disappeared like wax before the blast.

FIELD & LEITER'S MAGNIFICENT STORE,

second only in size and value of contents to one dry-goods house in the land, was already in flames. The streets were fast becoming crammed with vehicles conveying valuables, and the sidewalks were running over with jostling men and women, all in a dazed, wild strife for the salvation of self, friends, and property. The thieving horror had not yet broken out, and up to this time there had been a common, noble striving to aid the sufferers and stay the march of the demoniacal fire.

But now the sensation of weary despair, mingled with a grim acceptance of crushing fate, began to be noticed in the tones and doings of the populace. Liquor had flown freely, and from its primal nerving to heroism had passed to the usual inciting to recklessness and indifference. Thieves were beginning to ply their trade, and for once found more to steal than they could carry away; and express drivers and hackmen were charging atrocious prices ere they would consent to aid in removing goods from buildings thus far unconsumed. Hundreds of poor families were being rendered homeless, presenting pictures of squalid misery most pitiable. This was the first path that, like an immense windfall, mowed its way through the heart of the city to the North Division on the one hand and to the Lake on the other. Crackling and laughing demoniacally at the ruin and misery left behind, eager for more valuable prey, the flames sped on, taking in their course—the track continually widening from the causes mentioned above—Farwell Hall and the elegant stone structures surrounding it, and all the newspaper offices except that of the *Tribune*, leaving nothing behind but the grandest ruins the world ever saw. The reporters continued their work until what had been probable became a certainty—that *The Times* was doomed. It was then resolved to go to press at once, and, if possible, serve a portion of the subscribers, at least, with an account of the fearful calamity. The last words written were in the shape of a postscript, as follows:

"THE VERY LATEST—The entire business portion of the city is burning up, and *The Times* building is doomed."

The fire had already crossed Madison street, and it soon became apparent that the idea of issuing any copies of the paper must be abandoned. All efforts to that end ceased, and all endeavors were directed to the saving of as much as possible. It was too late, however, and comparatively little excepting the files were saved. The building caught fire in the upper story at about three o'clock, and fairly melted away under the intense heat to which it was subject. In half an hour nothing remained but a pile of smoking, smouldering débris.

The block bounded by Dearborn, Washington, State, and Madison streets was some little time in burning. Indeed, after the corner occupied by the Union Trust and Savings Institution had burned, it was believed that the vacant 150 feet front lot, created a short time before by the tearing down of the old Dearborn school, would save Mayo's corner and the St. Denis Hotel. But the fire, in spite of the terrible strength of the wind in the other direction, eventually contrived to beat up against the gale, and, by devouring the stores of Gossage and others, on the west side of State, and the book-houses of Griggs, Keene & Cooke, and the Western News Company, on the east side, to blister the St. Denis to the igniting point, and then McVicker's Theatre and the *Tribune* building formed the northern boundary of the South Division.

It was here that the few workers now left with courage enough to contest with miserable fortune made their final stand. The *Tribune* building was believed to be fire-proof, if any structure devised by man could be proof against such a combination of the elements as was now raging.

The Post-office had yielded to the assault and was only a smoldering ruin, and from away down to the devastated depot of the Illinois Central the flames had pushed back until they interlocked once more at the Custom-House with the fire that had torn its way from the Michigan Central Depot. Surrounded by the enemy on every quarter, and having held proudly up against the attack till long after daybreak, there was the same sad capitulations enacted here that had been the story of the entire night.

McVicker's yielded first, and was instantly a heap of brick and ashes, and the *Tribune* structure was not long in following, the walls of this latter structure, with those of the Custom-House, First National Bank, and Court-House, proving the most stubborn evidences of the worth of the architect's skill remaining in Chicago.

Up to this time the elegant and costly row of buildings on

Dearborn street, north of the Post-Office, had escaped. They included the two Honore structures, the Bigelow House, which was soon to have been opened, and the De Haven block, the latter extending from Quincy to Jackson street. The two blocks bounded by Monroe, State, Jackson, and Dearborn streets, that resting on Jackson street, including the Palmer House and the Academy of Design, were also intact. A new line of flame, however, had been formed some distance to the southward of the Armory and west of the Michigan Southern Depot, and was sweeping on in its mad, resistless career, and it was felt that the above-mentioned property was in the greatest peril.

The depot, a noble stone structure, upon which great reliance was placed for the safety of the adjacent property to the eastward, made but a feeble resistance, and soon, with a large number of passenger-cars inside, was in ruins. The large row of wooden tenements on Griswold street, fronting the depot on the east, succumbed at once, presenting a wall of fire of the length of the depot. It burned rapidly through to Third avenue, but at that point the wind, which had begun to show a changeableness it had not previously exhibited, veered to a point considerably east of south, in which quarter it remained for some time. Encouraged by this, a desperate fight was made on Third avenue, and for some minutes—minutes that seemed hours in the torturing alternations of hope and fear—the fiery monster was held at bay. The stone-yards on La Salle street also temporarily checked the progress of the fire south. Thousands of people occupying the large tract from Third avenue and Dearborn street to the Lake, watched the result of the battle that was to decide the fate of their homes with anxious countenances and bated breath. The wind benignly continued to blow from the same quarter, and the hopes that had been raised, slight at first, grew stronger. It was an awful crisis.

At no period in the history of that terrible day were more mo

PALMER HOUSE, STATE STREET.

THE SHEPARD BLOCK, DEARBORN STREET.

mentous interests trembling in the balance. The occupants of the Michigan-avenue palaces and the humble cottagers were there side by side, breathing supplications and agonizing prayers that their hearthstones might be spared. Many who read this were there; how futile the attempt to portray their feelings to those who were not.

Making a clean skip over the De Haven block, a shower of firebrands, hurled thither by a treacherous gust of wind, alighted on the roof of the Bigelow House, and that magnificent building was soon a seething furnace of flame, quickly followed by the two Honore buildings.

The one nearest the Bigelow Hotel was unfinished, but was rapidly approaching completion, and as a model of architectural beauty was hardly rivalled in the city.

From these buildings, as if maddened at their slight detention, the flames spread to the standing buildings west and southwest, with redoubled fury, enwrapping the block containing the Palmer House and Academy of Design, and that directly north, in an inconceivably short time.

The Palmer House was the tallest building in the city, being eight stories, three of which were comprised in its Mansard roof; and the scene of its demolition, which was more rapid than the account can be transmitted to paper, was inexpressibly grand. The march of the devouring element from this point to the Lake was uninterrupted, the intervening buildings, including many of the finest private residences in the city, melting away like the dry stubble of the prairie.

For some time after the ignition of the Bigelow House, the De Haven block stood unscathed, but, at last it, too, was forced to yield to the inevitable. It was a long three-story building, the opposite side of Dearborn street being occupied by a row of small wooden tenements. A stream was brought to bear upon these, and in the blistering heat three firemen, heroes every one, fully conscious of the tremendous interests committed to them, stood manfully at their posts. They did their work nobly and successfully. The De Haven block was levelled to the ground, and the whole row of wooden buildings had been perfectly protected. From a thousand parched throats the thankful ejaculation went up: "We are saved!" Delusive hope! One danger was averted only to be succeeded by others beyond the power of man to avert. The wind again suddenly turned to the southwest, carrying with it a baptism of fire which made it apparent that the whole remaining portion of the city north of Harrison street was doomed. Churches, palatial residences, everything was swept by the besom of destruction, an irresistible avalanche of flame.

In concert with the work of devastation just described, from the track of flame several blocks below, which had long before cut its way to the Lake, as if executing a well-devised military manœuvre, the fire had been steadily eating its way against the wind, the point of junction being at or near Adams street. From this it was evident that, even with the wind blowing a gale from the south, unless checked, the entire South Division was in danger. The supply of water had long before failed except from the basin, and more heroic treatment alone could save what remained of the city. It was at once and unhesitatingly determined upon, and then commenced the first systematic and thorough use of gunpowder as the only means of preventing the continuance of the work of ruin. It was conducted under the personal supervision of General Sheridan. Building after building was demolished, the reports of the successive explosions coming at intervals of a very few moments, and being plainly audible above the continuous din, each discharge announcing that at last the battle was being fought and won. The great fire which was to render Chicago forever memorable in the annals of history was ended in the South Division.

THE LAST BUILDING TO BURN

was "Terrace row," a palatial block of private residences on Michigan avenue, extending northward from Harrison street. Its destruction required two or three hours, as nothing remained in its rear to accelerate the work. About eighteen hours from the first discovery of the fire on De Koven street, the last wall of "Terrace row" fell. In the South Division, north of a diagonal line, reaching from the east end of Harrison street to Polk street bridge, there remained two buildings unharmed—one the large business block immediately north of Randolph street bridge, and the other an unfinished stone structure at the corner of Monroe and La Salle streets. The entire business portion of the city was obliterated. Two-thirds of the territorial area of the city was unscathed, but Chicago as a great business mart, the proud commercial centre of the growing West, was no more. Was ever devastation more complete?

Immense as is the burnt district in the South Division, for a single fortunate circumstance it might, and probably would, have been doubled. Immediately south of the Michigan Southern passenger depot was a long fire-proof warehouse; on the side fronting the fire there were but two windows, which afforded the only possible opportunity for the fire-fiend to effect a lodgment. These were successfully guarded by a small corps of men with pails. The building was saved, and with it undoubtedly the entire tract north of Twelfth street.

To complete the picture of ruin so vigorously painted already, we drop the *Times'* report here for a moment, and let another add a few touches with his gorgeous brush. The *N. Y. Tribune's* correspondent says: How can I give you an adequate conception of the vast and awful ruin which now occupies the entire site of the Chicago of a few years since? Standing at the Michigan Avenue Hotel, at the northeast corner of that avenue and Congress street, you look north along the Lake shore over nothing but ruins as far as the city extended in that direction, a distance of some six miles. A solitary grain elevator out on the pier at the mouth of the river is the only monument which remains on the Lake front. The eye utterly fails to take in the sweep of this field of ruin, even when you recall familiar knowledge of every foot of the ground. How *can* you make real hundreds upon two or three thousand acres of ashes, lime, and broken brick, where stood a day since a great city! Come back, then, to my spot of observation, the uninjured hotel just named. Directly before you was the large and elegant garden of J. Y. Scammon, and north of it a terrace of fine residences, among which were those of ex-Gov. Bross and Mr. Griggs, the well-known bookseller. All these went down before noon of yesterday, the fire spitefully beating back against a furious south wind, with a fierceness which made all South Chicago as fearful as if the hour of final doom had indeed struck. In several quarters during the morning there were amazing instances of this beating back of the fire, in consequence of the gustiness of the wind, and the ease with which the fire caught in all directions, in consequence of the excessive dryness of everything. The large empty corner occupied by Mr. Scammon's garden proved an opportunity to stop this on the Lake front; so Congress street became the southerly limit of the fire at the Lake front. This means a Lake front of ten blocks south of the river destroyed. Back from this front the solid business quarter of the city was built, eight blocks deep, every foot of which is down, with one or two slight exceptions on the extreme west of the district at the river bank. This is not all, either, that is down on the South Side. Going west from Michigan avenue, the southerly fire limit drops one block south to Harrison street, on Wabash avenue, and runs west on Harrison several blocks, and then on a diagonal southwest to the river and across, where, on the west side, in a tinder-field of dry lumber and exceedingly combustible buildings, an irresponsible cow kicked over the kerosene lamp which lighted all this disaster.

That unconcerned cow could not have chosen a point more admirably to the windward of the most solid and superb part of the city. It was at the close of a day of violent and really hot southwest wind, and that, too, after a month of most unusual dryness, when everything of wood, and especially everything of half-rotted wood, which abounds everywhere, was so perfectly dried that not petroleum itself could have made more entirely ready the destined victim of the fire-fiend. The danger, too, had come by stealth. The end of summer was really cold, though there was but little rain; but the latter half of September and the fatal first week of October brought constant, warm winds, under the pleasant softness of which field and forest and city became literally as dry as tinder. Chicago deceives any but a cautious eye. The ruin which defied the sea of fire most successfully is that of the First National Bank. On the site of this bank, less than four years ago, stood an old wooden house, so decayed as to be well-nigh ready to crumble into ruins. There is still a world of old pine in this condition in Chicago, where the original cheap structures are waiting until the lots are wanted at fancy prices, to cover with Athens marble, brick and iron. These vistas of decayed pine, dried to the condition of tinder, were the trains which fate had laid for firing our city. And every roof of the whole city, that even of the Water Works, which caught and burned before the great brewery near by was touched, had been put in perfect order for the swiftest and surest sweep of universal conflagration by the day and night steadiness of the southwest wind, and fairly heated for the match and the spark by the hot breath of Sunday's steady gale. And when the night of Sunday had closed in, without a vestige of moisture in the air, and fire broke out a little distance to windward of the costliest and closest square mile of Chicago, the end was as sure as if a fiend had prepared every inch of the devourer's path.

Half a dozen engines together, near the Court-House, had to be abandoned because of the rapidity with which the flames flew from point to point, minding no more about open spaces, streets, or squares than if they were carried over the distances between by so many trains of powder. One of the finest structures on State street, a great dry-goods house, seized in the rear, was seen to go down in barely fifteen minutes. The large hotels were bright spots in the burning, which raged from midnight to morning, and from morning to noon. The great book-stores, three standing side by side on State street, the finest single haunt of average book-buying in the country, and the store of S. C. Griggs & Co., exceptionally rich in all America in rare stock, were lapped by tongues of heat as many as the innumerable pages which shrivelled under the quick destruction, and all was gone. North and east of this point one solid mass of wholesale stocks, reaching to the depot and warehouses at the mouth of the river, crumbled into the maw of the easily-conquering doom. Taking in what lies outside of the district, ten blocks north and south by eight blocks east and west, a mile square of the very best of the city lies in ruins south of the short main trunk of the river, and between the Lake and the South Branch. This does not include the comparatively small district west of the South Branch, where the fire originated, and just north of which several blocks had been burned over on Saturday night.

The day of the fire was one of the worst which a dry and dusty city could experience. Beyond the limits of the fire was a frightful storm of dust and sand, blinding to the straining eyes of the hurrying throngs which filled the streets. It was a trifle of course compared with the other miseries, but it gave a dreadful added sense of the malignant character of the day. And now every wind that blows stirs a waste of ashes and lime, across which curious and sorrowing throngs tramp all day long, in and out among the remnants of brave buildings, over the charred pavements—never satisfied with gazing on a sight which perhaps may never be repeated. All accounts increase more and more the evidence of the most terrible intensity in the progress of the fire. The case of the Court-House, with the whole front of the block open on the south and the same on the north, suddenly bursting into a light flame, as if from oil easily ignited by intense heat, is as much in point as any. The fact was that the burning heat, which chipped the heaviest stone to such a singular extent, caused simultaneous combustion of large areas of exposed surface before any flames were actually communicated, or upon the first touch of flame at any one point. Among the tindery wooden buildings, which abounded especially on the north side, a rush of hot air—air that was almost red hot—would melt roof or walls as if they had been the lightest flummery. And these jets of heat went spitting about in the most capricious fashion, sometimes inexplicably avoiding an exposed corner, then returning to glean what remained. It was this in part which made so useless all efforts to head off or to stop the conflagration, though undoubtedly a more dreadful perplexity was to meet the shower of firebrands which were sweeping along on the heated gales. It was remarked on Sunday that pieces of burning pine fell on Saturday night two miles, or nearly that, from the fire of that night, and set fire to where they fell; and it was then said that it would seem as if a fire once under way in the city must sweep everything before it. The next twenty-four hours proved the justice of this apprehension.

The powers of the air defied interference, as soon as a sea of intense heat was created. On the south line of the burnt district the evidence is conclusive that the fire took all that was in its path, and took no more only from circumstances very little influenced by human intervention. The original fire burned east along the north line of the street which was its limit to the build ings of the Michigan Southern Railway, where the immensely long freight-houses, with the breadth of tracks west of them, proved a barrier which saved a large section of the city. Behind, or east of these freight-houses, is a row of peculiarly inflammable low houses. Happily the railroad buildings which were burned furnished less flying fire than that elsewhere, or the wind may have favored at the critical moment. At any rate, no fire took east of these freight-houses, while round the north end of the north one the line of conflagration went directly east along Harrison street to within one block of the Lake. On this block you still see where the work of demolition was commenced, but was suspended because the fire did not take hold of either the west or south sides of it. Along the line of Harrison street, mentioned just now, are two or three structures saved just as they stood, because the fire chanced to go round them. The easternmost of these is a church, north of which there was considerable vacant space, and west of which the houses were of brick, kindled from the rear and top, and burned out without very great intensity of conflagration. It becomes plain, therefore, that so much backing up of the fire as took place on Michigan avenue was only in conjunction with conflagration west of those blocks, which brought them under currents of fierce heat, and finally helped to destroy them.

Here we resume the thread of our former spectator's description of the fire in the North Division.

The four bridges on the main trunk of Chicago river fell an easy prey, but they were not needed to conduct the conflagration across, and speed it on its destroying way. The greatest number of easily combustible structures invited its progress in all directions, and so easily were new fires lighted far in advance of the general march of the destruction, that no regular line of fire front was preserved, nor did separate tongues of fiery advance, four or five of which existed most of the time, steadily hold their relative position. Now the burning terror would dart ahead a block or two in one place, and now in another, frequently giving less than time enough to the escaping population to put on necessary clothing. Great numbers, of course, were advised of the danger, and hurried their goods into the streets, to open squares, to the Lake shore, to any supposed place of safety—there to be burned, nevertheless, in the far greater number of cases. In all Chicago there were no finer private houses than great numbers of those here destroyed. The North Side was the earlier aristocratic quarter, and numerous elegant residences, with a rare charm of spacious grounds and fine shrubbery, maintained for this part of the city a New-England sort of charm not elsewhere to be found. All this was swept as if it had been a litter heap of tow and shavings.

The commencement of the fire on the North Side seems to have been at the Galena elevator, which is located on the north side of the main branch between State street and Rush street, the time when it first crossed over being about twenty minutes to six o'clock in the morning. Having once got a start to the north of the river, the fire rapidly progressed north, east, and west, the back fire west being unusually rapid. The corner of Rush and Illinois streets, three blocks beyond the elevator, where Judge Grant Goodrich resided, was soon reached.

The fire, then, as above intimated, progressed rapidly west, as well as north and east, first burning down the old Lake House, one of the oldest, if not the oldest brick hotel in Chicago. In its course west it also burned down, in addition to the other buildings, old St. James' Church, the oldest brick church in Chicago, which was occupied as a store-house. About this time, other portions of the North Side adjoining the river caught fire, and soon all North Water street, which was occupied by wholesale stores and large meat establishments, was in flames, the Galena depot, the Hough House on Wells street, and the Wheeler elevator west of Wells street, being also burned down. The bridges also were rapidly burned up, the flames from them helping to communicate the fire rapidly all along the north shore of the main branch. Not a bridge connecting the North Side with the South Side was left; Wells street bridge, Clark street bridge, State street bridge, Rush street bridge, all being burned.

The La Salle street tunnel also became impassable, the fire from the South Side rushing through it along the pedestrian walk, which was soon consumed, and filling the tunnel with smoke. At the mouth of the tunnel at the south end was found a dead dog, which had evidently met its death between a sheet of flame and a cloud of smoke issuing from the tunnel. The solid stone walls of the tunnel itself were cracked and chipped with the intense heat of the fire, the iron railings which protect the carriage approaches at each end being literally torn off from the walls and curved and bent into innumerable fantastic shapes by the fiery demon. Between Kinzie street and the river all was laid low and buried in a mass of undistinguishable ruins—wholesale houses, Uhlich's Hall, the Ewing block, the Galena depot, the offices of the Northwestern Company, at the corner of Wells and Kinzie streets, the Galena elevator, all were burned down in a miraculously short space of time. Between Kinzie and Illinois streets, from the North Branch to the Lake, nearly all was burned; among the prominent buildings consumed being the Revere House, on the northeast corner of Kinzie and Clark, the North Market Hall, one of the oldest buildings in Chicago, the Lake House, one of the oldest brick structures in the city, the mammoth reaper factory of McCormack & Co., a large sugar refinery, and an extensive coal yard; the last three establishments being located east of Rush street. The splendid new block, owned by McGee, on the corner of Michigan and Clark, was also burned down. A few fortunate buildings were left standing, but they only seemed to emphasize the ruins around them. These exceptions were about a block of buildings extending west from Market street to the North Branch, on the north side of Kinzie street, and a large brick building, occupied as a stove warehouse by Rathbone & Co., located to the south of Ogden slip, on the land which has been made between it and the slip, and which extends out into the Lake several hundred feet. A little to the east of the Rathbone building were several large piles of coal, which were burned up.

Between Illinois street and Chicago avenue the fire progressed with irrepressible fury and rapidity, soon enveloping the whole section, including in it both the most beautiful and the most forbidding portions of the North Division. On the west of Clark street and south of Chicago avenue was a section of the city densely populated; filled with buildinus occupied, many of them by two and three families; a region which in years gone by was noted for the disorderly character of its elections. Its only prominent features were a few churches, including the German Lu theran church, on the corner of La Salle and Ohio streets, and a Norwegian Lutheran church, built in 1855, on the corner of Superior and Franklin streets; the Kinzie school, a four-story brick building on Ohio street, between La Salle and Wells; the fine large structure known as the German House, dedicated last year, and containing one of the finest and best proportioned halls in the city. This portion of the city had, in fact, just begun to renovate itself; its streets were being raised and graded, and new buildings erected. East of Clark street to the Lake, between Illinois street and Chicago avenue, was the pride of the North Division. Its streets were bordered with rows of magnificent trees, beautiful gardens, elegant mansions, noble churches, all of which fell before the destroyer. Among the churches were the North Presbyterian church, an immense brick structure, on the corner of Indiana and Cass streets; a couple of frame churches on Dearborn street; the new St. James church, a beautiful Gothic stone structure, on the corner of Huron and Cass streets; and the vast structure of the Cathedral of the Holy Name, on the corner of State and Superior streets. Among the other prominent public buildings were the Catholic College of St. Mary of the Lake, occupying the whole block north of the Cathedral of the Holy Name; the Orphan's Home, conducted by Sisters of Mercy; the Historical Society's building on Ontario street, east of Clark, in which were kept, among many other valuable historical records, the original proclamation of emancipation by President Lincoln; and the North-side police station on Huron street, between Clark and Dearborn streets, a substantial and well-arranged building. Among the prominent residences were those of Mrs. Walter L. Newberry, whose grounds occupied the whole block bounded by Ontario, Rush, Pine, and Erie streets; that of Isaac N. Arnold, occupying the block north; that of McGee, occupying the block southwest of the Ogden block, etc. In short, this section of the North Division was full of beautiful residences and gardens.

Before tracing the progress of the fire further northward we may mention the burning of the water-works, and the curious, or rather incomprehensible manner in which it caught fire almost two hours before the time that the fire first reached the North Division across the main branch. As stated above, the Galena elevator at the edge of the main branch caught fire from the South Side at about 20 minutes to 6 o'clock. At about 20 minutes before 4 o'clock, a fire was discovered in the carpenter shop of Mr. Lill, built on piles above the shallow water of the Lake. The employés at the brewery immediately endeavored to extinguish the flames; but it was found impossible, and all the efforts of the men were confined to prevent their extension. Standing between the burning carpenter-shop and the water-works, extending northwest of the shop, stood one of Mr. Lill's book-keepers. Turning round toward the water-works, he exclaimed: "My God, the water-works are in flames!" This gentleman states positively that the flames from the water-works, when he first saw them, were issuing from the western portion of the pumping works, no flames being seen from the eastern portion of the grounds, which were occupied with coal sheds, etc. On the other hand, the employés at the water-works say that the fire commenced about half-past 3 o'clock in the morning; that it commenced in the eastern part of the water-works, and that it took fire from the shed. Another gentleman testifies that the carpenter-shop, or the cooper-shop, as he called it, was burned down before the fire commenced in the water-works, and that when the water-works were in full flame, the main body of Lill's brewery, with the exception of the carpenter-shop, was intact. The time of the commencement of the fire in Lill's carpenter-shop and the water-works, however, differs one hour; the last-named witness asserting that the water-works commenced burning at about half-past 2 or 3 o'clock. The gentleman referred to states that he had been to the Commissioners of Public Works several times to induce them to take precautions. But whatever may have been the origin of the fire at the water-works, it is certain that when it did commence the whole building was soon in flames, and in a few minutes the engineers had to rush out of the building to save their lives. The machinery was very considerably injured. The water-tower, however, to the west of the pumping works, was almost entirely uninjured.

Before relating the further progress of the flames northward, we must also notice the mingled scenes of sorrow and laughter, or tragedy and comedy, which were presented on what were once known as the sands—that part of the Lake shore which lies east of that portion of the North Side which has been described above. This sandy waste varies in width between one and two blocks, being the widest at the southern end near the river, where a frame building stood here and there before the fire. As soon as the fire broke out along the north side of the main river, and the rapidity of its progress showed that it would sweep the North Side or a considerable portion of it, all the inhabitants of the district described, lying east of State street—both rich and poor, both the tenants of the shanties and cottages which occupied North Water street, Michigan street, Illinois street, and the south end of St. Clair street, and the tenants of the aristocratic mansions north of this locality—fled to the Lake shore, carrying with them whatever they were able to carry in their hands, but little and but short opportunity being offered to do more. The scene was one of indescribable confusion, of horror and dismay, intermingled to the mere spectator with laughable incidents, which were, however, quickly drowned in the overwhelming horror which surrounded them all. Where the Lake shore or sands were narrow, and the burning buildings approached close to the Lake shore, despair reigned. The water was the apparent boundary of the place of refuge. The intense heat from the burning buildings, even the flames from them, reached the water and even stretched out over it, and the flying men, women, and children rushed into the Lake till nothing but their heads appeared above the surface of the water; but the fiery fiend was not satisfied. The hair was burned off the heads of many, while not a few never came out of the water alive. Many who stayed on the shore, where the space between the fire and water was a little wider, had the clothes burned from off their backs. The remnants of the sad scene presented a curious appearance on Monday. Scattered over the sands were broken chairs, shattered mirrors, drenched clothes without their owners, dresses, pants, coats, a motley array of clothing disowned. Boys wandered around picking out of the pockets of the deserted garments knives, change, etc.

Those again who lived west of Clark street in the district named, as soon as they saw that they must succumb to the advancing flames, after flying and moving north their goods from block to block, rushed across the bridges which, with one exception—that of the Chicago avenue bridge—remained standing. There was a grand emigration to the West Side of people and goods; of little children and big; of crying women and excited men; of broken furniture and cracked crockery; of wheelbarrows, buggies, one-horse teams, two-horse teams, heavy wagons, and light wagons—everything that could be saved.

But there was one bridge which proved unfaithful to its trust. Chicago avenue bridge appears to have caught fire from sparks before the main fire reached it. Thinking to be able to cross over this bridge, many people delayed their flight, hoping to save at least a part of their furniture before the flames reached their houses. But the delay was too long and the advance of the flames too rapid, and when they finally fled to the bridge it was too late. It was in flames. Under the approaches to the bridge the exhausted people tried to hide themselves from the flames, the stronger and less exhausted flying to the next bridge north—that at Division street. But the refuge under the bridge soon became a burning furnace. Those gathered under it soon saw the mistake they had made. The despairing ones stolidly stayed where they were, and were suffocated or burned to death. Those with hope still left ran out and attempted to fly north through the flames which were crossing the avenue. A few escaped, but with many it was only a death postponed for the space of a few minutes—burning garments, tottering footsteps, and then a fall to rise no more.

BORN ON THE STREET

As the fierce flames ran along the avenue, a woman ran out into the street, fell down, and gave birth to a child, but the birth soon became a death, and the mother and babe were soon lifeless bodies. In the mad hurry after each one's self, the mother and the child were deserted and left to their fate.

From the observation of many it would seem that the terror

and force of the conflagration on the North Side were aggravated by a fresh fire breaking out just north of Chicago avenue bridge at a time when the fire from the south had not advanced to within three or four blocks of Chicago avenue. It was this fire to the north that undoubtedly induced the weak and exhausted to take refuge under the approaches to the bridge, being unable to run around the fire to the north of the avenue, which was rapidly progressing both north and east. How many threw themselves into the river, with the vain hope of being able to cross the river or of being picked up, it is impossible to tell, but it is to be feared that in their mad and hopeless desperation many people in their flight from a death by fire, found a death by water.

SIXTEEN BURNED TO DEATH OR KILLED.

In a large blacksmith-shop, just south of the bridge, a number of workmen—stated to be sixteen—rushed into their burning building to save their tools, but the fire proved too much even for the sons of Vulcan. While catching up their tools, the walls of the building fell in and buried them in its burning ruins.

Perhaps the finest street running east and west in the North Division was Chicago avenue. Along its entire length, east of the river, it was filled with fine and costly buildings. During the present season alone several splendid buildings had been erected or were in process of erection. Among these were the building which was known, or to be known, as the Norwegian Hall, which contained, besides fifteen or sixteen stores, a large hall. The building had a marble front, and was nearly completed. To the east of this about two blocks, on the northwest corner of Clark street and Chicago avenue, was another fine marble front building almost completed. To the east of Clark street the avenue was filled with fine frame and brick residences. Among the residences on this street was that of the late Michael Diversey, the former partner of William Lill, and one of the earliest residents of

BURNING OF THE CH

OF COMMERCE.

Chicago, his house being perhaps the oldest residence of its size in the city. All these were burned from one end of the avenue to the other. Nothing was left but the water-works, themselves battered and torn by the devouring flames.

The surroundings of the water-works even were not without their tragedies. One of the firemen thinking, perhaps, that the heat of the approaching fire would not prove to be so intense and destructive as it actually was, crawled into a large water-pipe lying on the ground and was roasted to death. When fully awake to his mistake, probably all he saw at either end of his last refuge was a flame of fire.

North along Clark street, and on the branch tracks along Chicago avenue, Division street, Larrabee street, Sedgwick street, and Clybourne avenue, the horse-tracks were more or less injured; the tracks in some places being doubled up to a height of three feet. The tracks of the North-western road along North Water street, and extending between the government pier and the Ogden slip, were still more damaged, many of the ends of the rails being thrown eight or ten feet from their original position. In many sections of the track the rails have assumed a zigzag course.

At this time, between five and half-past five, the line of the fire as it progressed north was about a mile in width. Along the entire line the fire appeared as if attempting to see which portion could surpass the other in its march of destruction. To the east, near the Lake shore, were the large ale and lager-beer breweries of Sands, Hucks, Brandt, Bowman, Schmidt, Busch, Doyle, etc.; to the west, near the North Branch, was a densely inhabited district filled with wooden houses as dry as tinder. From the three, four, and five stories' height of the one, the sparks and burning charcoal from the wooden cupolas of the breweries were blown blocks northward, setting fire to the buildings on which they fell. On the west, the closely built wooden frame buildings, having no brick walls to temporarily stay

12

their progress, seemed to surrender instantaneously to the raging fire-fiend that did not crawl, but seemed to rush upon them with unrestrainable fury.

A TERRIBLE SCENE.

All seemed to be immersed in a hell of flame. No attempts were made to stem the progress of the fire. All that the tenants of the houses could do was to save a few of their household goods, and this, too, at the risk of their lives. The scene was rendered still more terrible and despairing by the fact that during the earlier stages of the fire thousands of the able-bodied men had rushed to the South Side to witness the fire there, not then dreaming that it would reach their own homes. Before the fire on the South Side, these fathers, brothers, and sons were gradually driven across the river, until the rapidity of the progress of the flames convinced them that their own families were in danger. Being at last convinced, they rushed in frantic haste to save what little they could. But they arrived at their homes, most of them, in an exhausted condition. They did their best, but the best was but little. All that many could do was to aid in saving the lives of their wives and children. With their all standing in their houses, many attempted impossible things, and rushed into burning buildings never to come out alive; for the wind rushed on in horrible fury, and seemed to envelop three or four houses at once in one fell swoop.

BETWEEN CHICAGO AVENUE AND NORTH AVENUE.

Until this densely populated district to the west of La Salle street, and between Chicago avenue and North avenue, had been wasted, there was no stay to the rapid progress of the fire. All that many people could do was to save themselves, and perhaps a few valuables that they could carry in their hands. A few, indeed, of those who saw beforehand that their homes would be burned down, even when the flames were half a mile off, saved, perhaps, half of their furniture; but many of these even were able to save but little. No conveyance could be found, in many cases, and piles of furniture were only saved from the house to be burned in the street. East of Dearborn street the scene was a parallel one; the homeless occupants of the houses in many cases rushed to the narrow beach which bounds this portion of the North Division on the east, and the same sufferings that occurred on the portion of the beach referred to south of this were repeated and aggravated by the narrowness of the beach. How many were killed, how many dangerously burned, it will be impossible to find out. Relatives and friends have not waited for the coroner, but have buried their own dead on their own responsibility, and no one person will ever know the names, or even the number, of the victims of the fire in the North Division. In the district mentioned, with the exception of La Salle street, Clark street, and Dearborn street, the population was densely packed. In many of the houses lived two or three families. To the east of it were large breweries, where, till the last moment, the employés worked to save the buildings, at last rushing to their own already burning buildings to save their families. Children, as is usual in poor districts, seemed to swarm around every building, and how many of these, left to their own care, infants, toddling children, little boys and girls, sank before the fire, it is impossible to estimate. Suffice it to say that hundreds have been missed who were seen at the fire, but never since.

The beautiful New England church went early in the day. Robert Collyer's stood defiant with its sturdy breadth and bigness, while behind and beyond it the conflagration did its will with everything else. There was some attempt to bring water in buckets from an open place, but it was not long before the vengeance which smote so mercilessly all around struck this noble monument also, and soon left the front and towers bereaved of all that made this one of the bravest and brightest spots in the whole city. In front of these two churches was Dearborn Park. North of this park a single residence was spared, almost capriciously and insolently. But from the wide scene of ruin, extending all the way across North Chicago, from the east bank of the North Branch to the Lake, the fury raged on to Lincoln Park, and far on between the park and the North Branch until North Chicago was almost completely blotted out.

On Dearborn street, diagonally opposite to the southwestern corner of Washington Park, was burned the New England Congregational church, one of the finest buildings of its kind in Chicago, and the most elaborately constructed of any ecclesiastical edifice in the city. The walls of the building stand. On the corner of Whiting and Dearborn streets, nearly opposite Washington Park, a block north of the last-named building, stood the beautiful edifice of Unity Unitarian church, of which Rev. Robert Collyer was pastor. The walls of this building also bravely withstood the advance of the flames; but it is to be feared that they will have to be rebuilt in order to secure a perfectly safe new structure. The whole length of Dearborn and La Salle streets, which from Chicago avenue to North avenue were two of the finest streets in the North Division, being lined with beautiful trees and splendid marble-front residences, were totally destroyed, not a house being left with the exception of that of Mahlon D. Ogden.

LINCOLN PARK AND OLD CITY CEMETERY.

These deserve special mention. Lincoln Park—the glory of the North Division—has been almost entirely preserved. But few trees have been injured except in the southeastern portion of the park, where the dead-house stood, and where a few trees are burned; the small-pox hospital to the east, on the Lake shore, being also destroyed. The grave-stone, or rather board memorials of the dead poor are many of them destroyed, and their relatives will know no more the place of rest of their kindred. The fences around the graves, the boards which have told to the wanderer their names, are all destroyed in the southern portion of the old cemetery. In the park itself many took refuge, though the great majority, as hereafter stated, fled to the prairies on the north-west.

North of North avenue no efforts whatever were made to stop the progress of the flames, with one exception, which will be hereafter mentioned. They followed out their course, the only means that prevented their progress both north and west being stretches of bare prairie, on which there was nothing to burn. Excepting on Clark and Wells streets, the houses were more or less separated from each other, occupying or being separated from each other by two or three lots, and often more. A small portion of the district north of North avenue and west of Wells street was thickly settled. At the corner of Linden and Hurlbut street stood the vast edifice of St. Michael's church. Its walls were left standing, but that was all. Its splendor is gone. A little church on the corner of Centre avenue and Church street, a branch of the New England church, was also burned, as also a German Methodist church on the corner of Sedgwick and Wisconsin streets; a little church on the corner of Clark and Menomonee, also the sub-police station on the corner of North avenue and Larrabee street.

At Fullerton avenue, a little over two and a half miles north of the river, the progress of the fire was finally stopped. A lull of the wind, between 2 and 4 o'clock on Tuesday morning, aided in the work of preventing the further progress of the flames northward; the only houses burned north of Fullerton avenue being Mr. John Huck's residence, and a building occupied by a Mr. Falk. Between the hours named, Mr. Huck's men turned out and beat out the sparks that came from the south as they fell on the ground. A slight rain falling at the same time, aided in the work.

During all this time, however, that the fire had been raging in the North Division, sometimes advancing directly northeast, sometimes progressing westward with a terrible back fire, people had been flying north and northwest until the few houses within reach in Lake View and beyond the limits were crowded full of refugees, and the flying population were compelled to take refuge on the open prairie. Here were gathered thousands of people—tired men, delicate women, children in arms without cover—without shelter of any kind; many indeed without clothes on their backs. Worse than all, here too were compelled to rest from their long-continued flight, the sick and the wounded.

The North-side horse-railroad stables were entirely consumed, and it is stated that over forty head of stock were burned up.

The boundaries of the fire in the North Division were as follows: With the exception of the few buildings mentioned above, the fire extended over all the North Division from the main branch to Division street, and from the North Branch to the Lake; very nearly seven hundred acres of territory. The fire left the North Branch at Division street, where it left a few houses standing along the side of the river. The back fire then extended to the river again, or to what is known as the North Branch canal, which connects the ends of a semicircle in the river, which bends over to the west. Following the canal or new channel of the river for a short distance, the fire then tended a little to the east as far as Halsted street, up which it extended to Clybourne avenue, the back fire extending along the avenue northwest to Blackhawk street and a little west until it reached Orchard street—a north and south street, excepting at its junction with the avenue, where it runs for about a block in a northeast direction. After reaching Orchard street, the fire proceeded north to Willard street, where it proceeded east along Howe street to Hurlbut street, across a couple of undivided blocks. Along Hurlbut street the fire proceeded north to Centre avenue, on which only three houses were burned down; the blocks around being nearly vacant. It then advanced up Hurlbut street to within about one hundred feet south of Fullerton avenue. In the meanwhile the fire had taken all east of this, with the exception of Lincoln Park. North of Fullerton avenue, the fire burned up only two houses; these being located east of Clark street. Here the progress of the fire was stayed in the manner stated above. C. Raggio's and two other houses on North Clark street, opposite the park, escaped destruction.

Here we part company with our guides, who have led us along the paths pursued by the hydra-headed monster, and turn again to hear the account of the

GRAVE OF THE FIRE,

from him who described to us its cradle.

Having seen the beginning of the fire, we thought it worth while to track it through its rise and its grandeur to its magnificent end after a glorious day's life. There is a very singular caprice of the fire in the North Division, equally remarkable with that in De Koven street. The house of Mr. Mahlon Ogden, a large frame building standing very near the street, is entirely untouched, while the entire region around it is laid bare. Even the church across the street, which stands entirely detached, is destroyed. The escape of the Ogden mansion is as complete and as mysterious as if it had worn an invisible coat of asbestos. The fire was no less singular in what it attacked than in what it spared. Just beyond this house, which would seem with its dry seasoned pine a most appetizing morsel for the fire-devil, there lies a green and tranquil grave-yard, with nothing in it which could attract a well-regulated fire. But this fiery tempest has swept in among these graves and tombstones, has sought out with an apparent disregard of conducting material, the humble wooden head-boards, and has even gnawed the marble in many places. The last expiring efforts of the flames were in the quiet German cemetery at the gate of Lincoln Park, by the shining beach of the Lake. It is here that hundreds of the hunted fugitives of the North Division, hotly chased by the fire, came to pass that first miserable night of hunger and cold. Loads of household goods were brought here, and dashed carelessly upon the ground. As the hard night wore on, and the cold wind came blowing in from the "unsalted sea," chilling the blood after the fever of the day, these unhappy people began to break up and burn the furniture they had saved, and brought so far with labor and pain. Everywhere you may see the traces of that wretched vigil of heart-breaking desperation. At one point there is a pile of half-burned picture-frames profusely gilded and elaborately carved, and at another there lie the scattered fragments of a richly inlaid cabinet. A library-chair has its back burned away and its upholstery wrinkled and singed with the watch-fire. But there are other and more revolting evidences of the misery which on that night gave many over into infernal guidance. I passed one modest grave, near the scene of a night-camp. A heart was carved upon the wooden tombstone by pious hands, and into this touching emblem a steel fork had been driven by some brutal fist. Above the outraged blazon were the tender words, *Ruhe Sauft* ("Sleep Softly").

The scenes witnessed in that quiet grave-yard during that night of horror were enough to appal the stoutest temperaments. A throng of half-maddened sufferers straggled through the grove looking for their friends and finding no one, oppressed by a weight of anxiety that caused them to neglect their physical discomforts. Delicate women came as they had escaped from death in thin fluttering night-clothes, blown about by the surly Autumn wind. Several were in a state which demanded the gentlest care and sympathy. Many little children were thrown into the crowd too young to speak their parents' names. And upon all, the crushing blow of an enormous and irremediable disaster had fallen, and rendered them for the moment incapable of any rational judgment. I heard of one company of German singers from a low concert saloon who flew out into the night with nothing but their tawdry evening dresses, who sat shivering and silent in a huddled group in the lee of a tombstone, their bare arms and shoulders blue and pinched, and the tinsel flowers in their hair shining with frost. They talked little, but sometimes they cheated their misery with songs, and it had a strange effect to hear in that gloomy and sorrow-stricken place the soft impurities of the Vienna muse, and the ringing and joyous jodel of the Tyrol. Near by, the fragments of a Methodist congregation had improvised a prayer-meeting, and the sound of psalms and supplication went up mingled with that worldly music to the deep and tolerant heavens.

The fire could get no hold on the green wood of Lincoln Park, and so gave it up and went furiously off to the left, and ate up all the pretty suburban houses on that side, and ended only when the wide prairie lay before it, with nothing more to burn. At the corner of Willow and Orchard streets the noble outline of the Newberry school bounds the line of devastation, as if to say that the future hope of Chicago, the power that shall yet rise superior to calamity, is Intelligence.

CHAPTER XXIII.

Thus ended what must be considered one of the most stupendous events of history, and the gorgeous descriptions above carry the reader, in imagination, onward from street to street, till darkness gathers upon the desolate scene, and the more desolate myriads who had been chased from their dwellings, and left roofless and almost penniless, many of them worse than beggars, because saddled with debts for property now hopelessly lost, and all securities utterly ruined. That night was the saddest ever experienced in our city—terribly gloomy for those who had not been burned out, and infinitely darker to the unfortunate. Everybody was thrown out of business, or had friends cast upon them for support or aid. The hungry were fed, the shelterless welcomed to a refuge, the naked clothed, and a general sharing of everything—an equal division—seemed going forward in every part of the saved district. Many people packed their goods and made arrangements to fly at the first alarm of new fires. Few slept soundly, even of the worn and weary. Children were in great distress, through the excitement of the day and the rumors that spread in wild profusion. The rain that fell was soothing to the mind and grateful to the eyes of those who were compelled to venture out the next day. Such dust had scarcely ever afflicted a people, and the smoke aggravated the visitation.

The presses were all lost, and there was an absence of any medium of reliable news. Correspondents are right in saying that "the wildest rumors were afloat, and people on the South Side were perfectly beside themselves with fear. The dead were multiplied into thousands; the fire was attributed to incendiaries; forty people had been burned in the Court-House; incendiaries had been caught in the act and thrown into the fire; vigilance committees had lynched others; men were dangling from lamp-posts everywhere; all the bank vaults had been burned out; the rest of the city was to be burned at night. The boldest robbery was still going on; organized gangs of thieves prowled through the streets laden with plunder. The police were worn out, and were worse than useless. Citizen patrols of the most ferocious character were firing off pistols everywhere. All along the northward progress of the fire there had whirled in uttermost confusion a throng of hurrying people, and of carts, wagons, carriages

—whatever could be drummed into the service to remove goods; and when night fell 75,000 to 100,000 people—north, west, and south—had either sought refuge with friends or were refugeless in the streets; and, added to all this, the city was wild with fear of what the night might bring forth; torches said to be ready to finish the destruction of the city; 1,500 thieves said to be organized for a raid of pillage upon the bank vaults, and whispers hoarsely breathed everywhere of fever and pestilence ready to fall upon a population left without water, with but short rations of food, with most insufficient shelter, and in the midst of loosened spirits of noxious evil stalking through the wide ruin; monsters of imagination evidently enough, and yet amply real to minds that could not possibly imagine a few hours before that any combination of effort could have burned to the ground the half that has fallen before the tumbling of one lamp into the litter of a stable."

If we dreaded the night, morning was, if possible, more dreadful still, for there lay the remnants of our lost city, and all around us were multitudes of dependent people and of wicked desperadoes. But the ground looked damp and the air was soft and mild, and the sun still shone in the heavens, reminding us of the ever-during mercy of Him in whose hands we were—"The Father of lights, with whom is no variableness or shadow of turning." It was well for us that our hands were so full of work for the miserable victims, for thus our own griefs were forgotten in the humane labors of relief, and our attention was diverted from those sickening ruins where lay the dead undiscovered, and the unopened smoking safes, and the wreck of all our city's greatness.

A ride over the burnt district from the little shanty to Lincoln Park, was more dismal than a walk through Pompeii, or an excursion among the wrecks of Paris, wrought by Communists from within, and Prussians from without. We leave a faithful observer to record what he saw in such a tramp.

Thursday, the third day after the fire, was clear, bright, and cloudless. The wind had died away, and I rode over the whole area of the disaster. There was no smoke or sign of remaining fire save in the great burning coal heaps along the river, or where mountains of smouldering grain were all that remained of the destroyed elevators. The fierceness of the flame had burned up everything combustible, and swept away the ashes as fast as consumed. The piles of crumbled masonry, hundreds of acres in extent, were even free from smoke stains. The streets were free enough to allow me to drive unimpeded. The Court-House is the most imposing ruin. Generally the larger structures are flat with the ground. The Sherman House *débris* are shapeless—almost level. So is all that remains of Field & Leiter's white marble store. The Pacific Hotel walls are one-third down, the interior totally burned out. The following costly buildings were designed to be fire-proof:—The Republic Life Insurance Company's building, Nixon's adjoining unfinished building, First National Bank, the Safe Depository, the *Tribune* building. Only Nixon's remains, it having been exposed to far less heat than the others. The rest are ruined. The late busy corners are almost undistinguishable, and old citizens contest the point as to whether this is Lake or Randolph, that Clark or Dearborn, until some familiar recovered landmark decides it. The only route to the North Division is across Lake street to the West Side, where we cross the North Branch at Indiana street, and drive northward three miles. We ride the whole distance on the raised grade of the Nicolson pavement, across a bare, treeless, vacant plain, and as we near Wright's Grove, we look southward and see from where we stand in our vehicle, the first and nearest unharmed structure, the Wabash avenue Methodist church at Harrison street, nearly four miles away. The elegant frame villa of Mahlon D. Ogden, in its wooden enclosure of an entire square, its graperies and wooden out-houses, is alone unharmed, an oasis in a wide desert. From the burned tract of nearly two hundred squares, every trace of combustion and combustible has disappeared. Even the turf burned up and its ashes blew away, leaving the naked soil.

The city will be rebuilt better than before. It will be a handsomer and a safer city than it could ever have been without this fire, but its purchase money strikes at the money centres of the world. Recuperation has already commenced, but it began in Chicago on Tuesday, in a city from which every public building, every newspaper, every power-press, all leading hotels, all but one wholesale store, eighteen churches, two great railway depot structures, six of its bridges, six large elevators, fifty vessels, and sixteen thousand dwellings had disappeared totally.

Using again the pen of the correspondent of the *New York Tribune*, we show what was transpiring by day, and how the scene appeared by night, as the time passed on. He is writing October 14th:—The town is beginning to fill with æsthetic sight-seers. The artists of the illustrated papers are seated at every coign of vantage, sketching for dear life against the closing of the mail. Photographers, alarmed by the prospect of speedy reconstruction, are training their cameras upon every unprotected point of picturesque ruin. They are sure of a ready sale of all the shadows they seize in these days. There has rarely been offered to the pitying admiration of men a collection of pictures of more poignant beauty. If one could divest himself of all feelings of sympathy and pain he could gain from these smoking squares the finest intellectual enjoyment. Monotonous as the gray stretch of desolation appears at first, the longer you look and linger the more this uniformity of character and color breaks up and reveals to you an infinite study of lines and forms. Of course, these ruins lack the consecration which has come with the course of ages to the splintered monoliths of Thebes and the gnawed plinths of Pæstum. But is there not an equal if not greater human interest in surveying these brand-new shards of a great city, and reflecting that the builders do not hide from our sympathies in the mists of immemorial time, but to-day live and breathe, think the same thoughts which found expression in these broken walls and melted columns, eat and drink and love and grieve and hope, and go on with work kindred to that which now has suddenly taken its place in the Past? Every one who has looked upon ruins has felt the keen, imperious desire to know what manner of men it was that built them and looked upon them when they were fresh in the sunshine of those older days. Half the joy and half the pain of travel is in this vain imagining. But here you look at these imposing wrecks, still Titanic and most impressive in a decay that already seems historical, and you reflect with a sudden feeling of surprise that you know by heart the sermon they are preaching. You are yourself a part of the life they symbolize, of the civilization which they express. You have heard the prayers and the oaths, the laughter and the cries, to the sound of which those walls went up. There is no unknown quantity in the problem they present. There it is—make of it what you will. If you come to nothing, do not blame time or history for the dust that is in your eyes.

Strolling through the town in the day-time, you see that it must have been a heat of singular intensity that melted down six miles of brick and mortar so soon into one undistinguishable mass. It took only about twelve hours to virtually finish the work; all that was done after that, was the after-wrath of the flame gleaning about the edges of the field it had reaped. But there has never been a fire which so completely attended to its business and slighted no part of its work. It seems like a mere figure of speech to speak of a quarter utterly destroyed. The phrase is always used about great fires, but usually means that all the houses are more or less damaged. In this case it is literally true.

Most of the houses are level heaps of calcined building material. The walls of the Custom-House are still standing; the Court-House wings refuse to fall. The fire-proof *Tribune* disdains surrender, though only a phantom house. A few heavily buttressed church towers wait also for the hammer of demolition. But with these exceptions, the central region of Chicago has ceased to exist. You can look through it to the far-off waste of the North Division. In many places the solid granite has cracked and peeled in great flakes, like stucco in the frost. The iron castings are partly melted and partly twisted into forms of startling grotesqueness. I have seen fluted columns, bell wires, gas and water pipes, wreathed and twisted among the smouldering ashes of a cellar like a coil of snakes of assorted sizes. Even the pretty gratings of the Safe Deposit Company, the best preserved of all, are fearfully warped and bent, like a character which has resisted temptation with a woful loss of temper.

These details we have been permitted to see for some days; for although the proprietors are eager to begin their work of reconstruction, the lack of water has thus far made it impossible to quench the smouldering flames. So that the light shimmer of the brooding heat hangs all day above the rubbish, and the air is full of the pungent odor of coals. When night comes a strange and beautiful transformation is wrought in the scene. Every evening since I have been here I have watched with increasing interest this marvellous and fascinating change. As the sun goes down in the prairie, and the night wind comes in from the Lake, this sleeping fire rouses and stirs in its slumber like a woman who shakes off the day's decorum, and flushes at the coming of her lover. The vast ignited coal-beds on the shore of the river throw red greetings to each other through the gathering shadows. The darkness slowly veils the lines of shattered walls, and one by one through the gloom twinkle out the delicate blue flames that spring from the anthracite coal-boxes of the burned mansions. They are so blue, and fine, and fragile, that they seem like forget-me-nots gemming the dusky field. They are very persistent though. They have been pouring tuns of water through the sidewalk upon one small deposit in front of Gov. Bross's residence, and yet at night it blooms as bluely and vigorously as if i were refreshed by the watering.

As the darkness deepens, the show increases in brilliancy, until, by a most lovely effect of reflection, the blaze from the unquenched fires strikes the clouds of smoke that hang over the city, and turns them a brilliant rose. The pillar of cloud becomes a pillar of fire, and all at once the dead lustre of this reflected light falls back upon the ruins and brings them out into pale and singular distinctness. It is not possible to imagine anything more terribly beautiful than this wild commerce of the fire and the darkness. From my window I see the whole sweep of the vast illumination. On the left a coal heap stretches beyond the river like a shore of fire ; a boat on this side is blackly painted athwart the blaze. The sky is flushed with the flame and mottled with driving clouds, and against it loom the ragged and torn walls of the Pacific Hotel, the sturdy arch of the First Presbyterian Church, and further to the right the broken outlines of the Court-House, far more reverend and graceful than ever in their forlorn incompleteness. All along the red horizon the coal heaps blaze and the sky is on fire, and the sharp angles of broken walls and the slim stems of black chimneys like minarets are drawn sharply on the crimson background. I do not know if it could be within the reach of painting to give any hint of the unutterable magic of this spectacle. No sunset was ever so rosy as that smoky sky. No frost-castle built on a window-pane out of a child's breath was ever more delicate than those fantastic ruins, flung like tattered lace against the drifting clouds. On the extreme right, just within the yellow blaze of the light that guards the breakwater, the great Central Elevator towers above the shore, shrugging its vast shoulders over the desolation, contem-

BURNING OF THE

OPERA-HOUSE.

plating its mirrored bulk by the flickering blaze of its fallen companion. After all this revel of form and color it is a relief to look to the east, where Lake Michigan lies in his night-dress of mist, and whispers as peacefully to the sandy beach as he did in the days before Columbus, and as he will in the days after Jonathan.

The crowning evil of all times of tumult and disaster is suspicion. We cannot burn witches now, nor tear out the tongues of Jews for imaginary crimes. But we can shoot old women for pumping petroleum if we are Parisians, and we can resurrect them in back alleys if we live in Chicago. That famous South-western verdict, which attributed a suicide to "accidence, incidence, and the acts of the incenduary," seems to have possessed the Chicago fancy; and though they do not positively hang or shoot their petroleum population, they say they do in their newspapers, and occasionally seize a shivering vagabond whom they find skulking on the sunny side of a barn, and drag him before General Sheridan for trial. Luckily this sagacious soldier has a cool head and an honest judgment, and insists on better evidence than poverty and dirt to hang a man, and the consequence is that not one case of incendiarism has been shown at headquarters. There have been two or three fires in regard to which the cry of incendiarism was promptly raised, but investigation at once made evident their accidental character. This general suspicion, however, has resulted in the establishment of an institution which is altogether laudable as long as the embers of the conflagration remain alive. A patrol of citizens has been formed in every block, and they all do sentry duty at stated hours. Every man out at night without cause finds it a little inconvenient to give repeated accounts of himself, and this of itself is promotive of the domestic virtues. The rule is certainly admirable in its application to that portion of the twilight population which always comes to the surface at such hours. In the day-time you may see them slouching about Wabash avenue, where their rascal faces and hang-dog air are never seen in ordinary times. It would certainly not be prudent to give the city up to them, and so at night they are kept in their own haunts. It is astonishing to see how simple and provincial Chicago has become. Standing sentry is positively the only recreation of men of the world. There are no clubs, no restaurants, no theatres, no libraries. There is no need of going out—if you go, a wall falls on you by way of warning. A little while ago, as I sat here writing, I heard a loud crash, and looking out, I saw that the high wall of Mr. Scammon's house had fallen. A furious gale was blowing from the south and roaring among the ruins. As I looked another wall came sprawling over the sidewalk. As the white dust rose and fled away with the wind, I heard a pitiful cry, "Help over dere! A man's got his leg broke." A dozen persons ran from the hotel and brought in a poor German who was watching the building, and had imprudently taken shelter from the wind under the wall. After he was safely bestowed I stood for a moment at the window looking westward at the fine arch of the Presbyterian Church, clearly and richly defined against the red glow of the sky. Full in my sight it tottered, parted with a dull report, and tumbled forward into the street. The gale increased in violence; the pale, shadowless light faded from the city as the wind drove away the illuminated clouds. The blackness of night, which had been hanging in the eastern horizon, swept in over the Lake to the town. The whistling wind was thick with lime-dust and sparks of fire. The blue flames of the anthracite burned more gayly, looking now like the witch watch-fires on some unusually tempestuous Walpurgis-night. A gentleman with a white cravat and a black face knocks, and requests, with the compliments of the authorities, that lights may be put out and windows closed. And so to bed, with a gale lashing the calm Lake into discontent, and the intermittent rattle of falling ruins, reminding one of an artillery battle between two absent minded armies.

CHAPTER XXIV.

In the Sacred Volume, the same incidents, scenes, and narratives are repeated under various forms, in order to give all shades of the important truths recorded, and impress all minds according to their different constitutions and conditions. It is necessary to read many histories in order to obtain just and adequate views of the course of events. One corrects another, or supplies what seems deficient in his story or estimate of things and men. In the accounts of some writers, the gas-works on the South Side exploded with noise and fury. Whereas the facts are these: When the fire seriously menaced the gas-works, to avoid an explosion, a sixteen-inch pipe was opened and the whole dischargod into the air. The wind carried it swiftly over the buildings, and the incendiary sparks set it afire, and in five minutes three squares were wrapped in a blaze. Thus everything conspired to give impetus to the work of destruction from first to last. All things seemed leagued in a fell conspiracy, and the efforts of man were almost powerless against the combined forces of nature, which wrought so eagerly together.

The *New York Independent* said "the great fire at Chicago need not have occurred if the firemen had been sober:" a statement either grossly unjust or frightfully significant. In order to do justice to the Department, we must let them be heard, and the verdict, based upon such evidence, will be more likely to accord with truth. Like the human face in its infinite variety of form and expression, every individual experience has some characteristic peculiarities. A gentleman telling his story, said to the writer: My house was as far from the fire, when I got home, as yonder brick building, a block away, one-twelfth of a mile, and yet I could not stand in the doorway, such was the violence of the heat at that distance. In speaking of his light insurance, he explained it, by observing that he did not consider his stock combustible, viz., marble for tombstones, mantels, and buildings; yet scarcely a whole piece survived the fierce heat, and his warehouse stood on the edge of the north side at the river bank.

A lovely Christian woman, who was in the heart of the burning fiery furnace, evidently realized the situation, at least in spirit, of the three worthies in Nebuchadnezzar's seven times hot oven, who had the form of the fourth with them, and so perished not, but triumphed through His grace. In describing her feelings as she fled, she said, she turned from flight and looked back upon the vast column of fire that swept adown the street burying all in destruction, and she thought of Paul's words —"as having nothing, yet possessing all things," and she seemed to herself, though stripped of everything and destitute at midnight, to be rich, because God was hers. My Father is better than His gifts, and He is still mine, blessed be His name! Her grief, though real, found its sanctifying grace, and out of all that burning she comes, as gold refined, shining and pure as a saint of God.

Many such true hearts were strengthened in their attachment to God. As a godly deacon said, I have my papers and my children—I am thankful. To him there was no such thing as despondency or gloom, for his treasures were laid up above the reach of the flames, and his hope did not consist in earthly prosperity, but in the mercy of Jesus Christ.

In sad contrast was the first utterance of a liberal minister as he opened his sermon among the ruins of his church edifice: I have nothing to thank God for. There can never occur such a crisis in any Christian's career, however dark; and adversity is the blessing most approved in the New Testament. Such seemed the prevalent view of the Christian people of Chicago. Mr. Moody, who saved from his library nothing except his Bible, not a scrap nor a book besides, was unchanged in the cheerful tone and temper which characterize his buoyant, believing heart. Said he, I asked myself, what shall I take? and I grabbed my Bible and ran out of my house.

Many men had their hair-breadth 'scapes and peculiar perils to encounter, either in rescuing their property, families, or neighbors.

Mr. George J. Read got together the firm's books and papers and put them in a bag to remove them to his own residence on the West Side, and offered men large sums to convey him and his valuables across the bridge. Finding time short and no one willing to aid him, he boldly proceeded to drag his load from the alley between Lake and Water streets; and, the fire drawing near, he chose Water street, and was making what haste he could, when a large mass of felt roofing came whirling down all ablaze and struck him fairly upon the chest. Quicker than thought he turned, so as to give the wind a chance to catch the burning mass, and send it flying away over the tops of the buildings across the street. By this sudden detaching of the incendiary felting from his person, he has no doubt he saved his life, as, in that hurricane, he would have been set on fire in an instant and perished there. He pursued his way amidst showers of fire and secured his precious treasure and reached his home in safety.

Mr. J. W. Goodspeed, the publisher, found himself encompassed with flames, in trying to get away from the store with his papers, which he fortunately took from the worthless safe, and, making a rush to break through, he was compelled to retire. Placing a handkerchief over his head and face, and measuring his distance, he leaped forward and reached a place of safety.

He tells how the wind poured the sparks down into the streets and narrow passages by which he and his father sought to make their way homeward from Lake near La Salle street, and whirled his chromos out of his arms through the air, almost prostrating them. They found an old cart back of their building, and loaded it with what few articles they could snatch from the clutches of the fire, and drew it some two miles in the night amidst the thronged avenues.

Mrs. Hobson, the milliner, carefully placed in a wagon her choicest goods, as many as she could collect at such a time, and, putting herself in the thills, drew her load down toward the Lake, where she hoped for safety. Stopping a moment to rest, she turned to her load—and it was gone; all had been stolen on the way, after her endeavor to save them. The powers of darkness seemed to be let loose to prey upon the people and turn human creatures into fiends.

A gentleman, who succeeded in getting a new carpet out of his dwelling, and removing it to a basement where he and his family took refuge, looked in vain for it the next morning. It was stolen. There was no mercy in the hearts of these plunderers.

A good deacon, trying to carry away his goods in wagons, saw a woman take up a valuable package and start off with her plunder, when he called to her and she laid it down. A moment after she repeated her attempt, and he laid hands on her. Again she took advantage of his momentary absence, to steal, and he, finding her obstinate, deliberately smote her with his fist, and she fell to the earth. This put an end to her depredations, and the church militant became the church triumphant.

A portion of the North Division was saved by Mr. Davis, who early saw that all was gone in the business portion of the town; and returned home to protect what little remained, his house, the shelter of his family. Procuring help, he dug three wells, and obtained water enough to wet the roof of his house and to keep carpets and blankets wet, by which all incipient fires from sparks were put out at once. He took a pail of water and a shovel and stationed himself where he could prevent the sidewalk and fence from burning. Being far out, the fire came to him late in the day. As flames would creep along the walk, he used sand and quenched them. Often the heat was so intense that he was obliged to wet his handkerchief from the pail, and breathe through that. He felt several times as if he must abandon his post, and allow his home to go down with the rest; but renewing his courage and moistening his face and hands, he continued to fight the fire till darkness set in on Monday night. While he still struggled with the devouring element, he felt a drop of rain fall on his cheek, the forerunner of the shower, and his grateful heart poured forth a shower of tears from his eyes. He could then retire and sleep with a sense of repose, and a consciousness that God had appeared for his deliverance.

Mr. Kimball, of the Michigan Central Railroad, was driven from his house towards morning, and fled to the beach, leaving choice mementos and collections. Many years ago, probably twenty-two, he was in India, and procured for a favorite aunt, who liked good coffee, a parcel of peculiar excellence. On a recent visit she gave him two pounds or more of this package of coffee, and he had determined that they would use it only on Sunday mornings for a luxury, as coffee like wine improves with age. That was burned too brown—probably scorched and spoiled. The birds were let out of their cages, and the books left to consume, and they seized what few things they could carry in bundles, and ran for dear life to the edge of the Lake. Here they stayed in a prison of fire and water, alternately wetting their faces and their handkerchiefs, through which alone they could breathe at times, and putting out fires that caught in their bundles from flying sparks. Seeing no other hope of rescue, Mr. K. and his wife made their way to the river, and stepped aboard the Alpena, which was in tow of another propeller, and rode out into the Lake three miles, where the boats anchored. There were sixty persons on board, and not a mouthful of food. The Lake was very rough, and, as a matter of course, it set them all cascading violently; from which condition they did not recover till an engineer came aboard and got up steam, and they were transferred to a large propeller that lay in the basin inside the dock by the light-house. Here they were generously provided with supper, at 11 o'clock at night, and tasted food for the first time since three the day before. They could not determine during that day, while they were riding at anchor, whether the whole city was burnt or not. They did see Terrace row on Michigan avenue in its conflagration, but the smoke was too dense and blinding over the water to allow any true knowledge of the extent of the destruction. Hundreds, and perhaps thousands, were similarly tortured by anxiety and doubt, until Tuesday morning, or late Monday evening.

A North-sider, worth a quarter of a million at the time of the fire, was glad to accept two pairs of blankets, as he said, "to keep the family warm." He had seven dollars in his pocket Saturday night, and spent two of that amount to pay a man for setting an article of furniture into the street, which was afterwards burned. His wife's and daughter's clothing on their wagon took fire and had to be abandoned. The latter became a mother that night in the basement to which they fled. His work-people clamored for their week's wages, or wished his assistance, as they were penniless. His safe was entirely lost. There was no bank open, and he was in straits such as press the life out of a proud man. How they survived till he could send into the country and make collections, and what they suffered, it were hard to tell.

A mother got separated from her two boys, and such agony as she experienced only mothers can realize. Through the bulletin in the church where the bureau of missing and found ones was kept, she learned that her jewels were safe in the little town of Austin, a few miles distant. Some farmer had picked them up, given them shelter, and reported them for the benefit of their mother if she were alive.

Mr. Holden reports, that when the throng was greatest about the First Congregational Church he saw a woman at a window beckoning earnestly to be admitted. Something in her appearance arrested him strongly, and he sent a policeman to bring her in through the crowd. In an hour from that time, perhaps, she stood by his side and explained that her husband, a German, was badly cut from his shoulder down to his waist, and had no attention. While she was telling her pitiful story, the poor woman fainted and fell to the floor, and was removed and cared for.

Some of the scenes that transpired about and in the fire were disgraceful beyond measure. The saloons were, many of them, thrown open, and men exhorted to free drinking needed but one invitation. Hundreds were soon dead drunk, or fighting and screaming; many thus fell victims to the flames, and some were dragged away by main force and rescued from roasting. Even respectable men, seeing that all was lost, sought to drown their misery by intoxication.

Would that more had been able to answer according to the hero of the following Chicago dialogue:—

"Well, Jim, are you burnt out?" Jim: "Not I; I don't drink."

We have too many whose very manhood is consumed by the "hot damnation," and stand like some of our blackened ruins, a mockery of poor humanity. Dr Goodwin tells us how the streets were here and there choked with the whiskey barrels rolled out of their hiding-places, and how they fairly ran, and were flooded with the infernal stuff. Why, there were quarters where, because of burst barrels and broken demijohns, the very air was drunk a square away. I remember down on Van Buren street, in one of the early hours of the fire, that while two or three of us were trying to help a poor widow save her little handful of stuff, we ran against a saloon-keeper hammering away furiously to tighten the hoops on a cask that had sprung a leak, and calling vigorously on the bystanders to help save his treasures; whereupon one of our Sunday-school boys mounted on a pile of barrels, and with a sly nod to me, set the spigot of a cider brandy-cask running; and I did not turn the spigot back, nor scold the boy!

But worse than this were the instances of theft and cold-blooded avarice which occurred and have come to light. One person was trying to remove valuable papers from an office and asked two firemen to help him, but they refused unless he paid them $50; the papers were destroyed. Drivers of express wagons have taken $100 and even $500 for an hour's use of their vehicles, in getting distressed people away from danger.

A book-keeper, engaged in conveying away the firm's records, fell fainting in the alley behind the store, overcome by exertion and suffocated by the smoke and dust. The shock restored him to consciousness, and upon attempting to rise he found himself unable to stand. Just then a man was passing, and he hailed him with a request for help. The wretch offered to assist for a hundred dollars. The fallen man said, "I have but ten, and I will give you that." For this amount he gave his arm to the poor sufferer, and saved his life. A girl carried her sewing-machine to four different points, and was forced from each by the advancing fiend. At last an expressman seized her treasure, and in spite of all her efforts drove away with it. Said the impoverished girl, "Do you wonder Chicago burned?" In front of a wholesale house the sidewalk was bloody from the punishment inflicted by the police upon sneak-thieves. Trunks were rifled after their owners had placed them out of reach of fire. They were broken open by dozens on the Lake shore, and the empty trunks tossed into the water. Pieces of broadcloth were torn into strips three yards long and distributed among a party who said, "These will make us each a good suit." Persons who saw and heard these things were powerless, and the confusion was so terrible that no one could look out for any one but himself, or interfere for the protection of others' property. It was a time when the worst forces of society were jubilant, and all the villains had free course. The Court-House jail had one hundred and sixty prisoners, and these were let loose to prey upon the people in the time of their helplessness and extremity. Such an event was a public calamity; but humanity would not permit the poor wretches to perish there, and no means were at hand to convey them to any other place of confinement.

One of our city papers thus deals with the oil-stone story:

The New York *Journal of Commerce* has swallowed the oil stone story; and assuming it as a fact that Chicago was built of stone heavily charged with petroleum, thus describes the process of destruction:

"An eye-witness of the process says he saw the flames cross streets and lick with long tongues at the stone buildings opposite. The latter, as they became intensely heated, emitted jets of gas, upon which the flames would catch and then go out again, repeating the operation a number of times, when—presto—the stone would apparently be in flames. This is precisely the action of fire on anthracite, as any one may see by watching a large lump of coal in his grate. Like coal, these stones were reduced to ashes."

That eye-witness had a lively imagination. We repeat that the only building in this city of any size built of the supposed oil-stone was the Second Presbyterian Church, and the walls of that building were not reduced to ashes, but stand conspicuously erect among the ruins of a hundred other buildings utterly destroyed. The foundation for this oil-stone theory is the following from a number of *Chambers' Journal:*

"In the neighborhood of Chicago there are enormous deposits of this oil-bearing limestone; some of the houses in the city are built of it, and after a while present a smeary appearance from exudation of the oil. The least thickness of the mass is thirty-five feet, and it has been estimated from experiment that each square mile of it contains seven and three-quarter million barrels, each of forty gallons, of petroleum."

Some years ago, when the oil-fever was at its height, and men were making fortunes in a week, some persons conceived the idea that the stone in an old quarry northwest of the city gave evidence of oil. If we mistake not, certain disembodied spirits encouraged the idea, and boring was begun. The oil-rock was perforated without getting a drop of oil; but the boring went on until at last they struck a vein of water in no wise tinctured with petroleum.

A countryman with a carpet-bag appeared the second week after the fire, and told his errand. He had a debt of five hundred dollars on his farm, and having heard of the great liberality of the Chicago people, how they took up collections of many thousands on a single Sabbath morning, he thought that they would be willing to pay off that mortgage for him, and thus enable him and his wife, as they were growing old, to live easy and take comfort the rest of their days. I suggested to him that the fire had impoverished us. Well, he said, he had thought of that, and had made up his mind, as he had some good apples, that he would donate to every person who gave him five dollars, a barrel of apples. Thus they would be helping him, and get something for themselves. Dinner was ready, and he sat down to a good meal; and after dining he entered into some account of his experience, and asked earnestly my opinion of certain heresies that were being promulgated in his neighborhood. Having run through all the subjects he could think of, he suggested that he should have to stay all night, and perhaps I could keep him, or send him to some of the benevolent people for a night's lodging. I intimated to him that every body was full, on account of the exodus of so many thousands from the burnt district to our quarter. Bethinking himself of another pastor, he started off to try and interest him, as I could give him little or no encouragement. It was doubtful whether he found the doctor in a mood to entertain his appeal for charity at that juncture. For charming simplicity and cool audacity this surpassed anything in my former experience.

How different the case of a noble man who came to his pastor for comfort and for nothing more: although he had been ruined, and his son had been driven away to another city for employment as an engraver, and his wife was in a distant city, he would not allow any appeal for assistance, as he had gone to work, though not a carpenter, as a foreman in re-erecting buildings on the desolated grounds. Won't you have a pair of boots? No; I can buy some. Nothing would he receive. He had been formerly burned out in Wisconsin, and had many times aided his unfortunate neighbors in similar troubles. He told how he had, early in Chicago's history, refused to invest his money in a block now worth half a million, and gone away up into Wisconsin, and there struggled and toiled, and finally lost everything.

A gentleman relates the following case of selfish, brutal meanness:—

In a church some blocks away, quite on the northwest verge of population, I found other examples of suffering. The first to greet me was a bright and brave German fellow, also a dry-goods clerk, who had rescued his wife and five children, and had saved plenty of good clothing and household stuff enough for tolerable comfort, only that he had no money and no chance of securing a house. He took little thought for himself, however, but showed me a family of ten—eight small children—the father and mother workers with the sewing-machine. They had owned a house and lot worth $4,000 or $5,000, with a debt of $700. The half-weekly payments for making up clothing had been their living. When the fire came, the two Singer sewing-machines were saved by burying them in the garden behind the house. Tuesday morning, on going to inspect, the man found ghouls just ready to make off with them. One of these saved appearances for the moment by offering to carry them to the owner's place of refuge, but on reaching this demanded $10, and took one of the $85 machines in lieu of payment. I am happy to say that two of Sheridan's bayonets are after that fellow, and that we have stern law for these extortions if the perpetrators are caught.

Another, a sufferer, states his bitter experience, and adds several interesting incidents:—

His residence was situated in the centre of the burnt district, and at an early hour was consumed. One of the first places to which he repaired was the Sherman House, in which he had friends. He found it on his arrival still untouched, but the guests were passing out in all directions.

Among other incidents he witnessed is one not the least strange of the many which have been told. A guest of the house, on his way from the West, had with him his invalid wife and children. In the hurry of the moment they were overlooked, and as the fire was rapidly encroaching on the building, he became frantic in his efforts to save his family. The conveyances around the hotel were all engaged, but by paying $1,000 he managed to secure an express wagon and thus escaped. On Wabash avenue the owner of one of its marble houses had his carriage and colored coachman drawn up at his door, preparatory to conveying his family to a place of refuge. Three ruffians on the look-out for plunder approached the carriage, and, jumping on to the seat, threw a sack over the head and shoulders of the coachman and hauled him to the ground. They rapidly drove away in the vehicle, leaving its owner to shift as well as he could without it.

Along lower Clark and State streets were located many livery stables. The horses were taken out at the first alarm and brought to what was thought to be a place of safety. Hundreds of them were gathered together in one inclosure. When the fire approached them they became strangely agitated, and their terror finally became so great that they broke from their fastenings, causing a general stampede. The scene was a frightful one. In their madness they trampled each other to death, and breaking loose among the crowds of fugitives, added not a little to the general alarm.

Going along Madison street, our informant was met by an excited individual, who was wildly shouting, "I knew they would do it!—I knew they would do it!"

On being asked to explain, he exclaimed, "The bloody Ku-Klux have done this, knowing us to have been extra loyal. They have burned our city, and it is useless for us to attempt to escape, for they will burn us up too!"

On lower Clark street, just below the Court-House, were some rows of splendid business houses. The upper portions were fitted up in furnished rooms, and, sad to say, were let to the less disreputable portion of the demi-monde.

Being steeped in the heavy slumber of vice, the fire had reached the lower part of the building before they were apprised of their awful danger. When they were roused from their lethargy, their terror was fearful. Appearing at the upper windows of the burning blocks, they found their communication almost cut off, and their screams were terrific. The staircases were still partly standing, and after great difficulty the girls were rescued from their perilous position. One young girl, an Italian, attracted the attention of all by her picturesque beauty, which was heightened by the tragic situation in which she was placed. Her hair, wildly flowing, reached almost to her feet, while the foreign expression of her features and the tragic *pose* of her attitude made her look like a tragedy queen. She was a striking illustration of the line,

Beauty unadorned, adorned the most.

Poor unfortunate! She looked fitter for a better life than the awful one she was pursuing. Who can say what treatment had driven her from her own sunny clime to our colder climate? Her looks were noble and striking, her bearing patient and courageous, and a feeling of intense relief was experienced by the spectators when she was rescued from the jaws of death.

Immediately before this incident occurred, a fearful scene was to be witnessed at the corner of Sherman street, about half a block west of La Salle, near the Michigan Southern Railroad depot. The street (which was a small one) was entirely occupied by bagnios, conspicuous among which was the corner one, run by a courtesan well known in Chicago as one of the worst characters that ever disgraced a city. Her name was Nelly Grant, otherwise known as Tipperary Nell, as that historic county had the honor of giving her birth. As usual the inmates on that fatal Sunday night were in a beastly state of intoxication. The fire crept upon them unperceived, and had it not been for a burly driver, the bully of Nelly, the inmates would have been burned in their beds. As it was the house had caught before any of them got out, and the screams, curses, and lamentations of the unfortunates were terrible to hear. "Nelly" herself was insensible from the effects of her potations, and her lover had to carry her out—no easy job, for she was not by any means what you would call a "light weight." He succeeded, however, in carrying her to a place of safety, and the remainder of the wretches were rescued without harm.

Going down Dearborn street our informant came to a gents' furnishing and jewelry store, which the fire was rapidly approaching. A crowd had gathered around, and the proprietor, unable to save his goods, said to them, "Take all you can, boys, for I can't

SCENE IN DEARBORN STREET WHEN TH

EACHED THE TREMONT HOUSE.

save anything." Several took wallets and filled them with valuables, but the police outside caused them to be delivered up, doubtless for the benefit of the relief fund.

A MUSEUM.

At Colonel Wood's Museum great preparations had been made for the production of "Divorce," but it has been indefinitely shelved till a new building is erected. The drama was one which would have exactly suited Chicago, as the city is celebrated for the ease and celerity by which the marriage tie can there be cut asunder.

The greatest contrasts were presented on all sides during the burning. Brave men were endeavoring to cheer downcast women with an appearance of light-heartedness which was far from real. Individual instances of gallantry on the part of women were not wanting, and our informant is in rapture with the coolness displayed by a widow, whose bravery extorted the admiration of all who beheld her. She had to cheer the spirits of some half dozen drooping maidens and guide them to a place of safety, which she did with perfect success. She was none of your "fair, fat, and forty" ones, but instead a young and pretty woman, and from all we can learn she will not long live in widowed blessedness, if any of her numerous admirers on the trying Monday can trace her.

The most ridiculous scenes ever mingled with the most terrible ones, and the spectacle of the effects that were being carried away was in many instances extremely amusing.

A lady who kept a boarding-house on Adams street struggled hard to get her stoves out at the risk of her life, and frantically abused her lodgers for defacing the walls of her house in carrying out their trunks. The flames were only half a block away at the time, and before she had ceased scolding her house had fallen in, nearly burying her in the ruins. By some the most selfish spirit was displayed. Next-door neighbors in many instances refused each other the slightest assistance, and much valuable property was thus lost that would otherwise have been saved. On the other hand, many whose homes escaped the conflagration acted with a large-hearted generosity, and freely shared their homes with all the sufferers they could accommodate. This spirit was particularly manifested by those whose losses had been greatest, and too much praise cannot be bestowed on conduct so noble. The sights to be witnessed on Tuesday were of the most heart-rending description, but as our correspondents have already narrated the most of the incidents seen by our informant we need not recapitulate them. One of them is, however, new. A mother who had lost her only child was wandering frantically among the ruins in search of her darling, and when she could discover no traces of it her reason fled, and she became a raving maniac. On Tuesday night the gentleman left the city for New York, and he presents a graphic picture of the excitement and suspense all along the line of the railroads. The train on leaving the depot was densely crowded, the aisles of the cars were filled with passengers, so that the wheels pounded with the weight, and two powerful engines were scarce sufficient to carry the convoy along. When it had got about three miles from the city a cry arose in the cars that the South Side was on fire, and a rush was made for the windows, from which a lurid glare could be perceived in the heavens over the lower part of Cottage Grove avenue.

A NEWSPAPER EXPLOIT.

The pluckiest thing we have heard of in connection with the conflagration is connected with the persistent issue of the Chicago *Evening Post.* That journal, like the others, and even more completely than the others, lost everything—building, presses, type, paper, material, and even the books. Two of the *Post* compositors, driven to the West District by the fire, found a little job-office, about Monday noon, open and completely deserted, the occupants having rushed to the fire then raging and seething like a hell across the city. One instantly wrote out an account of the fire as far as it had progressed, and the other put it in type, and they clapped above it the old familiar words, "*The Evening Post*," made it up in a page about six by eight inches, and exultantly printed it. So not one issue of that paper has failed.

It being announced that Rev. T. W. Goodspeed, of Quincy, Illinois, of the Vermont Street Church, who was present in Chicago at the time of the fire, and had witnessed many of its scenes and incidents, would give a narrative thereof at his church, an immense crowd was early in attendance, filling all the space in the building, while hundreds of others were unable to gain admittance. Mr. Goodspeed took no text, giving simply a narrative of what he saw. He commenced by saying:—

It was my fortune to be in Chicago when it was destroyed. I do not propose to give you a complete history of the conflagration. You are getting that from day to day through the newspapers. Many have said to me, "Tell us all you saw." This great calamity is in all hearts. We are not prepared to speak of or listen to anything else; and I have thought there was a sufficient reason for giving up this service to telling my congregation what I saw of this unparalleled conflagration. Sympathizing with this feeling, Mr. Priest has given up his service to be with us, as has also the congregation of the First Church. I fear you will be disappointed in listening to me, as I design to tell you only what came under my observation, and there were a thousand things I did not see.

The Chicago river runs directly west from the Lake almost a mile. It then branches north and south. That part of the city lying south of the main river, and east of the South Branch, is called the South Side. That part lying north of the main river, and east of the North Branch, is the North Side; and all west of the two branches the West Side. Each of these divisions is about one-third of the city.

You are aware that the great fire of Saturday night, which destroyed several blocks, was on the West Side, near the South Branch of the river. The fire of Sunday night and Monday began also on the West Side, near the scene of the other, destroying, with that, forty blocks on the West Side; swept across the South Branch, destroying a mile square of the South Side—the entire business portion of the city—crossed the river and laid in ruins almost the whole of the North Side, about 400 blocks.

Sunday evening I preached in the Second Baptist Church, which is nearly a mile west of the South Branch. We stopped in the study about half an hour after service, and started for my brother's home a few minutes after nine. It was then that we first saw the fire, a mile to the south-east. We continued to watch it from time to time till eleven o'clock, when, supposing it under control, we retired.

We were aroused a little before four in the morning. Hurrying on my clothes, I went out. The fire had got far up on the West Side of the South Branch, and had evidently crossed the river to the South Side, and was beyond all control. The wind was blowing fiercely from the south-west. The whole city was lighted up by the flames almost like day. As I hastened toward the river I noticed that the stars were all obscured as effectually as if the sun were shining, and the moon gave a feeble, sickly light. It was almost gray, altogether unlike itself.

As I proceeded the streets became more and more crowded. The whole West Side was gathering and crowding toward the river. I stopped to rouse my brother, but he had long been gone. A woman stopped me on Washington street and said: "My husband's place of business is destroyed, and we are ruined."

Reaching the river, I found that a large part of the South Side was still unharmed. Here I saw the massive blocks of the South Side in flames, and saw vessels being towed north to escape the fire. I followed the South Branch up to where it joined the North Branch and the main river, and looked down the latter to the Lake. Three or four blocks away the fire had crossed the river. Wells street bridge was burning. The spectacle was grand and awful beyond description. Great billows of flame swept clear across the river, while countless myriads of sparks and burning brands filled the air.

Proceeding, I crossed the Kinzie street bridge to the North Side. Here I met the fugitives—thousands of people, indeed, were going both ways — spectators to see, fugitives to escape. The streets were filled with merchandise and furniture. Women were everywhere guarding their household goods. The air was filled with a thousand noises. The screaming of the steamers, the whistle of the tugs, the cries of children, the shouting of men, the howling of the wind, the roar of the flames, the crash of falling buildings.

I went on as far as Wells street, and the wind was here a hurricane. The buildings on Water street and the south bank of the river caught, and almost instantly they were one vast volcano, throwing up great volumes of flame that were caught up and carried bodily across the stream. The river seemed a boiling caldron. We stood under the great elevator at the Wells street depot and saw on one of them a man wetting the roof. He had hose, and must have saturated the entire building with water, yet within fifteen minutes the building was aflame. I returned to the West Side. The fleeing people were carrying off articles of every description. Two men were wheeling away the Indian figure that had stood before their cigar store. One man was hurrying off with two whiskey bottles. I stopped again to look down the main river toward the Lake. The scene was even more magnificent and awful than before. This was indeed the grandest spectacle of all. The whole length of the river was then one broad sheet of fire.

With every fresh blast of wind great billows of fire would roll across toward the doomed North Side, as if filled with a mad desire to sweep it away in ruin. Then for a moment they would subside and show the three bridges wreathed in flames (the water apparently boiling underneath them), the black walls of the buildings on either side, and here and there tongues of flame shooting out from doors and windows and roofs. Then again two walls of fire, extending a mile away to the Lake, would flame up toward heaven for a moment, to be caught by the gale and tumbled in fiery ruin to the ground, or carried in great masses of fire to spread the conflagration. Going on from here I took my stand on Lake street bridge. The line of fire extended a mile or more down the South Branch. Several bridges had already been consumed. The great coal-yards were beginning to burn, and almost all the magnificent blocks of the South Side were in flames. From the slight elevation of the bridge, I could see almost two square miles of fire.

Looking toward the north-west, and seeing how directly toward the water-works the flames were rushing, it crossed my mind that they would be destroyed. I turned and hastened to my friend's house, a mile on the West Side, and immediately tried the water. I was too late, it would not run, and the great city of 300,000 people was without water.

Before seven o'clock I went to another friend's house and found him just returned from saving his books, and what merchandise he could. He had got into his place of business by the back way, and had been driven away by the swift demon of destruction. I went to another friend's house to inquire if his store was safe. He had visited the fire at half-past ten, and gone home confident it was under control. At three he had tried to reach his business place, and been driven back by the fire that raged between him and it. I got into his buggy with him and we started to find it. Reaching Twelfth street, which runs across the South Branch, a mile and a quarter south of the Court-House, we found the street crowded with people and vehicles, and all pressing toward the South Side. It was a little after seven o'clock, and of course daylight. We made our way to Wells or La Salle street, and tried to go up, but the flames stopped us. We went on to Wabash avenue, and found it to be so crowded as to be utterly impassable. We crossed to Michigan avenue, fell into the stream of travel, and worked our way up to the Michigan Avenue Hotel. My friend asked me to hold his horse five minutes, while he went to see what he could find. Left to myself, I had time to look about me. I despair of describing the scene to you. It beggars description. It was here that my friend Sawyer, who is with me in the desk, joined me; his clothes covered with dust, his hair filled with dust and cinders, his eyes red from smoke, his face black, so unlike himself that I hardly knew him. Michigan avenue was burning from within a block of where we stood a mile away to the river. The magnificent residences and great business houses were going up in flames and down in blackness before our eyes. Great volumes of smoke rolling away before the gale, concealed the North Side from view. But at every break or lift of the smoke, the great Central Depot could be seen all in flames. The fire was creeping away out on the piers, and had reached one of the immense elevators that stood near its end, and the flames were soon reaching up one hundred and fifty feet into the air. Every moment we expected to see the great Central Elevator, standing very near the burning one, fall before the conflagration that had devoured everything else in its path. But the wind seemed to veer suddenly to the south, and remained there an hour, and the great elevator was saved; with one exception, the only one on the South Side north of the line of fire. A steamer had reached the mouth of the river, but here the fire caught her, and I saw it run from one end to the other in little lines of light, and so over the rigging till the ship was all ablaze.

Meantime I was in the midst of the wildest confusion I had ever witnessed. The open space between Michigan avenue and the Lake was filled with every variety of household goods and merchandise. There must have been the furniture of a thousand families crowded into this narrow space. Rich and poor, white and black, were together. Over every pile of goods stood some one to guard it. Meantime other fugitives were every moment crowding into the already overcrowded space, and seeking room for their goods as well. Thousands of people pressed along the walks and filled the open spaces—some coming to see and others fleeing. The avenue was for hours one solid mass of teams. Up and down the street they pressed endlessly, going up empty and returning full. At length the press became so great that the street was completely blockaded, and the police began to turn the still on-coming multitude of vehicles backward. They chose the spot where I stood to accomplish this. Then began cursing and shouting; the teamsters insisting that they must go on, every one of them having valuable property just ahead; and the police insisting that to save men's lives they must turn back. The more determined teamsters went through in spite of the police, who were strangely inefficient. The more timid or reasonable tried to turn back in a street where there was hardly room to move forward. One backed into my buggy wheels as I crowded the sidewalk and waited; another ran into one of the shafts. Twenty feet ahead of me a horse tried to run away, starting directly toward me. He ran about ten feet and smashed two buggies. A rod to my left a driver ran against a buggy wheel and crushed it, regardless of the other's load. I grew more and more nervous, expecting every moment to have the horse and buggy ruined. Two hours and a half passed, and still I waited. I had plenty of time to look about me.

Every variety of vehicle passed me, loaded with every variety of article. I saw one of our former citizens, Mr. Pearson, carrying one end of a long glass case filled with his goods—hair done up in many forms. A dozen or twenty cows picked their way among the wagons. A woman found her way across the street, when there chanced to be an opening, leading a great black dog. The confusion was beyond all description. Up and down the Michigan Central track locomotives were constantly moving, drawing heavy trains, or alone, and, it seemed to me, blowing their unearthly whistles all the time. The fire-engines, a block away, added theirs, which were worse still. The voices of the police calling to the teamsters, the responses and often curses of the drivers, their impatient yells to one another, the cry of distressed citizens to the expressmen, the voices of the crowd, the roaring of the gale, the howling of the conflagration, the crackling of burning houses, the crash of falling walls, the ringing of bells, the shouts that greeted some new freak of the flames, and suddenly the sullen thunder that told us buildings were being blown up only a block away. The conflagration of the great day will hardly bring a confusion worse confounded.

The fire still made progress towards me, until the people in all the houses above and below me removed their goods and fled. Again came the thundering and shaking of the earth that accompanied the blowing up of a building. It seemed ominously near; I could see the fire on the Wabash Avenue Methodist Church, and was sure it was going, and that was behind me. At length the vast crowd, men and teams, precipitated themselves down the avenue like a falling avalanche, and the cry went up that the building on the corner just above us was to be blown up. Waiting no longer, I joined the fleeing multitude and made my way as fast as possible a block farther away. After three hours my friend returned; his coat gone; his face so black and his eyes so nearly put out, that for a moment I did not know him. He took his horse, to my great relief, and I proceeded up the avenue toward the Central Depot, to see what good I could do. On beyond Terrace row I went, and had the whole horrible scene before me. Not long, however, could I see it. The magnificent Terrace row was in flames, and the air was filled with smoke, and dust, and cinders, and live coals, and fagots of fire. The middle of this great row fell first, the ends following, covered in one black cloud of smoke, and ashes, and dust. It was almost past endurance.

Meanwhile the inflammable material in this narrow space caught fire in a hundred places. Beds, pillows, quilts, carpets, sofas, pianos, furniture, and it seemed to me that everything must be burned. With a small tea-chest I spent hours bringing water from the Lake, helping to extinguish numberless incipient fires which broke out continually among the heaps of goods. I returned home at three P.M., having had nothing to eat since six o'clock Sunday evening. Helping to carry a mirror up-stairs, I asked a woman on the way down to give me a drink from a full pail she carried, and she refused. In the evening, Monday evening, I took my station in the cupola of a four-story building to view the fire and watch, and for hours witnessed a scene which no language can describe.

In contrast with this calm and clear sketch of that memorable day by the young clergyman providentially in the city, we preserve

A WOMAN'S STORY OF THE FIRE.

Where shall I begin? How shall I tell the story that I have been living during these dreadful days? It's a dream, a nightmare, only so real that I tremble as I write, as though the whole thing might be brought to me again by merely telling of it.

We lived on the North Side, six blocks from the river—the newly-regenerated river, which used to be at once the riches and the despair of our city, but which had just been turned back by the splendid energy of the people to carry the sweet waters of Lake Michigan through all its noisome recesses. We were quiet people, like most of the North-siders, flattering ourselves that our comfortable wooden houses and sober, cheery, New England-looking streets were far preferable to the more rapid, blatant life of the South Side.

Well, on Sunday morning, October 8, Robert Collyer gave his people what we all felt to be a wonderful sermon on the text, "Think ye that those upon whom the tower of Siloam fell were sinners above all those who dwelt at Jerusalem?" and illustrated it by a picture of the present life, and our great cities, their grandeur, their wickedness, and the awful though strictly natural consequences of our insatiable pursuit of worldly prosperity, too often unchecked by principle; and instanced the many recent dreadful catastrophes as signs that not the Erie speculators alone, nor the contractors alone, nor the recognized sinners alone, but we, every man and woman of the United States, were responsible for these horrors, inasmuch as we did not work, fight, bleed, and die, if necessary, to establish such public opinion as should make them impossible.

I came out gazing about on our beautiful church, and hoping that not one stone of the dear church at home had been set or paid for by the rascality which our preacher so eloquently depicted as certain to bring ruin, material as well as spiritual; and so we pass the pleasant, bright day; some of us going down to the scene of the West Side fire of Saturday night, and espying, from a good distance, the unhappy losers of so much property. About half-past five in the evening our neighboring fire telegraph sent forth some little tintinnabulations, and we lazily wondered, as D—— played the piano, and I watered my ivy, what they were burning up now. At ten o'clock the fire bells were ringing constantly, and we went to bed regretting that there must be more property burning up on the West Side. Eleven o'clock—twelve o'clock—and I woke my sister, saying, "It's very singular; I never heard anything like the fires to-night. It seems as if the whole West Side must be afire. Poor people! I wonder whose carelessness set this agoing?" One o'clock—two o'clock—we get up and look out. "Great God! the fire has crossed the river from the south. Can there be any danger here?" And we looked out to see men hurrying by screaming and swearing, and the whole city to the south and west of us one vivid glare. "Where are the engines? Why don't we hear them as usual?" we asked each other, thoroughly puzzled, but even yet hardly personally frightened by the strange aspect of the brilliant and surging streets below. Then came a loud knocking at the back door, on Erie street—"Ladies, ladies, get up! Pack your trunks and prepare to leave your house; it may not be necessary, but it's well to be prepared!" It was a friend who had fought his way through the La Salle street tunnel to warn us that the city is on fire. We looked at each other with white faces. Well we might. In an inner room slept an invalid relative, the object of our ceaseless care and love, the victim of a terrible and recurring mental malady, which had already sapped much of his strength and life, and rendered quiet and absence of excitement the first prescription of his physicians. Must we call the invalid? and if we did, in the midst of this fearful glare and turmoil, what would be the result? We determined to wait till the last minute, and threw some valuables into a trunk, while we anxiously watched the ever-approaching flame and tumult.

Then there came a strange sound in the air, which stilled, or seemed to still for a moment, the surging crowd. "Was it thunder?" we asked. No, the sky was clear and full of stars, and we shuddered as we felt, but did not say, it was a tremendous explosion of gunpowder. By this time the blazing sparks and bits of burning wood, which we had been fearfully watching, were fast becoming an unintermitting fire of burning hail, and another shower of blows on the door warned us that there was not a moment to be lost. "Call E——" (the invalid); "do not let him stay a minute, and I will try to save our poor little birds!" My sister flew to wake up our precious charge, and I ran down stairs, repeating to myself to make me remember, "birds, deeds, silver, jewelry, silk dresses," as the order in which we would try to save our property, if it came to the worst.

As I passed through our pretty parlors, how my heart ached. Here the remnant of my father's library, a copy of a Bible printed in 1637, on one table; on another, my dear Mrs. Browning, in five volumes, the gift of a lost friend. What should I take? What should I leave? I alternately loaded myself with gift after gift, and dashed them down in despair. Lovely pictures and statuettes, left by a kind friend for the embellishment of our little rooms, and which had turned them into a bower of beauty—must they be *left*? At last I stopped before our darling, a sweet and tender picture of Beatrice Cenci going to execution, which looked down at me, through the dismal red glare which was already filling the rooms, with a saintly and weird sweetness that seemed to have something wistful in it. I thought, "I will save this, if I die for it;" but my poor parrot called my name and asked for a peanut, and I could no more have left him than if he had been a baby. But, could I carry that huge cage? No, indeed; so I reluctantly took my poor little canary, who was painfully fluttering about and wondering at the disturbance, and, kissing him, opened the front door and set him free—only to smother, I fear. But it was the best I could do for him if I wished to save my parrot, who had a prior right to be considered one of the family, if sixteen years of incessant chatter may be supposed to establish such a right.

What a sight our usually pretty quiet street presented! As far as I could see, a horrible wall—a surging, struggling, encroaching wall—like a vast surface of grimacing demons, came pressing up the street—a wall of fire, ever nearer and nearer, steadily advancing upon our midnight helplessness. Was there no wagon, no carriage, in which we could coax our poor E——, and take him away from these maddening sights? Truck after truck, indeed, passed by, but filled with loads of people and goods. Carriages rushed past drawn by struggling and foaming horses, and lined with white, scared faces. A truck loaded with goods dashed up the street, and, as I looked, flames burst out from the sides, and it burned to ashes in front of our door. No hope, no help for property; what we could not carry in our hands we must lose. So, forcing my reluctant parrot into the canary bird's cage, I took the cage under one arm and a little bag, hurriedly prepared, under the other, just as my sister appeared with E——, who, thank God, was calm and self-possessed. At last the good friend who had warned us appeared, and, leaving all his own things, insisted on helping my sister to save ours, and he and she started on, dragging a Saratoga trunk. They were obliged to abandon it at the second corner, however, and walk on, leaving me to follow with E——. "Come, E——, let us go," said I. "Go where? I am not going. What is the use?" he answered, and he stood with his arms folded as if he were interested merely as a curious spectator. I urged, I begged, I cried, I went on my knees. He would not stir, but proposed going back into the house. This I prevented by entreaties, and I besought him to fly as others were doing; but no. A kind of apathetic despair had seized him, and he stood like a rock while the flames swept nearer and nearer, and my entreaties, and even my appeals to him to save me, were utterly in vain. Hotter and hotter grew the pavement, wilder the cries of the crowd, and my silk and cotton clothing began to smoke in spots. I felt beside myself, and, seizing

E——, tried to drag him away. Alas! what could my woman's strength do? There followed another shout, a wild push back, a falling wall, and I was half a block away and E—— was gone. "O God, pity those poor worms of the dust, and crush them not utterly!" was my prayer.

How I passed the rest of that cruel Sunday night I scarcely know. Wandering, staring, blindly carrying along my poor parrot, who was too tired to make a sound, I seemed to be in a dream. Starting north to get help, running back as near to the flame as I could in the vain hope of finding E——, bitterly reproaching myself that I had ever left him an instant, I passed three hours of which I can hardly give any account. I know that as I turned wildly back once toward Dearborn street, I saw the beautiful Episcopal Church of St. James in flames. But they came on all sides, licking the marble buttresses one by one, and leaving charred or blackened masses where there had been white marble before. But the most wonderful sight of all was the white, shining church tower, from which, as I looked, burst tongues of fire, and which burnt as though all dross of earth were indeed to be purified away from God's house forever. As the tower came crashing down, the bells with one accord pealed forth that grand old German hymn, "All good souls praise the Lord." I almost seemed to hear them, and to see a shadowy Nicholas striking the startled metal for the last time with his brave old hands. "If this is right, if it can be right, make me think so," groaned my soul, and the souls of many weeping women that night, as they fled homeless and lost through that Pandemonium of flame and tumult.

Constantly faces that I knew flashed across me, but they were always in a dream, all blackened and discolored, and with an expression that I never saw before. "Why, C——, is this you?" some frightened voice would exclaim, and a kind hand would touch my disordered hair, from which the hat had long since fallen off, and some one, only a little less distracted, would whisper hopefully a word about E——; that he might not be lost, that the actual presence of flame would arouse him, and so on; and I loved them for saying so, and tried to believe them. Very little selfishness and no violence did I see there. Neighbors stopped to recognize neighbors, and many a word was exchanged which brought comfort to despairing hearts. "Have you seen my wife and children?" would be asked, and the answer given: "Yes, they are safe at Lake View by this time." "Won't you look out for my baby?" (or Willie or Johnny, as the case might be). Out would come tablets or papers, or names or inquiries would be noted down, even by the man who was making almost superhuman efforts to save a few goods from his burning house. Some friend—it was days before I knew who—took my parrot and forced a little bottle of tea and a bag of crackers into my hand as I wandered, and I was enough myself to give it to a friend, whom I found almost fainting with heat and fatigue, and who declared that nectar and ambrosia never tasted better. At last I found myself opposite Unity Church. Dear Unity! will her little circle of devoted ones ever come together again, and worship sometimes, and work for the poor sometimes, and sing and play in her beautiful under-parlors sometimes, and love each other always? I know not, but I know that I wept and beat my hands together, and raged hopelessly, when I saw that the beautiful homes on the west side of Dearborn street were gone, and the Ogden Public School was one bright blaze, while the graceful and noble Congregational Church, next to Mr. Collyer's Church, had caught fire. Nothing could save our pride and joy—our darling for which we had made such efforts in money and labor two short years ago, that the fame of Chicago munificence rang anew on our account through the civilized world.

I was grieving enough, Heaven knows, over my private woes; but I awoke to new miseries when I saw our pastor's great heart,

SCENE AT THE JUNCTION OF THE CHICAGO RIVER.—THE FLAMES (

'ATE WITH THE SHIPPING AND DESTROY THE GRAIN ELEVATORS.

which had sustained the fainting spirits of so many, freely give way to lamentations and tears as his precious library, the slow accumulation of twenty laborious and economical years, fell and flamed into nothingness in that awful fire. I turned away heart-sick, and resumed my miserable search after the face which I now felt almost sure I should never see again. A new sight soon struck my eye. What in the world was that dark, lurid, purplish ball that hung before me, constantly changing its appearance, like some fiendish face making grimaces at our misery? I looked and looked, and turned away, and looked again. May I never see the sun, the cheerful daily herald of comfort and peace, look like that again! It looked devilish, and I pinched myself to see if I was not losing my senses. It did not seem ten minutes since I had seen the little, almost crescent moon, look out cold, quiet, and pitiless, through a rift in the smoke-cloud, from the deep blue of the sky.

Two dear children, whom I had taught peacefully on Friday in our cheerful school-room on Chicago avenue, met me, crying, "Oh! have you seen mother? We have lost her." This appeal brought me to myself. I felt that I had something else to do than wander and grieve; so I persuaded the lost lambs to go with me to a friend on La Salle street, where I felt sure we should find help and comfort, and which everybody supposed would be safe. Indeed, a very curious and rather absurd feature of this calamity was that nobody thought his house would burn till he saw it blazing, and also felt perfectly sure that this was the last of it, and that he and his family would be safe a little further up; so the North-siders never began to pack up till the fire crossed the river, and then the lower ones moved about to Erie street, six squares from the river, then stopped. Then they were driven by the flames another half-dozen streets, losing generally half of what they saved the first time; then to Division street, then to Lincoln Park, where heaps and heaps of ashes are all that

15

remain to-day of thousands of dollars' worth of eatables and furniture.

Exhausted and almost fainting, weeping and sorely distressed, I finally landed in a friendly house, far up on La Salle street. As I stepped inside the door E—— appeared, quiet, composed, and almost indifferent. Burnt? Oh, no; he was all right. Did I suppose he was fool enough to stay and be burned? There was D——, too, if I wanted to see her, in the parlor. Did I feel reverently thankful? Ask yourself.

C——.

We recall Byron's lines in Childe Harold, although the situation is inverted:—

> "Oh! who could guess if ever more should meet those mutual eyes,
> Since upon night so sweet such awful morn could rise!"

The night here was "awful" and the morn "sweet."

We give another leaf from personal experiences of painful interest. The narrator was a lodger in the St. James Hotel, and says:—

I was awakened about three A. M. by some one pounding upon my door, and after springing from my bed, discovered that the whole city was in flames. I hastily put on my clothing, and going into the corridor I saw a crowd of men, women, and children clustered about the door. Returning to my room, I gathered my goods quickly into boxes, and carried them down to the sidewalk. Hearing a shout, I seized a satchel and a small trunk, and rushed out. As I reached the door, I saw some men coolly loading my boxes into a wagon. I called to them, but they laughed and drove away. The street was full of people with bundles of every description on their backs. I pushed at once for the West End. Neither Michigan nor Wabash avenues were then on fire, and I rushed down the former. The hot air almost burned my face. The smoke was stifling me, and my clothes were covered with ashes and cinders. As I passed along the avenue, I looked up each street to the west to see where the fire headed me off in that direction. I had the fire behind me on the north, and the Lake was on my left. My object was to try and get to the west side of the city, near Union Park, where I knew a gentleman named Mason. The Lake was on my left, the city on fire behind me, and as I passed along Wabash avenue I could see the fire raging furiously on West street, at the head of Lake, Randolph, Madison, Monroe, Congress, Adams, Jackson, and Van Buren streets, and away to the south as far as Fourteenth street. Here for the first time I saw a clear passage to the west. When I reached this point I was utterly wearied out, and I sat on my trunk in the street. In a few moments I saw a man pass by, and I asked him to give me a hand with my trunk. He said he would, and we walked up Fourteenth street. After going a short distance I saw that I could not carry the trunk any further, and I told the man who was assisting me that I must give out. He urged me on, and after going about a block I saw a man standing at his own door, looking in the direction of the fire. I told him that I had been burned out, and that I was so wearied I could carry my trunk no further. I asked him for permission to put it in his yard until I should be able to convey it to a safe place. He gladly consented, and between us we took the trunk into the yard. He and I then returned in the direction of the town. In the mean time the fire had reached the great business quarter, and most of the streets from South Water street to the river were in flames. After waiting until the progress of the fire was arrested, I made my way across Twelfth street bridge, which was then the only one standing on the south, to Union Park, on the western side of the city. Here I found my friend's house, and was joyfully and hospitably received. I was so wearied out that I remained there asleep all day on Monday. Such a gale never raged before as that which blew from the southwest in Chicago during the night of Sunday and the morning of Monday.

The only fragment of literature saved from the immense stock of the Western News Company was this

CURIOUS MEMORIAL.

A single leaf of a quarto Bible, charred around the edges. It contained the first chapter of the Lamentations of Jeremiah, which opens with the following words:—"How doth the city sit solitary that was full of people! how is she become as a widow! she that was great among the nations, and princess among the provinces, how is she become tributary! She weepeth sore in the night, and her tears are on her cheeks: among all her lovers she hath none to comfort her." It was a singular circumstance, that Rev. Mr. Walker, of Connecticut, upon hearing of the catastrophe at Chicago, preached from this text, not knowing that this was all that remained of the store in which his son was a clerk. There was only a general correspondence in the actual experience of our city to that of the city bewailed by the prophet, for we were not solitary nor widowed, neither did we become tributary. There was sore weeping, but our lovers did rise up and comfort us with solid comfort. The relic hunters were extremely busy, and some of them coined money by the sale of their commodities to strangers and citizens who wished to retain some small remembrance of the powerful heat that melted everything in its progress. Glass two inches thick fell before it in streams. A gentleman found in the ruins a lot of dolls melted and run together. He called them fire-proof babies. In a store furnished with paints, oils, and glass, the specimens were elegant. Glass, in masses, was tinted with brilliant colors of every hue. So great a variety of curiosities was never found unless in old Pompeii, where the ashes preserved objects in a more perfect state. It seemed sad to see the merchant princes succeeded by little boys, whose stands were upon the corners where the heaviest business transactions occurred, or the most elegant goods had been displayed. It gives an air of romance to many spots, to remember that here and there men struggled for life, in the dark hours, and surrendered to the foe. Take, for instance, a scene like this which is vividly sketched by the *Tribune* :—

While Madison street west of Dearborn, and the west side of Dearborn, were all ablaze, the spectators saw the lurid light appear in the rear windows of Speed's Block. Presently a man who had apparently taken time to dress himself leisurely appeared on the extension built up to the second story of two of the stores. He coolly looked down the thirty feet between him and the ground, while the excited crowd first cried "jump!" and then some of them more considerately looked for a ladder. A long plank was presently found and answered the same as a ladder, and it was placed at once against the building, down which the man soon after slid. But while these preparations were going on there suddenly appeared another man at a fourth-story window of the building below, which had no projection, but was flush from the top to the ground—four stories and a basement. His escape by the stairway was evidently cut off, and he looked despairingly down the fifty feet between him and the ground. The crowd grew almost frantic at the sight, for it was only a choice of deaths before him —by fire or by being crushed to death by the fall. Senseless cries of "jump! jump!" went up from the crowd—senseless, but full of sympathy, for the sight was absolutely agonizing. Then for a minute or two he disappeared, perhaps even less, but it seemed so long a time that the supposition was that he had fallen, suffocated with the smoke and heat. But no, he appears again. First he throws out a bed; then some bedclothes, apparently; why, probably even he does not know. Again he looks down the dead, sheer wall of fifty feet below him. He hesitates—and well he may —as he turns again and looks behind him. Then he mounts to the window-sill. His whole form appears naked to the shirt, and his white limbs gleam against the dark wall in the bright light as he swings himself below the window. Somehow--how, none can tell—he drops and catches upon the top of the window below him of the third story. He looks and drops again, and seizes the frame with his hands, and his gleaming body once more straightens and hangs prone downward, and then drops instantly and accurately upon the window-sill of the third story. A shout, more of joy than applause, goes up from the breathless crowd, and those who had turned away their heads, not bearing to look upon him as he seemed about to drop to sudden and certain death, glanced up at him once more with a ray of hope at this daring and skilful feat. Into this window he crept to look, probably for a stairway, but appeared again presently, for here only was the only avenue of escape, desperate and hopeless as it was. Once more he dropped his body, hanging by his hands. The crowd screamed, and waved to him to swing himself over the projection from which the other man had just been rescued. He tried to do this, and vibrated like a pendulum from side to side, but could not reach far enough to throw himself upon the roof. Then he hung by one hand, and looked down; raising the other hand, he took a fresh hold, and swung from side to side once more to reach the roof. In vain; again he hung motionless by one hand, and slowly turned his head over his shoulder and gazed into the abyss below him. Then gathering himself up he let go his hold, and for a second a gleam of white shot down full forty feet, to the foundation of the basement. Of course it killed him. He was taken to a drug store near by, and died in ten minutes.

But by far the saddest case here was that of a beautiful and refined woman, known in art and operatic circles, whose husband is missing, and who escaped herself in only a night wrapper; was driven to distraction by the terrors of the wild flight, and was picked up in Lincoln Park in a state of more than half insanity. In the direst need of care from her own sex, ready to die almost from extreme exhaustion, and wandering in mind most of the time, she had had last night only the nursing and help which two men could give, and now lay on a pallet upon the church floor, directly behind the rear pew on one side. A young woman cared for her during the day, but at night female imagination lent partial insanity too great terrors, and care which should have fallen to womanly sympathy, devolved on the rude, though kind and skilled hands of men. The man whose brave and clear head gave him chief charge had had experience in a hospital; but it was pitiful that womanly protection should not be at hand, and that the couch of such a sufferer should not be tenderly spread under a private roof. Unhappily, the entire length of burnt Chicago intervened between all these sufferers, on the North Side, and that part of the city where suitable care could have been secured for them.

The most disastrous event in the horrible whole seemed, for a time, to be the destruction of the books of record in the Court-House; but it is found, on examination, that the loss is by no means irreparable. Many of the essential books are safe, though in one case—that of Messrs. Shortall & Hoard—the rescue was a marvellous achievement. This firm was located in Rooms 1, 9, and 10 Larmon Block, northeast corner of Washington and Clark streets.

The following account of the way in which the books were extricated was taken *verbatim* from Mr. John G. Shortall, senior member of the firm, by a *Tribune* reporter:—

I had just come home from church, and had been sitting in my house, No. 852 Prairie avenue, and was going to bed. I looked out of my north window and noticed a very bright light in the sky. I had been, from some unaccountable cause, quite apprehensive in regard to fire for some time previous; and, on noticing the light, determined to go to the fire, although it was not in the direction of my office.

I met a friend on the cars who was also going to the fire, and we crossed Twelfth street bridge, and got up to the side of the fire on Canal street, and followed it up from block to block to Adams street. We then got on Van Buren street bridge, and watched the progress of the flames for probably an hour and a half. I then had no idea that the fire would cross the river, and I argued with myself several times whether I had not better go home, but kept on staying watching the fire; and, while standing on Van Buren street bridge, I noticed a new body of flame—I should think there was an intervening space of fully half a mile untouched by fire. This new fire broke out, as it seemed to me then, in the vicinity of South Water street and Fifth avenue. When I saw this new light, I started for my office in Larmon Block immediately.

On reaching the office I found, as I apprehended, great danger existing from the awnings, which were outside the building, the embers dropping down very thickly on the roofs of the buildings, and on the fronts, and signs, and awnings. I ran upstairs, got into the office and tried to cut away the awnings in front of our building and that of the building adjoining; but, owing to the absence of anything adequate, I had to give that up, and simply press them close to the wall, that the embers might drop off them, and not be caught in them. Even then I scarcely believed it possible that the Larmon Block could take fire, and I requested the men in the upper portion of the building with buckets of water, to put out any embers that might fall there and endanger the building. In another half hour I felt more apprehensive, and went on the street to find an express wagon. This must have been an hour and a half before the building actually burned. I stopped probably fifteen different trucks and express wagons, offering them any pay to work for me in saving the books. Seven of them at least I engaged, one after another, they faithfully promising me that they would come back when they had carried the load and done the work in which they were engaged; but no one came back. At this juncture I met a friend, Mr. Nye, who was look ing out, as I was, for the danger. I told him I needed him, and he answered me promptly that he was at my service. We both watched some time longer for express wagons, but could find none. At last, when the Court-House cupola took fire I told my friend that we must have an express wagon within the next five minutes or we were utterly lost. He stood on Clark street and I on Washington, determined to take the first expressman we could find. The first one happened to come along on his side. He seized the reins with one hand, and taking a revolver from his pocket with the other, "persuaded" the expressman to haul up to the sidewalk, notwithstanding his cursing and swearing. When I came back from my unsuccessful watch, I found the expressman there, and my friend, handing the lines and revolver to me, went upstairs to help our employés, who were then in the office, to carry down the volumes. We got round with the wagon to the Washington street entrance, and, after filling the wagon, found that we had but about one-quarter of our property in it. Just at that critical moment I saw a two-horse truck drive up to where I was superintending the packing of the books, and my friend Joe Stockton, whose face was so covered with smut and dust that I did not recognize him until he spoke, turned over the truck and driver to me, with the remark: "I think, John, this is just what you need." I never felt so relieved or so thankful for anything as I did at his appearance with that substantial aid at that moment. We unpacked our impressed expressman immediately and set him adrift with $5 in his pocket for his five minutes' work, and commenced to pile our property on friend Stockton's truck. Meanwhile the flames were roaring and surging around us. Six of our boys were carrying down the volumes as rapidly as they could, and I, standing on the truck, was stowing away the books economically as to space. About that time they told me the Court-House bell fell down.

It must have been about two o'clock. I never heard the bell fall,

I was so excited. Toward the last, when we had got our indices all down safely, and we were trying to save other valuable papers and books, many of which we did save, it was stated that Smith & Nixon's Building was about to be blown up. Our truck was headed toward that building. The sky was filled with burning embers, which were falling around us thickly. As soon, I think, as the information was given that that building was to be blown up, the crowd rushed past us down Washington street, toward the Lake, terribly excited, shouting and warning everybody away. My driver was very nervous, and, on one pretext or another, would start his horses up for a rod or so, swearing that he would not be blown up for us or for the whole country; but I succeeded in stopping him eight or ten times during the excitement. In the mean time, our men were coming down the stairs laden with our property and returning as rapidly as they could. I was standing on the books, packing them in the truck, and the embers were flying on them, and I picked them off as they fell and threw them into the street, until, a rod at a time, we reached the corner of Dearborn and Washington. Messrs. Fuller and Handy were the last to leave the office, and they did not leave until Buck & Rayner's drug store was on fire. The store, as we believed, was full of chemicals and explosive matter. At that time the Court-House was a mass of flames, and our own building was burning, and other buildings in the immediate vicinity entirely destroyed. Three of us then started with the truck for my house, which we reached about three o'clock that morning. I had our property unloaded and placed securely within; and, after giving the driver and others some refreshments, I started again for the fire to see what aid I could give other sufferers.

There are three abstract firms who have saved portions of their books. Our own firm and Chase Brothers & Co. have saved their indices, digests of records, judgment dockets, and tax-sale records complete, together with many valuable memoranda, and probably 130,000 pages of copies of abstracts and examinations of titles, which are sufficient, we believe, with the aid of proper legislation, to establish the title to every tract of land in the city of Chicago and Cook County. Messrs. Jones & Sellers, I am informed, have saved their books of original entries, but have lost their indices. They have also, I understand, saved many volumes of copies of abstracts made. All these valuable documents, in the absence of the records themselves, are a firm security for titles to real estate in the city and county, and are sufficient to prevent any iniquity being done. Without them we should have to return to the tomahawk, pre-emption, and possession.

THE POST-OFFICE CAT.

A sketch of the doings of the Post-office in connection with the fire would not be complete without a notice of the office cat. She (or he) had been once before burned out, and was therefore, in a measure, prepared for this calamity. On the night of the fire the cat was present and assisted in the removal, though she did not go herself. When the work of removing the safes was in progress, the tearing away of a portion of the ruin revealed the faithful public servant in a pail partially filled with water. She had rented this as temporary quarters, and apparently enjoyed the cool shelter which it afforded. From her position it appeared impossible that she could have gone away and returned after the fire, and so she may be set down as the only living being who passed Sunday night and Monday in the burnt district.

A little before two o'clock on Monday morning, when the fire was raging, G. W. Wood, Assistant Superintendent of Railway Mail Service and Special Agent of the Post-office Department, arrived at the Post-office, convinced that the building would go. He was, of course, aware of the responsibilities which he would incur in removing anything from the office; but chose to disregard the requirements of red tape in the interests of the citizens who would suffer by the loss of their letters. When remonstrated with by a gentleman connected with the customs, he defined his position by saying that they might wait for an act of Congress if they chose, but he should do his best to save what he could. On this declaration of principles he proceeded to load everything portable into wagons belonging to the department. The force of men was large, the transportation ample, and the direction vigorous, the result being that every letter, both registered and common, in all the boxes and compartments of the office was hastily dumped into sacks and removed out of reach of the fire. All the mails in the building were rescued, with the single exception of a small one which came over the Fort Wayne road, and which, owing to the fact that it was four hours late, no one knew anything of. The registers and other matters belonging to the office, including the furniture in one of the private rooms, were also loaded in the wagons, and the whole was taken up town. Some of it had to be moved a second time, its first station having been on Harrison street; but everything eventually brought up in safety. On the evening of Monday, Mr. Wood telegraphed to Postmaster-General Creswell what had been done, and on Tuesday was instructed by that officer to spare no expense to carry on the mail service as well as before.

The work of getting out the safes belonging to the office from the ruins was undertaken a day or two after the fire had passed over, and the result was, in the main, satisfactory. There were some $60,000 worth of postage-stamps on hand, and these were rendered useless. Though not totally destroyed, they were so badly charred as to render their use impossible. They were forwarded to Washington immediately. The most valuable contents of the safes—the books of accounts—were found uninjured and in perfect order. The only exception was the cash-book, which, through some inadvertence, was left in a desk.

A gentleman says of the suffering:—

"I have just made a personal inspection of the condition of North Chicago. I entered by the northernmost of the bridges on the North Branch, in order to see first what was left. On my way to this bridge I came upon a young man, in the open lot, sitting on the ground by the side of a box of bread. Inquiring if he had gone into business as a baker, I found that he had been to some freight cars near by to procure some supplies, that his box contained meat and apples as well as bread, and that he was resting on account of feebleness produced by exposure after the terrible exhaustion of Monday. I shouldered his box and went with him just over the bridge to the temporary refuge which he had found. Besides himself there were his wife, three young children, and a widowed aunt with eight children, the eldest a girl. He had had a good situation as a clerk in one of our leading dry-goods houses in State street, and, with his aunt, owned four small houses, the rent of two of which was the sole dependence of the widow and her eight children. The fire took all they had, except the clothes in which they escaped, and about fifty dollars in money, which the young clerk had just invested in boards to build a shanty on their lots in which to house the double family of thirteen. The chance of obtaining employment for the man seemed fair. He slept out on the prairie the night after the fire, and was nearly helpless the next day from fatigue and severe chills. Probably great numbers laid that night the foundation of ague or consumption. The Sunday night had been very warm, and Monday, until toward midnight, was so mild as to make sitting out not quite uncomfortable for a well person. But a sharp change occurred about midnight, rain came, with violent and very chilly winds, under which even the robust suffered severely. Those who had some covering found the wind too much for them, and many lacked even the chance to shield their wearied bodies from the blast, and their little ones from the chill unfriendliness of the dropping skies. The rain was not drenching, nor was the wind near to freezing, but both were just at the point which makes excessive discomfort to the hardy, and to the enfeebled is the touch of distant but certain death."

A lady from St. Louis found in her rounds of mercy a mother and her daughter under a sidewalk. The latter had been confined there, and her babe was in such distress that the little creature's eyes protruded upon its cheeks. They were instantly provided for, but the little one could not survive the shock. The world into which it came was too hot just then, and not the "cold world" of poetry and despair.

I am told by the physicians here that as many as five hundred cases of premature birth have been reported, and the many helpless mothers who gave birth to children along the Lake can be numbered by scores. I can only weep as I hear this terrible tale. One told me last night is almost too much for human heart to bear. The daughter-in-law of a clergyman here gave birth to a child in the flight along the shore, and was separated from the family, and neither mother nor child have been found. Another —a lady in the Sherman House—was carried out in the arms of her husband, the new-born babe clasped to her breast, and both died in the father's arms before reaching a place of safety. The poor man, crazed with grief, was last seen along the shore of the Lake, with his dead across his shoulder. Again, I heard of a fine-looking woman in a night-dress being seen wandering along the Lake shore with twin babes, all of whom have died without recognition, and been buried by the city. These are but a few among the many awful horrors of that night.

FEARFUL ADVENTURES.

A graphic writer, not wholly reliable, however, says he had been watching the fire for hours, till at length it began to approach his boarding-house on the avenue, when he became seriously alarmed for the inmates, many of whom were helpless women, and among them Rosa D'Erina, the Irish prima donna, who had just conclud ed a series of entertainments.

I had previous to this made no efforts to save my effects, and it was now too late, as I found the balcony of the house (a wooden one) on fire when I got down. The women were panic-stricken, and seemed utterly incapable of action, but we succeeded, amid great difficulty, in rescuing them from danger, and, along with them, we wended our way towards the Lake shore. But my feelings were so much excited I could not remain long in any one place, and I again went citywards. I walked along Adams street, which had up to that time escaped, and found the Academy of Design, situated corner of Adams and Dearborn streets, still untouched. The Palmer House, on State street, a little lower down, also stood, and for a moment a feeling of hope sprang up in my breast that something might be spared even then. But this was a short-lived feeling. The Honoré Block, in process of erection by the father-in-law of Potter Palmer, caught, and now the Post-office, which had acted as a barrier against the progress of the flames eastward, was in imminent danger, and the district around seemed abandoned to destruction. The utmost exertions were made to save the mails, papers, and valuable contents of the office, and in a great measure, I believe, they were successful. Finally the Post-office caught; but being a very solid and substantial structure, it withstood the fire longer than any other which I had seen. Its interior was completely gutted, but the walls remain, and on Tuesday they served me as a guide through the ruins. Familiar as I was with the city, I could not otherwise have found my way. No chance now remained for the Academy of Design, and we mournfully watched the rapidity with which the fatal element was surely encircling it. It was filled with valuable paintings, among others Rothermel's great picture of "The Battle of Gettysburg," which had been on exhibition for some weeks past. I heard the picture was taken out in safety, but it was considerably damaged, and the largest portion of the remaining paintings had to be abandoned to the fire. The building was not fire-proof, and, being built with more regard to display than utility, was quickly and effectually consumed. The walls toppled over with the heat, and fell with an awful crash, and it is feared many perished in the ruins who had recklessly ventured too near. There were a large number of valuable stores in this district, and they with their contents were completely destroyed. I had a narrow escape for my life just then, and even now I can scarcely realize how great the risk was. I had, in my eagerness, gone too near the Honoré Block, when a falling timber struck me on the forehead and felled me to the ground. I was completely stunned for a moment, but the love of life was strong, and I struggled up, minus a hat, a loss which I soon replaced, as there were hundreds of them flying through the streets minus heads. I picked one up which made my appearance more picturesque than flattering. The east side of State street, towards the bridge, was now in flames, and the block of buildings north of Field, Leiter & Co.'s extensive wholesale store (the largest in the West) was being rapidly consumed. The employés of the house, some five hundred in number, had been busily engaged all the night in removing the most valuable portion of the goods to a place of safety; but the heat had become so intense that they beat a retreat and abandoned the immense stock to its fate. The building itself belonged to Potter Palmer, and was the finest business house in the city. It could not be valued at less than a million and a half of dollars, only part of which was covered by insurance. It caught in the roof, and in an instant was enveloped in flames. Gunpowder had previously been placed in the basement, but the fire was long in reaching it. At length a terrific explosion apprised the spectators that the end had come. The fragments were scattered around for blocks, wounding and maiming many persons, and shaking the foundations of the solid earth on which we stood.

A FAMILY TERRIBLY PERISH ON THE ROOF OF

N VIEW OF THE MULTITUDE BELOW.

When I opened my eyes again (which I had closed on account of the flying sparks) all I could perceive was a smoking heap of ruins. Immediately adjoining Field & Leiter's stood the booksellers' block of Chicago, in which the largest book trade in the West was transacted. It remained intact at six o'clock, when everything around it had been burned; but the fire, which by this time had made its way east to Wabash avenue, ignited it in the rear, and it burned up like tinder, the stock, of course, being perfectly inflammable. The loss must have been immense, as, the book trade being unusually active, a tremendous stock was on hand. Previous to this the water-works had been burned, and the city was now without light or water. The water-works were, in the opinion of the Chicago people, the finest in the world; but whether this be so or not, they were magnificent structures of their class, fitted up with all the modern improvements, and in perfect working order. When they burned, the Fire Department, which had never rendered much service, practically ceased to exist, and seven of the engines were abandoned to the fire.

The panorama was now awfully grand and magnificent, and presented a most imposing spectacle to those who had coolness enough left to appreciate the vividness of the scene. On every side, far as the eye could reach, the forking flames were shooting up, jumping entire blocks with the rapidity of lightning, filling the air with burning timbers, seizing fragments and inflammable material of every kind. The crash of falling buildings would at swift intervals drown all other sounds, and almost blind the spectators with dense masses of smoke, causing for a moment a darkness that could be felt. The entire northeast part of the city had now been consumed as far as the river, and the interest became concentrated in the southern part, towards which the fire was cleaving its resistless way. It was daylight, but you could not distinguish the difference between day and night, as the streets presented the same appearance, and the atmosphere and heavens looked as they had done for hours previous. Lower State street, south of Jackson, was one vivid blaze, the ruined structures belching forth whole columns of fire and smoke, and in the midst I perceived the Palmer House still standing.

I was astonished that it had held out so long. From its great height one would imagine that it would be one of the structures most liable to go, but it stood longer than others supposed to be completely fireproof. Its entire contents had been saved, and Mr. Palmer remained in the building to the last. I was told by a gentleman that Mr. Palmer's wife, who was waiting outside the building for her husband, had been struck in the building with a blazing fragment and severely burned. I did not see this myself, but I have no doubt of its veracity, as occurrences like this were innumerable. When the hotel finally caught, it rapidly burned up, and communicated the flames to St. Paul's Universalist Church, situated on Wabash avenue, which in its turn carried them further on. Adjoining the hotel were a great number of saloons and some of the most disreputable bagnios in the city, and when the dwellings caught it was horrifying to see the rascally proprietors selling liquor in the front part of their premises, and the rear on fire. Many of them met the fate they so richly deserved, the buildings falling on them before they could manage to escape. The burning district was now abandoned by all who valued their lives, and all who could reach the Lake shore, where they hoped to be safe. Subsequent events will prove how futile were their hopes.

But I am anticipating the order of my narrative. Wabash avenue, adjoining the Palmer House, was principally built of marble blocks, which were used for the better class of boarding-houses. Just above, where the private residences commenced, was located the Farwell Block, occupied by Farwell, Hamlin & Hale, and other prominent merchants. It had been destroyed about a year ago, in the great fire; but had been rebuilt at an enormous cost, and made more magnificent than it had been at any previous period. I thought it might have been saved, but it was not to be; it seemed as if all that was valuable, costly, and noble must be sacrificed to the relentless and conquering element, which was "monarch of all it surveyed." I did not witness the burning of this block, but I was told that it burned up with a rapidity that was perfectly terrific. On the opposite side the boarding-houses commenced, and in one of them, No. 159, a lady was burned to death before she could be rescued. The cross streets intersecting those running north and south were everywhere igniting, and I saw that everything was going to be swept clean to the Lake. I had by this time found the ladies of our party, and a few of us set to work to erect a kind of breastwork as a protection against the blazing fragments, which were falling thickly around. The scene on the Lake shore was awful in the extreme. Hundreds and thousands of people had carried what effects they had saved down there, in the hope of safety; but the last hopes they entertained were gone when they perceived Michigan avenue, the last street east facing the Lake, ablaze in several places. The terror and agony became intense; women were wildly screaming; young girls, with dishevelled hair and apparel all awry, could with difficulty be prevented from throwing themselves into the Lake. Children were seeking lost parents, and parents lost children; wives their husbands, and husbands their wives. Strong men fainted with the agony of despair; while high above all could be heard the brutal cries of wretches, who, maddened with strong drink, which was flowing like water, seemed bent on rapine and pillage in the midst of the universal dismay. I think history has never recorded a scene so full of all the elements of terror and dismay; for my part, the remembrance of it shall haunt me as long as I live. The breakwater became crowded with fugitives, and the trains of cars which were being taken from the Great Central depot must have caused numerous accidents.

Many had got along the edge of the Lake and immersed themselves in the water up to their waists, in the frenzy of the moment; and even here they were not for a moment safe. My position was immediately opposite Adams street, to which the fire had not yet come. On the corner opposite stood the magnificent residence of Mr. Honoré, the father-in-law of Potter Palmer. The next building was a Swedenborgian church, one of the strongest stone edifices I ever saw. So strong was it that I was certain the flames could not penetrate it. They did, however, and now our danger was great, as the fire was directly opposite to where we stood. We made all the precautions possible to save our lives, as we knew when the fire should pass a given point we would be comparatively safe. We confiscated two large carpets which we found on the ground, and immersing them in water, placed them on the tops of chairs, and got the five women of our party under them. We had not a moment to lose, as the fragments from the church and Mr. Honoré's house were rapidly coming in our direction. The heat was so intense that the carpets immediately dried up, but we had pails, and as fast as they dried we wet them again from the Lake. I had, on leaving my residence the night before, put on my overcoat, and I was congratulating myself all the time on my forethought; but even this had to go, as it took fire on my person, and I had to be baptized over again, adopting the plan of total immersion, for I jumped into the Lake. My companion's pants were on fire in several places, and he had to do likewise, so we were both in at the same time. We got out again, I minus my coat and he with his pants in a tattered condition. The heat was terrific; my face was literally scorched, and my eyes I thought would melt out of my head, but I was mercifully preserved, and I weathered the storm until the fire passed. When the smoke had cleared a little I looked north, and what attracted my attention first was the Pullman building, which had just caught fire. It burned to the ground in thirty minutes, and the great Central Depot, one of the ornaments of the city, adjoining Pullman's, was soon a living flame. It was occupied by the Illinois Central, Michigan Central, and Chicago, Burlington & Quincy Railroads, and was the head-quarters of the former and latter companies. Just beside were the two large elevators, in which were stored millions of bushels of grain; and taken as it was, all in all, it was one of the most valuable parts of the city. The depot burned up, and along with it hundreds of cars of every description; but the elevators remained standing entire when the depot had been utterly consumed. I believed them, as did every one else, to be safe; but a new and unperceived danger soon attracted our attention and convinced us that we had been too premature in our suppositions. There was a large quantity of shipping anchored at the mouth of the Lake, waiting to be laden with grain from the elevators. No one thought of these, as it was supposed they had made their escape out; but it was not so, and the tall masts catching fire from the sparks, communicated with the nearest elevator and set it instantly ablaze. It made a terrific fire, [illegible] burned the entire day; but, strange to say, its companion, just beside, escaped uninjured, and is preserved, with its immense stock of grain.

The fire companies from other towns were beginning to arrive, but owing to the scarcity of water they could accomplish but little. In one district they did good service, however, as they had a supply of water from the river, and, owing to their exertions, the fire did not spread to the West Side. All the stores of goods which were piled up on the banks of the Lake had become ignited, and the entire ground was one sheet of fire; yet, in the midst of all, fiends in the shape of men were pursuing their hellish trade. Whiskey barrels, which had been rolled down, were burst open, and men, and even women eagerly drank the fire in liquid form.

Free fights were, in numerous instances, indulged in, and the ruffians, rolling over each other, were burned and trampled to death. We had succeeded (after paying a fabulous price for an express wagon) in removing our women from the scene, and they made their escape to the southern part of the city, where they were for the time being safe. I remained, as I was determined to see the last of the spectacle, and I made my way, along with a printer on the *Evening Journal*, to the West Side. The task was one of no ordinary difficulty and danger, as walls were everywhere falling, and the ground was strewn with burning embers. We found the bridges gone, and could not tell how we were to cross the river until we met an attaché of the *Journal*, who told us we could cross at Madison street, a portion of the bridge still standing. I was surprised to find the Pittsburg and Fort Wayne depot still unburnt, when the stronger built ones had perished; but the fire on Saturday night had cleaned a space around it, and this, probably, accounted for its preservation. On reaching Canal street, we found that the *Evening Journal* had, with commendable enterprise, secured the Interior Printing House, and had already commenced to get up an extra, which they issued the same evening. My companion was called upon to work, and I was now left alone, and I went on towards the Galena depot of the Chicago and Northwestern Railroad, which I could not find, it having been long before consumed. The freight houses attached to it had also been destroyed, so that wherever I went nothing but ruin, complete and awful, met my gaze. The fire seemed to have spent itself on the South Side, and as the North Side then seemed safe, I thought I would go and endeavor to find a place of refuge, which I did far down Indiana avenue. I was so blinded with the smoke and scorched with the flames that the two ladies of the house could not recognize me, and it was with difficulty I gained admittance. I could find no water; so I took a couple of pails and started for the Lake, which was more than a mile away. I could scarcely drag my legs after me; but I managed to get back, and, as it was now pretty late, I retired to bed, but not to sleep. The heavens were more lurid than they had been the night before, and I was at a loss to imagine the reason of this, little dreaming that the whole North Side was being rapidly consumed. I rose early and went north, and I shall never forget the sights I witnessed on that terrible Tuesday. When Twelfth street bridge was reached, the roads leading out of the city were perfectly blocked with people, hurrying away, while vehicles of every shape and description were engaged in carrying their effects. Many fires were still blazing, and around Madison and Washington streets immense coal heaps were burning, the heat from which was very acceptable, as the morning was bitterly cold. Crowds were gathered reading

THE MAYOR'S PROCLAMATION,

the first which had been issued, and a universal gloom seemed to have settled on all faces.

We interrupt the story here to give this document, a full account of its origin being furnished in the next division of this work. Strange as it may appear, and the fact illustrates the completeness of the ruin, it was not for hours that a press could be found on which to print the proclamation.

The following proclamation was issued, and gave confidence :—

"WHEREAS, in the Providence of God, to whose will we humbly submit, a terrible calamity has befallen our city, which demands of us our best efforts for the preservation of order and the relief of the suffering,

"BE IT KNOWN that the faith and credit of the city of Chicago is hereby pledged for the necessary expenses for the relief of the suffering. Public order will be preserved. The Police, and Special Police now being appointed, will be responsible for the maintenance of the peace and the protection of property. All officers and men of the Fire Department and Health Department will act as Special Policemen without further notice. The Mayor and Comptroller will give vouchers for all supplies furnished by the different Relief Committees. The head-quarters of the City Government will be at the Congregational Church, corner of West Washington and Ann streets. All persons are warned against any acts tending to endanger property. All persons caught in any depredation will be immediately arrested.

"With the help of God, order and peace and private property shall be preserved. The City Government and committees of citizens pledge themselves to the community to protect them, and prepare the way for a restoration of public and private welfare.

"It is believed the fire has spent its force, and all will soon be well.

"R. B. MASON, *Mayor.*
GEORGE TAYLOR, *Comptroller.*
(By R. B. MASON.)
CHARLES C. P. HOLDEN,
President Common Council.
T. B. BROWN,
President Board of Police.

"CHICAGO, *October* 9, 1871."

We allow our reporter to continue his sad tale:—

The ground was so hot as to be almost unfit to walk upon, and in passing over it I thought of the torture of olden days, when wretches were forced to walk on heated metals as an ordeal for their real or fancied crimes. Dead bodies were being everywhere picked up. In one group I saw as many as thirty-eight corpses which had been gathered together for interment. I went over the ruins of my former abode, on Wabash avenue, and, while doing so, stumbled over something which I at first supposed to be a charred timber, but a nearer investigation proved it, to my horror, to be a dead body. The head, arms, and legs were gone; nothing remained but the trunk. Who it was, whether one of the inmates or a stranger, I could not learn; but it was there, and I felt a sickening sensation creep over me I could not control.

The Mayor had issued another proclamation directing the closing of the saloons, but no attention was paid to it; the police made no endeavor to enforce it, and in nine cases out of ten were themselves intoxicated. Even had they been closed, there was plenty of whiskey left; barrels of it were to be found in all quarters, and they certainly were freely broached. I never saw so many under the influence of drink, and where the roughs came from I cannot imagine. Full as Chicago was with them, I had never believed she contained so many as I saw on the streets on Tuesday. Many had come in from other cities; every train was bringing its contingent, and many began to look anxiously for the military, as it was feared the ruffians would complete the destruction by setting fire to what remained. All the available citizens were enrolled as special constables and invested with extraordinary authority, and they did all that men could do to subdue the disorderly element; but it was beyond their power to do so effectually, and outrages and rapine were hourly on the increase. As the day wore on the exodus from the city increased, the railroads furnishing free accommodation. All who had friends outside were leaving in hundreds, and long trains were leaving in rapid succession, carrying their loads of living freight to all points. As the lines had been burned up near the scene of the fire, passengers for the East and West had to go out to Twenty-second street to get their trains, and the rush and jam around these places was something terrific. I went to the temporary depot of the Michigan Southern and saw the afternoon train leave. The most pitiable sights were to be witnessed among the heart-broken refugees. I saw one woman, who had lost her child in the fire, seized (just before the train left) with strong convulsions, so severe that a medical man present said it was im possible for her to live. She was carried to the nearest dwelling, but I did not succeed in ascertaining her subsequent fate.

This was no isolated instance. Scenes like this were so numerous, that after a time they ceased to cause any surprise. The train, which consisted of seventeen coaches, slowly steamed away, and in the annals of travelling a more sorrow-stricken multitude was never carried by a company.

Thus abruptly breaking off from this narrator, we introduce our readers to one of the professors in the Academy of Design, Mr. Alvah Bradish, who writes the following letter concerning this institution and its recent destruction :—

To the Editor of the Chicago Tribune:

SIR: Among the more recent and cherished institutions of Chicago that fell a victim to the late fire, none will be more missed than the Academy of Design. The artists of Chicago had been organized for several years, and were steadily advancing the cause of fine art. Within the year they had planned, and seen growing up under their fostering care, a most beautiful edifice, almost wholly devoted to art purposes. It was situated on Adams street, between State and Dearborn, near by the Palmer House, and three blocks south of the Crosby Opera House. The academy building had been constructed especially to meet the wants of the artists. It comprised eighteen studios, all of which had been engaged before the building was finished. It was, indeed, a beautiful home for the arts, and for those who were making art a profession. The gallery was spacious. For fine proportion, for a true elevation and clear light, it was not surpassed in this country. The lecture room was ample, and the handsomest in the city; the reception room and studios, the stair-cases, approaches, school-rooms, were all fitted up in a style of elegance that speedily won the popular favor. The Academy had been thus founded by an enlightened body of artists, who were animated by the true ambition of adding the glory of art-culture to the other distinctions of the Garden City. These artists are mostly men of reading and culture. They foresaw the three essential conditions of a permanent and beneficent institution of fine art—an Academy of Design founded on principles that would insure growth, durability, and popular favor; *schools*, life and antique, for the thorough discipline of students; a *gallery*, open at all times for the display of the best works, pictures, and statuary; and *lectures*, both special and general. No academy can stand long without the full recognition of these three conditions; and they had been abundantly discussed, recognized, and established. During the past twelve months—the brief existence of their beautiful home—the artists had given numerous public receptions, and had varied their collection of pictures. They had offered to the public many rare works of the best modern masters, both American and European. Already the Academy had become the centre of art-attraction and art-culture in Chicago. Her schools, conducted by competent professors, had attracted a large number of pupils. Applications from the country and from the city were numerous for the coming winter. A centralized home had united the artists. They were working in good faith. Members of Council had, with unselfish enthusiasm, devoted a large portion of their time and thoughts to a wise administration of their trust. The younger members were making rapid progress in their studies, and felt the influence of a generous competition and the example of such rare works as the gallery contained, always open to their inspection. Already the Academy owned some good pictures; some others had been generously given. The Scammon collection of antiques was the gift of a gentleman of taste—an example that would soon have been followed by others. An art library was in contemplation. The artists were proud of their success. I can declare that no institution in Chicago had so speedily won such general favor. Its influence on public taste, and on the life, labor, and future of the artists was so manifest and so admirable, that it was universally recognized and acknowledged.

Mr. Rothermel's great picture, the Battle of Gettysburg, ordered by the State of Pennsylvania, had been on exhibition for two months. It was an immense canvas, sixteen feet by thirty-three, and was drawing crowds of admirers to the gallery. The attendance had been on the increase for two weeks past, when, on the Saturday previous to the destruction of a great part of the city, the visitors numbered one thousand. The coming winter would have witnessed one of the rarest exhibitions ever seen west of New York. Many pictures had arrived. The schools, thoroughly organized, would have been full; special lectures would have been enlarged and continued; and the course on the theory and history of the fine arts, first opened at the inauguration of the Academy, would have been given during the season. The leading artists were preparing pictures for the coming reception in November. It should be observed that Mr. Potter Palmer was putting up an edifice next to the Academy, with an iron front, of an elegant design, to be constructed especially for art purposes, studios, music rooms, etc. Most of these had been already taken. The struggle which the artists had thus made in the noble cause of art in Chicago would be crowned with success; for this new building would be opened through to the halls of the Academy proper,—thus concentrating the entire art interest and artistic genius of Chicago on Adams street. At this time there were a great many valuable works of art in the gallery and scattered through the studios—Drury's large and precious collection; Ford's beautiful Ohio wood scenes; Deihl's careful studies and designs; Jenks' conscientious labors; Elkin's world of Rocky Mountain studies; Bradish's popular "Leather Stocking," his full-length portrait of the late Douglass Houghton, and numerous smaller works; Pine's attractive group of children; James

Gookin's charming "Fairy Wedding," a gift to the Academy. Cogswell's studio contained some of his best portraits. Reed & Son's studio was crowded with pictures and studies. Pebble's studio contained numerous works of high promise. Other young artists, or students, occupied rooms and pursued their studies in the building; so that, with these hundreds of pictures and out-door studies, the Academy was emphatically the centre of art-interest and the cherished home of the artists.

On that memorable morning, the ninth of October, that witnessed the most dreadful conflagration of modern times, some of the artists were at the Academy by one o'clock. The great fire had only reached Clark street at two o'clock. The artists were not yet alarmed. At three o'clock the fire had advanced greatly northward on La Salle and Clark streets; the wind was sweeping through the streets, and carrying the fierce element towards the Chamber of Commerce and the Court-House. By four o'clock the great Pacific Hotel and the Rock Island Railroad depot were enveloped in flames. Would the new Bigelow Hotel and the Honoré Block be saved? The artists, gathering on Adams street, waited in painful suspense. Would the wind, now more terrific and pitiless than ever, lull for a moment, or would it veer a degree north, and thus save all this portion of the city? These thoughts flashed through our brains or quivered on our lips. Soon the Pacific Hotel—a magnificent structure, and nearly finished—was in ashes. The forked flames, made irresistible by the hurricane of wind, had struck the Bigelow Block, standing on Dearborn street, and wrapped it in a red winding sheet in a moment. The atmosphere was filled with brands, cinders, combustibles, all on fire, careering through the air. The splendid Honoré Block was seized by the devouring element, the unfinished roof furnishing the ready kindling, and these two stately blocks—the pride and ornament of a new street—faced with new marble, five or six stories high, were all enveloped in a few moments; were penetrated and swept by the fire fiend. Though we knew that on Adams and Quincy, Monroe and Madison streets, west, to the river, the finest structures had sunk before the blast of fire, we still clung to some hope. But the wind was on the increase, if possible. The writer stood for an hour close by, and witnessed the approach of the awful tornado, advancing rapidly and with irresistible strides north, but with less violence east, and at times hesitating to cross a broad street or strike a new victim. But what power could resist this hurricane of fire that came, as it were, in isolated sheets of flame through the air? The interior of these two noble structures were like appalling volcanoes that swallowed, from moment to moment, heavy timbers, walls, columns, as they fell inward. It was a sublime sight. Before this awful conflagration, in which already some of the most beautiful and costly structures of the city had melted like soft metal, the artists stood helpless in their anguish, but still hoping, praying, that they and their cherished home might be spared. A slight change in the wind would do this, for as yet not a building east of Dearborn street had been touched. The Academy was still safe; the eastern walls of the two noble blocks, though all luminous with interior fires, were still standing. Especially the Honoré Block, with its colonnade of white marble still firm, seemed to offer a solid bulwark to defend the more eastern portion of this part of the city. So intense was the heat of these edifices, all on fire from the pavement to their roofs, that the artists and groups that pressed forward toward Dearborn street to witness the sublime spectacle were obliged suddenly to retire and cover their faces. The south end of the Honoré Block, struck and torn by the blast, would give way. It bent, swayed, and surged for a moment, and finally twisting round, as it were, by the insatiable embrace, toppled over, stayed a second, then fell, with three upper columnar stories, carrying roof and cornice, crushing over into Adams street, shaking the earth for many rods about. Then shot up from the wreck a column of flame, through black smoke and cinders, that lit up the Palmer Hotel and threw a ghastly light on the façade of the Academy. In half an houı these volcanic fires had perceptibly decreased, and the artists were greatly encouraged.

But soon the Bigelow Block became the centre of tragic interest; for here the fires, sweeping over from the Pacific Hotel, now in hopeless ruin, seizing every intervening building and every combustible object in its way, had acquired a vehemence and violence most appalling. Now seemed the moment of greatest danger; for the Bigelow was directly west of our block. Between us was one brick five-story building; the others were low wooden tenements. They were like ovens, but covered by a hose in the hands of two colored men, who, with unsurpassed heroism, stood their ground. For a long time, by moistening the sides and roofs of these two buildings, the fires were kept at bay. They might burst into flame at any moment! Now the lofty walls of the Bigelow Hotel were all aglow with the fire inside, that seemed to crackle and roar with a triumphant sound as everything was devoured; the windows and archways belching forth tongues of red and white flame that reached nearly across Dearborn street. But, even up to this moment, when we saw the walls of the Bigelow and Honoré Blocks still standing firm, though greatly shattered, the artists took courage. These walls, that had risen like a dream of beauty under the eye of their architect, who stood now in our midst, seemed to offer a solid bulwark to the advancing enemy. Indeed, there was almost a shout of gladness heard from the group of artists that gathered in front of these torn and shattered battlements. There was a moment—one short moment—of congratulation and joy. It was five o'clock —not quite daylight. The wild ocean of fire had gone far off northeast. The awful destruction, the ruin, the dreadful havoc that followed that fierce march, cannot be told. We did not dream of its extent; we might hope some beneficent power would arrest its progress. We could hear the crackling of flames, the hurricane that scourged every street—that sent the fierce fiend through whole blocks; we could hear the distant roar, overpowering like an ocean-symphony, all near sounds. This sublime roar went moaning, like a storm at sea, through all the beautiful structures on Washington and the dense blocks north to the river. Who shall describe the swift horror that suddenly overwhelmed all those beautiful homes on the North Side? Happily, at that moment, we could not know of the dreadful scourge that was passing two miles north of us.

In the mean time, by six o'clock, in the face of so imminent a danger, the artists had taken measures to save such pictures as could be reached. All the smaller pictures in the gallery had been cut or torn from their stretchers. Some of the artists were too far away to be present. Some, living on the West Side, were cut off by the intervening fire. Up to half-past six, even, there was hope for us; but, before seven, some of the artists had gone several blocks south on State street, to make observations. The fires were advancing directly across Dearborn, along Jackson—the wind unchanged, and blowing with all its untamed violence, and rolling an ocean of fire over whole blocks of wooden dwellings, devouring everything it touched. No human power could save now the blocks south to Van Buren street, and we had become directly in range of this new danger; for no abatement could be seen, but, if anything, more fierce, more insatiable, this hated tornado carried whole roofs, planks, windows, all on fire, directly over the intervening tenements. And the Palmer House stood in range of this fiery storm, and the Academy but twenty feet from its walls, and overtopped by its stately Mansard roof. What pen shall depict the scene that appeared to our view? Every street and alley crowded with crazed, helpless fugitives; Adams and State, Quincy and Jackson, Van Buren and Wabash, one living,

WHERE THE FIRE BEGAN.

POST-OFFICE

LAND-OFFICE, ILLINOIS CENTRAL R.R.

REPUBLIC LIFE INSURANCE COMPANY.

MASONIC TEMPLE,

SOME OF THE RUINED BUILDINGS IN CI

STOM-HOUSE.

CHAMBER OF COMMERCE AND COURT-HOUSE.

N STREET.

CROSBY'S DISTILLERY.

FIRST NATIONAL BANK.

PHOTOGRAPHED BY THOMAS T. SWEENEY.

moving, screaming mass; helpless families; decrepit old age; infants on pillows in the streets; sidewalks crowded with furniture, chests, glasses, bedding, horses, wagons—all in confusion, without order, without kindness to neighbor, and none to direct or advise, but all fleeing from the brands and cinders that filled the atmosphere; rushing from block to block, weighed down by household goods; driven from house to house, till they reached the Lake shore beyond Michigan avenue, where hundreds of loads had been left or thrown on the sands. A few hours after this everything along the water's edge here was on fire,—the poor, desperate owners escaping only with their lives.

The artists had stood bravely by their beautiful temple, ready to aid if, by any chance, hope could come through any efforts or sacrifice of theirs. Up to this hour when the flames crossed Dearborn street, the Palmer House and the group of buildings near by could be saved; but when word was brought that State street was threatened south of us, all hope was abandoned, and the artists were obliged to look for personal safety. In the mean time, long before this hour—by seven o'clock—Mr. Reed, our Secretary, had given orders to have Rothermel's great battle-piece taken from its stretcher and saved from the approaching flames. There was ample time for this, though, in taking it down, it has suffered serious injury. Its great weight required several men to carry it out, and, in a bent, broken condition, it was taken to the steps of Trinity Church, Jackson street, and afterward to the university building, four miles south. Its subsequent fortunes for two weeks, to the time it was delivered to the distinguished artist who had designed it, may be given to the public by Mr. P. F. Reed, in whose charge it was. The Academy had a policy on it of $30,000. Such of the other pictures as were not carried by hand were placed on carriages and wagons. These were tied together, and, under the guidance of one of the artists, were moved by hand, by slow degrees, through the dense crowd, through Adams street and Michigan avenue, often blocked and arrested by opposing teams, and the suffering, crazed fugitives, but from time to time making progress, until, after infinite difficulty, the precious loads reached Harmon court, out of danger. By eight o'clock the wide area from Harrison street south, and Dearborn street west to the Lake, was all threatened with destruction which a few hours after witnessed. The writer of this, as he saw the five or six vehicles loaded with their precious freight of pictures, frames, books, trunks, and boxes belonging to artists and others, did not feel too sure they could make their way through such a confused mass of human beings in a state of indescribable excitement and frenzy. When the cortége passed the superb block known as Terrace row, facing the Lake, little did he think that, within two hours, all those beautiful homes would be levelled to the earth. Here lived Governor Bross, Mr. Griggs, Mr. Scammon, and other gentlemen of wealth and culture. The block was much admired for its stately grandeur. The next day its location could hardly be identified,—a shapeless mass of undistinguishable, smoking ruins.

It might be nine o'clock, and the Palmer House was still untouched. An imposing edifice, surrounded by an ocean of fire, its lofty three-storied Mansard roof, with five stories beneath it, rose supreme over all other buildings near by. But, soon after this hour, from pavement to roof it was one sheet of flame. Its walls swayed and trembled as the wind roared against its projecting portico, its windows and doorways belching forth to the north long spikes of red flame, forked, like ten thousand serpents, reaching out and lapping the walls of the Academy building as in horrid derision. The hotel thus covered with a sheet of flame,—its interior all red and dazzling with inextinguishable fires,—the walls of the Academy, only a few feet off, were heated, and the lower windows and doorways penetrated by an element as irresistible as fate. Was there any hope now left for the academy? Soon through its broken windows, down through its noble expanse of skylight, came the whirlwind of flames and murky elements, down-crushed timbers and walls, staircases, pictures, casts, —all the precious works that filled the studios of absent artists, —now all on fire, and adding intensity and grandeur to the whirling volcano of the interior,—a blackened, burned mass of art ruins for one moment, then shot up a sharp, dazzling spire of red flame, far into the impending smoke-cloud that rolled like a pall over the expiring structure, as though to proclaim a savage triumph over the fond hopes and labors of genius.

Thus perished the Chicago Academy of Design.

From this letter it may be seen how widely the blow smote; and yet, even farther than many think, were these tidings like deep wounds piercing. In the studio of one of our Chicago artists in Rome, we sat and heard the future of the Academy discussed with enthusiasm. Mr. Leonard W. Volk, who stands pre-eminent in sculpture, and was President of the Academy, and had purchased several valuable works for its use, has been cut to the heart by this loss. Far from home, and among a foreign people, this great sculptor has wept over the disaster which has come upon his own fortunes, and upon the career of his cherished institution.

A gentleman who had been presented with an expensive watch, went abroad a few months since, and left this valued gift in his safe, in a fire-proof building. Doubtless he wishes he had even exposed it to all the dangers of a foreign tour, now that it has been so thoroughly destroyed. One who has given his life to the examination of shell-fish, and had collected the materials of a scientific work on conchology of special value, and expected an appropriation from Government for the publication of his researches, has not a scratch of the pen nor the minutest shell left out of the conflagration. Such losses can never be replaced. In a great city, where everything was done by the representatives of all nations, there is an almost infinite variety of loss, from the toy all along up to the Medical College with its collections of a quarter of a century of existence.

"The lamentable tragedy at the Historical Society building is the darkest episode of this day. The people in the vicinity of this edifice, confident of its strength, gathered their most valued possessions and crowded the cellars in assurance of perfect safety. Among them were citizens of note, the venerable Col. Stone and wife, Mr. and Mrs. Able and two daughters, Mr. and Mrs. Carpenter, Dr. Leai and family, with several others not so well known. While the frightened group were moving a trunk, the librarian caught sight of a flame, and shcuting to the rest, rushed from the fatal place. The others, at least twenty in number, were not seen to emerge, and there is no doubt that they perished, as the building was soon tottering in utter wreck. The original copy of the Lincoln Emancipation Proclamation perished among the most cherished memorials of this Society.

"Death came to the crowds in the open air as well as in the buildings. A great following of ruffians, emboldened by the absence of the police and half maddened with liquor, assaulted several saloons on the verge of the fire, and held the ground against the advancing flame. When the moment of need came they were too drunk to get away. In this portion the fire came on with such incredible rapidity, that mothers threw their children down from the windows and then leaped down after them. Throughout the day and night every foot of advance was a complete surprise. In Chicago avenue, a noble thoroughfare one hundred feet wide, the people were confident of escape, and took little or no precaution. Here, as on Wabash avenue, when the fire did come, panic aided the devastation. Thoughtless women piled mattrasses and fragile goods in the street, and the dropping sparks took but an instant to make the avenue a glowing pathway of fire. The side streets were built wholly of wood, and the thin walls burned like shavings. This region, over by the Lake and the great Lincoln Park, seemed to offer safety. So a great rush was made for the park, and the refugees made themselves comfortable in the delusion of security. After ravaging to the limits of the city, with the wind dead against it, the fire caught the dried grasses, ran along the fences, and in a moment covered in a burning glory the Catholic Cemetery and the grassy stretches of the great park. The marbles over the graves cracked and baked, and fell in glowing embers on the hot turf. Flames shot up from the resting-places of the dead; and the living fugitives, screaming with horror, made for a moment the ghastliest spectacle that ever fell upon living eyes. The receiving vault, solidly built and shrouded in foliage, fell under the terrific flame, and the dead burst from their coffins as the fire tore through the walls of the frightful charnel-house. In the broad light of to-day the place is the most ghastly I ever saw, not even Cold Harbor exceeding it in awful suggestiveness. Above the graves charred stones stand grim sentinels of the dead, no more memorials of anything but disaster. Every inscription has disappeared, and even the dead are robbed by the flames. The park turned into a wilderness of fire, the crowds doubled backward and made for the avenues leading westward and to the south, to reach which they must cross the river. Many of the bridges were in flames—the rest were already choked with the heavy wagons which, tearing the way through, cruelly aggravated the distress of the thousands of foot-sore women and weary men. Fully 30,000 people were afoot in this quarter, and this mass densely wedged into barricaded streets between trampling horses, kept up a ceaseless stream far into the night. With the night new volumes of flame shot out on the air, and new crowds were hurled among the flying masses. There was no hope of saving the city: the struggle was simply for life. Half-clad women fled moaning through the streets, and at this time, it is asserted, robberies were perpetrated in some of the remote private residences. A vast throng reached the prairie, and sunk exhausted on the ground; the air was filled with a torrid heat, and even at this great distance immense particles of cinders fell in showers. The dreadful agony of separated families came to add its horrors to the calamity. Babies were found alone in the multitude, and countless little people crept about crying wildly for their parents. A blessed rain came down slowly, and the fire, stayed in its advance, rolled backward and flamed up with greater fierceness in the immense coal piles in the very centre of the town. Then a new agony came upon the people. The only untouched portion of the town was brilliantly illuminated, and for a time it seemed as though not a roof was to be left in the great city.

"The first victims were the poorer classes, and as they were driven from their burning homes they hurried with the goods they had been able to save (or to steal) to the eastern and southern parts of the city, as if with an instinct that the fire must fall back before the stone and brick palaces of the rich. Thus the lower end of Wabash avenue became choked with the débris of disaster and flight. Cursing men, shrieking women, and terrified horses stumbled over the streets and sidewalks, pursued by the tempest of flame and the scorching blast of heat which swept on from the centre of the city. For one awful moment the whirlwind rushed through the beautiful avenue; but, happily, at Congress street its ravages were stayed. How shall any one forget that extraordinary scene, where the horrible and the ludicrous, the mournful and the grotesque, mingled like the visions of a nightmare? Ladies half-clad, but loaded with heavy burdens, rushed madly from those luxurious houses, and joined the hideous throng of the struggling poor, inextricably entangled with wagons and horses, and trampled by thieves and outcasts. Some had just put on all their finery to save it. Many were almost naked. Not a few carried infants nursing at the breast, and a great many were hugging lap-dogs. Tipsy men, fantastically clad, made ribald jokes upon the fugitives. Families who had been lucky enough to get trucks to cart away their valuables and bric-à-brac, sat disheartened on top of the load. Parties interrupted in the midst of a carouse ran madly about, too drunk to knew what it all means. All the while the motley throng pushed frantically southward. The weak were thrown down by the press and trodden under foot. For hours and hours the panic hegira continued, pushing out towards the prairie. From Monday morning at daylight the fear was for life, not for property. In this dire extremity the greed of man added to the horror of the scene. Drivers of carts and carriages crowded over from the divisions of the city presumed to be safe, and demanded outrageous rates for the slightest services. Yet it is to be said to the credit of human nature that hundreds of honest men turned out heartily to aid their more unfortunate neighbors. In all the horror of this southward pressure there was a continual stream of curious people from the distant regions crowding eagerly forward to see the vast illumination. The counter-currents, as they met, caused frightful mishaps and confusion. Men and women, maddened by the red terror behind, fought ferociously for a pathway to safety. Near each church vast masses were assembled with a sort of assurance of safety in those sacred precincts. Presently rumor came that it had been resolved to fight fire with fire. Laird Collier's church was to be blown up, and the dense crowd in the vicinity broke frantically for a new refuge.

"Late in the morning the people of the North Division were involved within the sudden horror of fire and death. A great crowd had assembled at one of the avenues leading to the burning region, where the close approach of the fire moved the bridgemen to turn the draw. The move was of not the slightest avail. The fire lapped the slender wood-work in the vicinity, leaped lightly from bank to bank, and before the bewildered people could make a movement toward safety they were helplessly environed by raging walls of fire, the Lake rolled lazily beyond them, and with one impulse the great crowd made for its shelter, and buried themselves in sand and water. This scene was simultaneous with the Wabash avenue stampede."

In illustration of the excitement that robbed some of their senses, and made them do the thing they did not care to do, and leave undone what they ought to have done, we mention the case of a lady who gathered her silver into a basket to place it in her husband's safe, as they could scarcely bear it away with them without danger of losing it. When she came to the moment of depositing the valuables, she took instead of the silver a pin-cushion, worth half a dollar, placed it carefully inside, closed the safe, and ran out of the house. The safe preserved everything it contained, and the lady now possesses her pin-cushion as a relic of the Great Fire. Truly it must be considered a costly reminder of the agony and fright of that dreadful morning.

We have heard of people becoming so upset in such a moment as to throw mirrors out of the window and carry cook-stoves down stairs with particular care. In such heat it was difficult to keep cool. Men entered their stores in the rear, and before they could open their safes they were driven out of the front door by the pursuing flames. It became then a race for life, and sometimes the fire proved too swift for the unfortunate fugitives. Horses grew frantic, and refused to move until a blanket or robe was wrapped about their heads to hide the fearful glare.

A mother, escaping with her babe clasped to her bosom, suddenly plunged from the darkened staircase into the blaze of the approaching fire. Her darling, terrified and shocked by the quick flood of light, and partaking the mother's alarm, made one quivering motion and died in her arms. This was worse than loss of home. What a burden did that mother bear through the horrors of that conflagration!

A business man, who had seen his buildings and machinery sink into ashes, and a prosperous business disappear in an hour, was summoned a few weeks afterwards to bury a new-born babe. He was a strong man, to whom tears were strangers But when he communicated the sad news to his pastor, he exclaimed in the midst of sobs and weeping, "Oh, this is our first great sorrow. The loss of property is nothing; but our little one is gone, and I feel so sorry for my poor wife."

A business man, watching by the couch of his dying wife, knew that his books and papers were all burning; but he stirred not from her side, and ere the embers were cold amidst the ruins of his marble store, he saw the remains of his companion lowered into the grave. Everything seemed to combine to crush him, but he bore up like a Christian hero.

A clerk of the Court, who must be a man of kind heart, since a merciful man is merciful to his beast, put his cats in a bag, and tied a string around the neck of his dog, and thus laden sought safety in swift flight. Another clerk of the Court, having put all things in order for removal, was about to leave his house, when his little rat-and-tan dog sprang from his perch and clasped his legs around the neck of his master, and there clung like a child, and thus was saved. He too perceived the danger, and loved his life too well to be sacrificed without a struggle. Doubtless dumb animals felt the horrors of that woful night as well as human beings.

As an instance of the sagacity of the dog, so often observed and justly celebrated, a gentleman fleeing before the flood of fire ran down a street across which the flames were already pouring in torrents, when his faithful dog began to bark and jump up upon him, and hinder his advance in the fatal direction. The master at length perceived the animal's purpose, and stopped to take a view of the course before him, when he was able to discern the danger of further progress, and turned in time to escape by another way. In a bank vault under one of the great buildings that fell before the blast of heat, a mouse was discovered safe and lively, without the smell of fire on it. This relic may hope to become one of the "lions."

The President of the Illinois Central Railroad, arriving early upon the scene, found that he could not reach his family on the North Side by the bridges, and, after arranging for the safety of freight-cars, books, papers, and other property, he employed a tug to convey him down the river, out into the Lake, and so along the shore till he could gain a landing, and thus access to his wife and children. But the fearful smoke and heat made the attempt a failure, and he returned bewildered and almost crazed by anxiety and the horrors of the time. He put every machinery in motion for the purpose of ascertaining the fate of those dear to him, and on Tuesday, at four in the afternoon, he learned that they were all safely housed in Evanston. How many happy meetings like this occurred within that mournful week of the fire!

A gentleman, living near the corner of La Salle and Madison, started early at the commencement of the fire to relieve his brother-in-law, near whose home it began. At midnight he hastened back, fearing the progress of the devastating element, to provide means of escape for his own household. When he arrived within two blocks of his late dwelling, all was gone in smoke and flame. They had been in the very central line of fire, and now where were his wife and four children? Scouring Madi son street, he at last discovered his wife seated on her trunk in a doorway, and disconsolate as ever woman was. His joy upon seeing her was swept away by the information that the children had gone north by La Salle, while she went east down Madison. The eldest, a girl of eighteen, had taken in charge the three boys and two trunks full of clothing, and sought escape or protection with a friend at the mouth of the tunnel leading to the North Side. Leaving his wife in a place of supposed safety, the anxious father engaged a man to go round the blocks where they might be expected to have fled for refuge, while he also sought for them where he hoped they might be. He was compelled to return, baffled and disappointed. Removing his wife and the trunk still further, he stayed and fought the fire till the water-supply failed, and then they joined the procession marching along the avenue southward out of the range of the fire, now rising again into uncontrollable fury. The crowd seemed orderly, solemn, and composed. Ladies of wealth and position were blackened by soot and dust, many of them dragged trunks by their handkerchiefs fastened into the handles, or carried bundles or boxes. All were intent on saving their lives and something besides from the general wreck and ruin.

The purpose of those whose fortunes we are describing was to gain the West Side by way of Twelfth street bridge, and then to seek a refuge with old friends. There they hoped to meet the children if they were yet alive. Twelve mortal hours elapsed before this worthy couple rested under the hospitable roof of their friends on Park avenue. Their children were not there. Their hopes were dashed, and whither to turn they knew not. The father, almost frantic, returned to the scene of desolation, hurried from place to place, made inquiries of all his friends, and got no tidings of his lost ones. At night he turned homewards with a heavy heart. But upon reaching the threshold, there were the gleaming faces of his loved children. Two hours after he left the house of his friends they had appeared, bag and baggage. Their story was one of romantic interest. When the alarm of approaching peril roused them, the women wakened the other members of the family. With great difficulty they got their colored servant sufficiently wide awake to realize the situation. They at once resolved to save their best clothing, and the boys were dressed up in their Sunday best. They loaded themselves down with whatever apparel they could get on their persons. The mother wore away several skirts, and both were arrayed in their finest silks. They also filled three trunks, and throwing their beds over the piano to save it from water in case the engines should deluge the house, they bade adieu to their home. The mother took one direction and the children the other, in hopes that between the two routes one would prove to be safe. They did not then apprehend the magnitude of the danger, nor conceive that the gigantic blocks could be melted by the flood that was sweeping across the city, driven by the hurricane. The young lady bethought herself of the sewing-machine, and found two men willing to aid her in its removal. Back she went with her noble helpers. One of these had a wooden arm, which he lost without knowing it at the time, in aiding her to save the machine. Driven from the refuge she had hoped to be secure, at the mouth of the tunnel, she found a milk-wagon, and got herself and the boys and their rescued property conveyed to the North Side. The driver proposed to stop at his residence, believing it to be out of danger. But the young woman said no, and induced him to convey them still further. When he finally returned to his home, after they had been disposed of, he saw only its smoking embers. So fast had the demon wrought!

Supposing themselves secure in their distant retreat, they began to think of father and mother. Soon, however, the tidings came that their refuge was threatened, and they were about to load up for a retreat still further north, when the heroine bethought herself of her West Side friends, and hiring a dray, she packed the goods upon it, and for ten dollars she and the children and property were conveyed to their asylum, which they reached some time in the afternoon. And at night the family were reunited, glad and thankful, even though they were homeless and almost like beggars, upon the verge of winter.

A white-haired Scotch lady, who was taken from the fiery furnace, aud barely saved, said that her father's picture, an oil painting, and her mother's Bible, were consumed, and her eyes became moist and her voice choked, as she added, "These are the things that trouble me most." Choice mementoes of those dear to her heart, never to be replaced, were more precious than jewels and velvets. Oh, the diabolical energy of this fiend, which spared nothing sacred, nothing cherished, and smote, with human bodies, the idols of the heart, and reduced to ashes fondest memorials of the past! Bridal gifts presented to those who were about to become brides, and nuptial offerings half a century old, were all melted and dissolved without mercy. And some who had expected to approach the altar in gorgeous array, stood up in calico, and were adorned with paper flowers. Doubtless they were as happy in these simple fixings as if they had been peers of Solomon in all his glory. Yet some courage was necessary on the part of those who plighted their faith and took upon them the yoke of matrimony amidst the ruin of their fortunes and prospects. And common sacrifices and struggles will knit them into closer and tenderer fellowship.

Among the peculiar losses by this fire were heir-looms long held in families as sacred treasures, and never to be restored. Their value was inestimable to those who had them in charge. A man of gray hairs, describing to his pastor the events of that fearful morning when they were hurried out to escape personal injury, said that they seized in their haste things least valuable, and left other articles that money could not buy or replace. "There was my father's picture, the only one owned by any of the family relations. It was forgotten and lost." As he uttered these words his voice faltered, and he broke down in tears.

A German musician of splendid abilities, who had lately come from his fatherland with his wife and five children, was driven to the prairie, where they lay out two nights exposed to the autumn blasts and dews without protection. His loss of personal effects was almost entire, and beggary stared him in the face; but kind friends sought him out and relieved their necessities with abundant supplies. There was one thing no hand of mercy and charity could return. He had brought with him a violin three hundred years old, for which Ole Bull had offered the family three thousand dollars, and been refused. It was a darling of the artist's heart, and when he feared lest it would suffer harm in the flight out of the flames, he resolved to bury it in the yard, and did so. Ordinarily such a precaution would avail much, even if the earth was but slightly piled above it. But, alas! the precious wood was consumed by the fierce heat, and he found, upon returning for his treasure, only charred remnants. Who can ever describe or enumerate the losses of this kind in such a sweeping, all-consuming conflagration, which allowed so little time for reflection or action?

A gentleman who owned a choice library ordered the expressmen to load up with books. When another team came for its load, the question was, What shall we bring out? The answer came, "Books!" And so he saved his whole collection, and has them intact, while all else was lost.

Other men employed all the hands they could find to roll out their liquor-casks and save this fiery fluid, whose ruinous effects are worse than those of flame, because they burn up men's souls, and involve them in other evils than those which end with time.

Some ladies resolved to secure their best clothing, and accordingly dressed themselves up in silks and velvets and jewelry, even putting on several skirts and dresses in order to carry away as much as possible by their only means of conveyance.

It was the only thing possible to many to remove their families, and then they were "saved, yet so as by fire." One man brought from an upper story his aged mother, and left her standing upon the sidewalk, while he hastened back for his sick wife.

Upon reaching the rendezvous, the poor man missed his mother. The flame and smoke and confusion were so great that he had but a moment to search for her, and was obliged to fly and leave the spot. He never looked upon that venerable form again. She was lost, and perished.

A gentleman in one instance was coming down the steps of his house, in perfect safety for the moment, as he supposed, when a vast sheet of flame whirled down over the whole building, striking him to the ground, and only not making an end of him because it was lifted up for a moment by a gust of fresh air, under cover of which he staggered away. A saddle-horse just left unhitched before the door dropped in his tracks with no attempt to get away, and died almost instantly. A house-owner went for a wagon and assistants, expecting to have ample time to remove all his goods; when the wagon was procured he found that it was hopeless to attempt so much; then he made up several bundles, only to find that the larger of these must be left behind; then the bundles first carried out were set on fire by the shower of sparks in the street, and the last man coming out was smitten down, as I have related, on the very steps; so that the party not only did not save their goods, but barely escaped with their lives. Remembering that in very many cases the getting away the family was similarly interrupted, some idea may be formed of the terrible fashion in which people were surprised and almost swallowed up. In the case of a family particularly known to me, the lady looked out of her window to get a glimpse of what she had heard of as a fire two miles off, and before she could summon her household and get on her clothes her house was in flames. She got away herself half dressed, with but a wrapper hastily snatched, as she hurried her little ones into the street. This was before day on Monday; but during that forenoon of flying terrors, great numbers had equal difficulty in getting out, after discovering imminent danger where it was supposed no danger existed. I have learned definitely since my last, that Robert Collyer and his family made their first removal to his church, then a second to the house of a friend several blocks west and a little north, where there was supposed to be no danger, and thence they were driven in a short time to find refuge eventually in the son's cottage, on the remote edge of the city, at a point where something was spared.

CHILD'S RELIC.

A child of seven years went, at five in the morning, to her church, which was likely to be burned, and looked for some article which she might save. Her younger sister stole away with her, and they both fixed on the communion-service as the most valuable and precious thing they could carry. This had been purchased by special contributions, and was sacred in the children's eyes. The plates and cups were taken charge of by the eldest, and the flagon by the youngest. Out into that cloud of smoke and dust these heroines marched in that early twilight, and they faced it four hours—the youngest, meanwhile, having lost her burden, and become separated from her sister. Three days after the fire the father found the eldest child, and she still clung to her treasure, and would give it to no one but her minister. Such an instance of pious love and devotion to the sanctuary has hardly an equal in the annals of time. Both these dear girls were dearer than ever to the father's heart, and we trust God himself looked on them with a smile.

The greater part of the fire in the North Division occurred after daylight on Monday, and the spectacle presented in that quarter was such as would be presented by a community fleeing before an invading army. Every vehicle that could be got was hurrying from the burning district loaded with people and their goods. Light buggies, barouches, carts, and express-wagons were mingled indiscriminately, and laden with an indescribable variety of articles. Others were hurrying to the scene from curiosity, or

ST. PAUL'S CHURCH—UNIVERSALIST.

CHURCH OF THE HOI

METHODIST CHURCH BLOCK.

FIRST PRESBYTE

SOME OF THE RUINED CHURCHES OF C

-ROMAN CATHOLIC.

ST. JAMES'S CHURCH—EPISCOPAL.

:H—SOUTH SIDE.

SECOND PRESBYTERIAN CHURCH.

PHOTOGRAPHED BY WILLIAM SHAW, CHICAGO.

to complete the work of rescuing friends and property before the monster could destroy them.

People crowded the walks, leading children or pet dogs, carrying plants in pots, iron kettles not worth ten cents, or some valueless article seized in the excitement; many looked dolefully upon the lurid clouds, still far away, and wondered whether they and their homes were in danger; and others looked as though they had spent the night in a coal-pit or a fiery furnace. There was such "hurrying to and fro" as the world seldom sees, with univer sal agony and distress.

A gentleman on the train with several of our merchants going home from New York, says:

A wretched cripple came into the train with a doggerel petition asking for aid to put him on his legs again. "Just our affair," they laughed; "we're all cripples together;" but they showered the "stamps" upon him, which he received with all the surly discourtesy of his race. Then they began to ask each other where they would put up, facetiously mentioning the burned hotels.

"Is the Pacific open?" asked one.

"Yes, at the top," said another, and the jest was highly relished.

At Laporte a man came on board, of whom one of the passengers asked: "How about my house?"

"Burned," was the reply. The next question consisted merely of a searching glance, and the answer was, "She's all right at our father's; we got your papers out of the safe this morning; they are all right, too."

"Well," said the merchant coolly, "when a man has his wife and his papers, what more does he want?"

A CHICAGO MAN'S GOOD FORTUNE.

The first man I met on leaving the train was the Hon. L. Swett. I asked if he was one of the few fortunates. He smiled and nodded. I congratulated him on the safety of his house. "Oh! that's another matter; my house is gone, but my wife and children were saved." This spirit is too common to be remarked, yet when you compare it with what you see among other people, it seems very admirable.

A druggist came to me one day in Madrid, half insane because he had bought a soda-fountain which he could not work. He tore his hair, bit his fingers, and called down maledictions on his birthday, because he saw $300 in danger of being lost. "My ducats and my daughter!" A Chicago man is very fond of his daughter, if he has one; if not, he is equally fond of his neighbor's daughter. As for ducats, he likes the gaining of them remarkably well; but when he loses them, he thinks much less of them than of those he intends to gain.

The gloomiest man on the train was the representative of a great New York house, which had large credits in Chicago.

Potter Palmer left home on Sunday night worth many millions; despatches reached him at every station on his way East, and every despatch announced the loss of a fortune. But he did not tear his hair, nor did he speak disrespectfully of the day he was born. He doubtless thought very vigorously how he was to go to work to get back those millions.

A lady who resided on the North Side, thus gives her experience: I do not speak or write of this terrible event as one who has only listened to the report that flew from lip to lip, but as one who stood face to face with death, and counted the leaden-footed hours as they dragged by their endless length, and prayed for the coming of the morn—that morning whose dawn was to reveal only more clearly than the lurid glare of the flames had done, how wide-spread was the ruin that the fire had wrought. I had attended service at St. James, and returning, retired early to rest. At twelve I was awakened by some of the boarders in the house coming in from the fire, and passing through the hall up to their rooms in the story above mine. My room was on the south side of the house, and the light shone through the shutters and fell in red bars on the opposite wall. I sprang up and looked from the window. The fire was a mile or more away, but the roar of the flames and the crash of falling buildings could be plainly heard, while a meteoric shower of sparks filled the air, and cinders fell like snow around us. The wind was blowing very hard, and constantly increasing. At one o'clock footsteps were heard hurrying through the house, doors opened and shut, and anxious faces peered out to ask, "What of the night." At two o'clock a message was sent the round of the rooms, "Pack your trunks." The fire was rapidly nearing us. At three, the order came to "bring out the baggage"—human strength was in vain—human power was as a thing of nought—human ingenuity or courage was powerless before that angel of destruction whose red torch lay at our doors. The fire could neither be controlled nor checked. The gas, already burning low, went out, and with a terrible, oppressive sense of the impending danger, we went outside the door, and sat down on our trunks, where they were piled awaiting transportation to a place of safety. Our house, 364 and 366, was in the centre of the last block of buildings at the east end of Ohio street. Beyond that was a space of unimproved ground about two blocks in extent, then a large lumber yard, then the beach and the Lake. Opposite us was a fine block of buildings, consisting in part of the residence of H. M. Miller, the well-known jeweller, and a large first-class private boarding-house. Like our own, it was the last block, and beyond it the unimproved ground spread down to the water. Upon this space our trunks were placed, but the heat and the falling cinders soon drove us down to the beach, to the very water's edge.

There we again sat down, only, as it proved, to wait the coming of the hungry flames. At five o'clock, all that was left of what had been our pleasant home, was a heap of iron, brick, and ashes; and even while we congratulated ourselves upon our personal safety, and jested lightly about a "tent on the beach" for a temporary local residence, the cry was heard, "The lumber-yard is on fire!" It was only too true. Like flashes of lightning from the breast of some purple cloud, fire leaped forth red-tongued from a score of points, then a broad sheet of dancing flames and flying cinders; and in a moment more the heat from the dry, seasoned pine lumber was intolerable. No pen can do justice to the scene that ensued. No imagination has power to picture the sickening details. No tongue can convey to another an idea of its horror. As far as we could see to the north, the beach was covered with goods of every description. The household gods of the rich and the poor lay side by side, and the millionaire and laborer sat down together to guard them. If *death* is a leveller, what less can be said of a calamity like this? The lady who yesterday rolled by in her carriage with her coachman in livery, or who held her silken robes daintily aside, while some child of poverty crept humbly by in rags, hushed her own bitter lament to speak soothing and gentle, but groundless words of hope and encouragement to the homeless wretch by her side. The sparks fell amidst the piles of bedding and clothing around us; fire broke out in every direction, and we were compelled to abandon everything, and fly as fast as our weariness would permit, toward the North Pier. What hand guided our flight, only the heart that is stayed upon its Maker, knows;—surely it was not reason, for that seemed to have utterly forsaken the mass of humanity that fled, amidst groans, and tears, and curses, and prayers, the neighing of frantic horses, the lowing of frightened cattle, the yelping of dogs, and the cries of cats that were half consumed by the fire while they yet lived. Neither the weakness of age nor the helplessness of infancy were sacred in that hour when all were desperate. Suddenly, while we pressed on in our mad wild flight, a shriek—a woman's shriek—freighted with inexpressible agony, rang out on the air, rising above the Babel-like confusion that surrounded us, and, looking back, I saw a sight that chilled my blood, even in that moment when our terror was so intense as almost to preclude the possibility of another sensation. A pair of powerful horses rendered uncontrollable by the heat and smoke and confusion, had thrown down a boy of six or seven years of age, and the heavily-laden dray to which they were attached passed over his head, killing him almost instantly. The mother sprang forward and caught up her child, and, with the mangled and bleeding head pressed to her bosom, gave expression to her sorrow in most heart-rending cries, that rose, shriek upon shriek, as she staggered on with her lifeless burden. Scorched by the intense heat, suffocated by the dense smoke, blinded by the sand and ashes and cinders, the crowd pressed on. Alas for him or her who fell by the way! There was a cry, a groan, and the tidal wave of humanity swept on, and all was over.

Our flight was stopped at last by the river. Kind hearts had devised means to aid us, and kind hands drew from shore to shore a dry dock laden with its living freight. I crossed with the first, and climbed from the dock to a schooner, thence to the shore, and then, over piles of hewn timber, over heaps of stone, and bricks, and rubbish,—how, is known but to Him who has promised that "as thy day is, so shall thy strength be." Three or four steamers lay moored at the North Pier that had come into port during the night, and our party went on board the Alpena, while others went on board the Morning Star and the Corona, and as many as could be were taken off by schooners and tugs, but yet the majority were left upon the beach. Soon the flames spread to the shipping; several schooners were burned, then the flames were seen bursting from the windows of the steamer Navarino, and the miserable refugees who had sought shelter here fled panic-stricken from this new danger. For a time it seemed that our own boat must share the same fate, for she was aground, with her fires out, and only the almost superhuman efforts of her officers saved her. At nine o'clock we were safely anchored in the Lake, and the doomed city was hidden from our sight by the pall of smoke that enveloped it. We secured a state-room, and the three ladies and two children who made our party crowded into the berths, where we tried in vain to rest our throbbing temples and weary limbs. The day wore slowly by, and as the gray shadows of the early dusk crept over us, I went out on the deck to take, what it seemed then, must be my last look at Chicago. The long, low stretch of shore lay spread out before us, and as far as the eye could reach was an almost unbroken line of lurid, cruel fire. To the north and to the south the flames leaped, and swayed, and surged like hungry fiends. The wind still blew a perfect tornado, and, in spite of two anchors, our boat rocked to and fro on the wild waters, like a spirit that could not rest. One long look of sorrow and despair; one long look of bitter, unavailing regret for her fate; one long, sad, unspoken look of farewell to the Queen of the West, that peerless city that was being tried as by fire, and I turned to enter the cabin, when a group attracted my attention. In the centre was a woman who, under other circumstances, must have been very beautiful, crouching upon the floor, with her white hands fast locked together. Her great brown eyes were tearless, but eloquent with their dumb woe, and ever and anon moans burst from the quivering lips that spoke no word of the sorrow that had almost unseated her reason. They told me she was the mother of three little children; the youngest a babe of a few weeks old.

Her husband had gone out in the night and had not returned, and when the fire drove her from her home, she started down the beach with the crowd, a little nurse girl, herself, a mere child of a dozen or fourteen years, assisting in the care of the children. One of them had fallen, and being injured, she had put her babe in the nurse's arms to be able to better assist the child when the crowd pressed forward, and before she could recover herself she was parted from her helpless little brood. Back and forth through the throng she had run, calling aloud for them to come to her, until exhausted, when, she could not tell how, she had come upon the boat. Frantic with suspense as to the fate of her husband and children, she paced the cabin through the long cold night, and her moans and the sullen plash of the waves, as they broke against the boat, mingled with my dreams, as in imagination I lived over again the scenes of that terrible night and day. At midnight sufficient rain had fallen to subdue the fires, already partially exhausted, and when the bleak, cold morning broke we looked upon a scene of desolation such as never was seen before in the New World. We had partaken of no food since our late Sabbath dinner, the Alpena having no stores on board, and indeed the excitement had stimulated us to that extent that it is improbable that even the nectar or ambrosia of the gods would have tempted us to break our fast, or that the royal banquets of Cleopatra would have provoked a thought of hunger; but now a sickening faintness crept over us, and we were weak and worn. At eleven o'clock Captain Samuel Shannon, of the propeller "Toledo," came into port and visited the Alpena, and learning the facts, invited us, with a seaman's proverbial generosity, to come upon his boat and eat a warm breakfast, to which he had the satisfaction of seeing full justice done. At noon we left the boat and once more trod the streets of that city whose wealth, and prosperity, and luxuriant growth had been the pride of the world, as well as the marvel of the age. But now, shorn of her glory by one fell blow, she sat, a queen indeed, but a queen whose emblems of royalty were broken, whose robes of Tyrian purple trailed in the dust, whose shapely limbs were swathed in sackcloth, whose feet were buried in the ashes of her ruined palaces. Yet with all our hearts we did her homage, for the world of earth and air and water were her empire, and her throne was as enduring as the blue Lake that lay before her.

A wholesale grocer, residing on the North Side, was absent from the city. His wife, a delicate woman, finding the flames suddenly upon her house, snatched up a silver cake-basket and a valuable little clock, took one of her two children in her arms and another by the hand, and fled. As she sped before the pursuing fire, she found her strength failing, and begged the driver of a passing express wagon, lightly laden, to help her in her extremity. He would for the clock. She submitted to the exaction, was carried three blocks, and then forced to get down. The cake-basket bought her another ride of about the same distance, and then she was forced to finish her flight on foot, her means of satisfying the rapacity of drivers being exhausted. Finally, more dead than alive, she reached a place of safety.

On Monday evening a knot of men, from 35 to 40 years of age, stood on Michigan avenue, watching the fire as it fought its way southward in the teeth of the wind. They were looking grimy and dejected enough, until another, a broad-shouldered man of middle height, a face that might have belonged to one of the Cheeryble brothers, shining through the overspreading dust and soot, approached them, and clapping one of their number on the shoulder, exclaimed cheerfully: "Well, James, we are all gone together. Last night I was worth a hundred thousand, and so were you. Now where are we?" "Gone," returned James. Then followed an interchange, from which it appeared that the members of the group were young merchants worth from $50,000 to $150,000. After this, said the first speaker, "Well, Jim, I have a home left, and my family are safe; I have a barrel of flour, some bushels of potatoes, and other provisions laid in for the winter; and now, Jim, I'm going to fill my house to-night with these poor fellows," turning to the sidewalks crowded with fleeing poor, "chuck full from cellar to garret!" The blaze of the conflagration revealed something worth seeing in that man's breast Possibly the road to his heart may have been chcked with rubbish before. If so, the fire had burned it clear, till it shone like one of the streets of burnished gold which he will one day walk.

A woman living on Ontario street, between Market and Franklin, brought out her two children, aged five and seven, safely, and then went for a baby. The children followed her back, and none came out alive.

The Quinn brothers went into their house while it was untouched by the fire to secure some clothing, but in getting out had to jump through the windows.

Mr. Malcomb, who died about two hours before the fire reached his residence, was burned almost beyond recognition.

A story is related of the proprietor of St. Caroline's Court, a hotel on the West Side of Chicago, illustrative of General Sheridan's idea of the eternal fitness of things. The General called at the hotel and inquired the price of board. "Six dollars per day," was the reply. "The price before the fire?" inquired the General. "Two dollars and a half." General Sheridan replied that he would run that hotel himself, and at $2.50 per day. He placed an orderly in charge, and at once put a stop to exorbitant charges.

The following curious incident is well authenticated: Mrs. ——, the housekeeper of a prominent hotel, had made up her mind to leave the city a few days before the fire. She had not drawn her salary for some time, and it amounted to $1,000. On Saturday this amount was handed to her by the proprietor. The boarders at the same time got up a testimonial, amounting to $150, and presented her with the money that evening. She deposited the greenbacks under the carpet in a corner of her room. When the fire was raging, Mrs. —— rushed into her room and succeeded in saving a favorite canary-bird. But she forgot all about the money.

The son of Mayor Mason, of Chicago, is worthy of Chicago and of his large-hearted sire. Everything was swept away except his wedding presents, which were at the house of his father. This house was saved. He sold them to Tiffany & Co. for $5,000. With this money he will now re-establish himself, opening a stove store for the time being in the basement of his father's elegant residence. The young man shows the real Chicago pluck.

A locomotive engineer was on his freight-train, forty miles from the city, when he heard the fire was raging on Michigan avenue. He said, "I asked permission to go on with my train, and was forbidden ; I put on steam, and they put down the brakes, but I pulled my train as near to the depot as I could, and left it in charge of the fireman. I hurt nobody and did no harm to anything ; I went straight to the place where I left my family, and dragged out their bones. When I came back to my situation they told me I was discharged, and I am now homeless and helpless."

Men were desperate, and deemed almost anything justifiable. One who saw that he could not escape, opened his veins that he might not know the horrors of death by fire. Another, probably rendered insane by losses and terror, was found with his throat cut from ear to ear. Men who were laboring to rescue their books and papers from the peril, were so involved in the mazes of the fire, that they tried several streets before they were able to escape, and then suffered serious inconveniences or injury in the final struggle that saved them. One, in trying to gather a few things from his room, fell suffocated, and, recovering presence of mind, crawled to the window, and calling on men to catch him, leaped from the second story, and was able to rejoin his family. A fireman brought a two-year old child to a lady, which was snatched out of the upper story of a lofty building in the heart of the fire. The little thing was scorched and singed, and when asked, "Where is papa?" he answered, "Gone to church."

"Where is mamma?" "Gone to church." So unexpected was the fire, that the parents had not time to find their darling after church. Some 300 were caged up near the river, and taken off by the steamer that lay close at hand. Others, hurried out of their home and cut off from egress by any street, fled to the Lake shore, and as the furious element closed around them they were pressed into the water, and kept themselves for hours by dipping their heads into the cool element. Children were immersed repeatedly, in order to keep them from being scorched, and many came from their wet refuges more dead than alive. A family who had spent several years abroad, and collected many valuable works of art and souvenirs of their journeys, were driven from one place to another, and finally took refuge in a stable. The proprietor begged them to take his carriage and drive it off to save it. In this they escaped several miles to a place of safety, having nothing left but what they wore upon their persons.

A man at the corner of Division and Brandt streets had apparently secured his household goods in an open lot; but the flames mercilessly attacked his effects, and seeing there was no further chance of saving them, he knelt down and offered a brief prayer, after which he arose, clasped his hands in wild despair, and looking to heaven, exclaimed, "God help me now," and was soon lost to view in the dense smoke through which he endeavored to make his escape.

Mr. Kerfoot gives the following graphic account of his escape from the fire with his wife and children: "Being the owner of a horse and carriage which I used to go to and fro from my business, when I became satisfied that my house would soon be enveloped, I brought my horse and carriage before the house, and placed my wife and children in it. There was then no room for me, so I mounted the back of the animal and acted as postilion. While driving through the flame and smoke which enveloped us on all hands, I came across a gentleman who had his wife in a buggy, and was between the thills hauling it himself. I shouted to him to hitch his carriage on behind mine, which he did, and then got in beside his wife. I then drove forward as fast as I could, for the flames were raging around us. After proceeding a short distance, another gentleman was found standing beside the street, with a carriage, waiting for a horse, which was not likely to come I directed him to fasten on behind the second carriage, which he did, and in this way we whipped up and got out of the way of the flames with our wives and children, thank God."

A remarkable instance of courage and presence of mind is told of Mr. E. I. Tinkham, of the Second National Bank. On Monday morning, before the fire had reached that building, Mr. Tinkham went to the safe and succeeded in getting out $600,000. This pile of greenbacks he packed into a common trunk, and hired a colored man for $1,000 to convey it to the Milwaukee depot. Fearing to be recognized in connection with the precious load, Mr. Tinkham followed the man for a time at some distance, but soon lost sight of him. He was then overtaken by the fire-storm, and was driven toward the Lake on the South Side. Here, after passing through several narrow escapes from suffocation, he succeeded in working his way, by some means, to a tug-boat, and got round to the Milwaukee depot, where he found the colored man waiting for him, with the trunk, according to promise. Mr. Tinkham paid the man the $1,000, and started with the trunk for Milwaukee. The money was safely deposited in Marshall & Illsley's bank, of that city.

Mr. Nathaniel Bacon, of Niles, Michigan, student-at-law with Messrs. Tenney, McClellan & Tenney, at No. 120 Washington street, slept in their office. On waking at about one o'clock, and seeing the Court-House on fire, he saw that the office, which was immediately opposite, would surely go. Judging that one of the safes in the office would not prove fire-proof, he promptly emptied the contents of his trunk on the floor of the doomed building, and, filling it with the interior contents of the safe—books, valuable papers, money, etc.—shouldered the trunk and carried it to a place of safety on Twenty-second street, losing thereby all his own clothing and effects except what he had on. That young man is a hero.

In the midst of all that was sad and terrible, there was an occasional gleam of the humorous.

One merchant, who found his safe and its contents destroyed, quietly remarked that there was no blame attached to the safe; that it was of chilled iron, and would have stood, but that the fire had taken the *chill all out.*

A firm of painters on Madison street, bulletin their removal as follows, on a sign-board erected like a guide-board upon the ruins of their old establishment:—

MOORE & GOE,
House and Sign Painters,
Removed to 111 Desplaines st.
Capital, $000,000.30.

An editor of a daily paper has received several poetical effusions suggested by the late disaster; but he declines them all, on the ground that it is wasteful to print anything which requires every line with a capital, when capital is as scarce as it is now in Chicago.

A bride, who entered the holy married state on Tuesday evening, determined to do so in a calico dress, in deference both to the proprieties and the necessities of the occasion. But she desired that her *toilette de chambre* should be, if possible, on a more gorgeous scale. Being destitute of a *robe de nuit* of suitable elegance, she sent out to several neighbors of her temporary hostess to borrow such a garment, stipulating that it must be a *fine one*. So peculiar is the feminine nature, however, that her modest request excited no enthusiasm in her behalf among the ladies to whom it came. This is not a joke.

A sign-board stuck in the ruins of a building on Madison street, reads: "Owing to circumstances over which we had no control, we have removed," etc.

CHICAGO, October 12, 1871.

To the Editor of the *Chicago Evening Journal*:—

The attention of Chicagoans is called to the 8th chapter of Deuteronomy, and the clergy of the city are respectfully requested to take the same for a text on Sunday morning next.

MERCHANT.

One of our merchants, reported insane, was heard from at New York—where he had gone to bury a sister—in the following noble manner:—

Mrs. Potter Palmer:

I have particulars of fire. Am perfectly reconciled to our losses. We shall not be embarrassed. Have an abundance left. Be cheerful, and do all possible for sufferers. Will return by first train after funeral. POTTER PALMER.

The fugitives from our city were good, bad, and indifferent. The men of pluck and value to us generally stood by the wreck to restore the town. Many truly unfortunate could do no better than to leave for a time. Some found the place too hot for them. Among these may be reckoned the villain who thus ignobly perished in Ohio, where he had gone to retrieve his fortunes. Says the Lima *Gazette*:—

The fire in Chicago has begun to make itself felt in the rural districts. Additions are daily made to the population of the country towns. These additions consist generally of men with scarred faces and sinister looks, who are looking around for some opening in the way of business and trade.

On last Saturday, October 28, one of these enterprising unfortunates visited some of our farmers in Amanda, German, and Marion Townships, in this county. He was in the horse trade. Wherever he went he wanted to buy horses. All day Saturday was consumed in fruitless attempts to buy a horse. Night found him in Marion Township, about three miles west of Elida.

Between seven and eight o'clock in the evening he entered the house of Andrew Stever. Stever is about sixty years old, a bachelor, and has the reputation of owning considerable of this world's goods. He lives alone in a small log cabin, is the owner of the farm on which the house is situated, and has resided here for twenty years past. He has the reputation of being a peaceable, quiet, inoffensive man.

After entering the house of Stever, our horse-buyer from the burnt district introduced the subject of horses, and proposed to buy one from Stever. Stever had none to sell. He then inquired for matches, and requested Stever to furnish him with some. This Stever proceeded to do. The matches had scarcely passed from his hand before the stranger drew from his pocket a revolver, and, presenting it at Stever, asked him "if he saw that?" Stever replied that he did, but that "this was no place for it." Stever in the mean time had observed that the features of his visitor were disguised by daubing mud in his moustache and whiskers, which were of not more than a week's growth. Stever, therefore, by way of precaution (and which precaution had also probably been quickened by the sight of the revolver), opened the blade of a pocket-knife, and kept it in his hand—his hand in his pocket.

A motion on the part of the stranger to present his revolver was the signal on which Stever acted. Grasping the hand that held the pistol with his left hand, he told the man he must leave the house. A terrible struggle ensued. Tables were turned over and broken, and everything movable in the house was displaced. Stever kept his hold upon the pistol-arm, while the stranger strove to beat him over the head with the pistol as severely as was possible under the death grip of Stever. While this was going forward, Stever continued with his right hand to ply the knife. This he continued to do, although he was under, to so good purpose, that, to use his own language, he made him "grunt." His hold upon Stever relaxed, when Stever rose from the floor, the stranger rising with him. On getting to their feet, the stranger reeled and fell in the portal of the door, when Stever jumped over him and ran to a neighbor—a Mr. Carr. With Mr. Carr he returned to the house, where they found the nocturnal visitor where he fell. He gave one or two gasps after they got to the house and was dead. The Coroner's inquest on Sunday developed the following facts: Before entering the house he visited the stable and procured a bridle. This, with his hat, overcoat, and shawl, he left near a stack of straw. On his person were found six watches, two revolvers, one single-barrelled pistol, and $82.50 in money. His arms were tattooed with India ink. He was apparently about forty-five years old, with as forbidding features as one seldom sees. There was nothing on his person to mark who he was or whence he came. The Coroner's jury examined Stever, and the body of the unknown was disposed of by the Coroner.

A policeman in New York City found four women and a child standing on the corner of Chambers and West streets. In answer to his inquiry they told him that they had just arrived from Chicago by the eight o'clock train, and, being entirely destitute, they did not know what to do. The officer took them to the station-house, and Sergeant John J. Fitzgerald, who was in charge,

NEW ENGLAND CHURCH—CONGREGATIONAL.

ST. JOSEPH'S PR-

RUINS OF THE BIGELOW HOUSE.

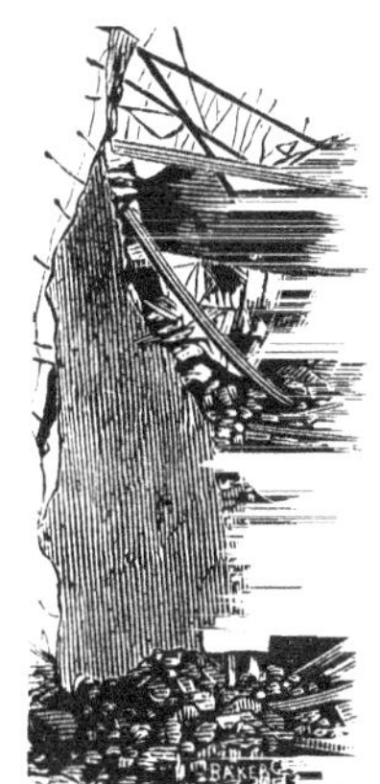

RUINS OF THE ILLIN

RUINS OF THE PACIFIC HOTEL.

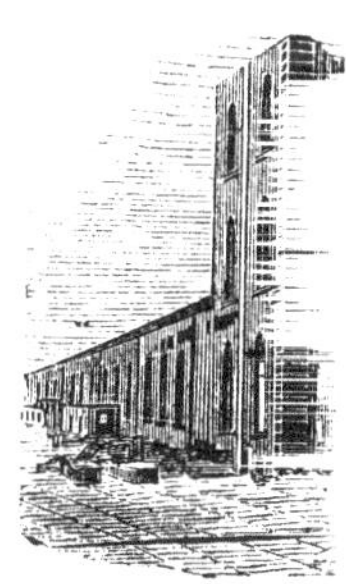

RUINS OF THE

:RMAN CATHOLIC.

UNITY CHURCH—DR. COLLYER.

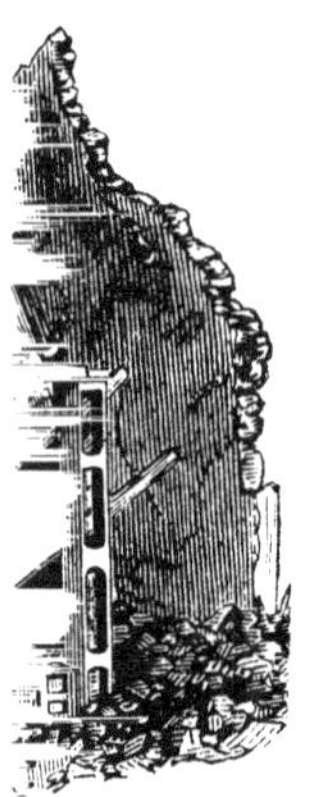

'AL LAND OFFICE

RUINS OF THE METHODIST CHURCH BLOCK

'ON DEPOT.

RUINS OF SAND'S BREWERY.

examined the case. Finding the women were just what they had represented themselves to be, sufferers by the disaster in Chicago, he made every effort in his power to accommodate them the best way he could for a short time in the station-house. He then sent men out to the neighboring houses to state the case of the poor people. Assistance soon came in the person of Mr. N. Huggins, proprietor of the Cosmopolitan Hotel, who desired the sergeant to send the women over to his house, and they should have everything they needed until the proper authorities came to look after them. The women were then sent to the Cosmopolitan, and gave their names as Lina Mylo, Minnie Ditzler, Annie Fris, and Bridget Mahon and child. They were sent up stairs, were properly cared for, and, being tired from the hard ships they had lately undergone, they all retired except Annie Fris, who made the following statement to a *Herald* reporter of the scenes through which she had just passed:—

My father was a silversmith on State street, and lived in the house with my mother. I wanted to learn to cook, so I went out to the house of a young friend of mine to get taught. My father wanted to bring her into the house, but I did not want that, as I preferred to go to where she lived. He tried to keep me at home, and bought me a piano for $1,000, and I had only just taken two lessons on it when all was burned.

I am the only child my father and mother had in this country. We belonged to Medo, in Bohemia, where I have a sister married now. On Sunday night, about nine o'clock, I went to bed, and had been asleep for about an hour, when the other girl woke me, crying fire. I jumped up and rushed to the windows, but everything all around where I could see was in a great big blaze. I pulled on something, and all ran down the street to save my father and mother, but when I got within about half a block of them the fire was all in the house, and father was hanging out of the window, stretching his hands out to me, calling to me to help him, but I could do nothing. Then I turned to go back to my friend's house, but some men had come along the street, and they threw bottles of kerosene and matches into the place until everything was on fire. I don't remember what occurred after that, I was so frightened. When I saw my poor father burn up before me, and heard my mother shrieking out to me, and I could do nothing for them, I would have rushed into the house and died with them, but some men picked me up, threw me into a carriage, and took me away out to the other side of the city. I was on the college grounds with hundreds of other people. I did not know any one there, and no one knew me. I have no relatives in this country anywhere. I was two days in that place without anything to eat but some little bits of bread that a lady gave me. I did not want to eat. I was so distressed about my family, and having nowhere nor any one to go to, I went into the woods with all the other people, when the fire came to us, and there we had nothing scarcely for three days. We had to sleep on the grass when we did sleep, but that was very little, as we had too much trouble to think of it. On Friday I left Chicago because I did not know what to do. Some ladies gave me a pass to New York in a church, and I came on here. It made me so sad and sick to remain in Chicago that I thought I would rather go anywhere than stay there. Two of these ladies who are with me promised to take me with them, as they have some friends here, but they are very poor themselves, and I don't know what they will do. My father had some money in the bank, but I don't know in what bank, or how much it was, so that I suppose that is gone too. I am just fourteen years of age, and I have nothing in the world but just what is on me. I think if I could get back home to my own country I might get something. I don't know what to do. I have scarcely thought about it yet, for my poor father and mother they did everything for me. All the people were very kind to me since I left Chicago. I got something to eat at Buffalo, and then the people on the train gave us something as we came along. The police were unusually kind to us when we came here, and it makes up a little to us to find so much charity and feeling in the people.

Miss Fris is an interesting-looking young lady; she speaks English freely; and as soon as the present grief of her loss and the bewilderment of the strange situation she finds herself in wear off, would prove a great acquisition to many a private familv in some position, as she is willing to work.

Our young city had many charitable institutions, and among them, on the extreme north, was the Half Orphan Asylum, where much good was being done by a few benevolent ladies, in relieving these unfortunate children whose natural protectors were unable or unwilling to care for them in a tender and humane manner. There were some seventy odd children in the asylum at the time of the fire, including about a dozen infants, and when it became evident to the matron that the building would have to be vacated, she at once made preparations and looked about for the safest means of removing her charge of little children. The North Side line of omnibuses had removed some half dozen of their vehicles to the neighborhood of the Asylum for safety; and Mrs. Hobson at once sent a gentleman to secure the services of a few omnibuses in which to remove the children to a place of safety. His efforts were fruitless, and he returned to the asylum with the intelligence that the omnibuses could not be had, as the persons in charge would not allow them to be used. Mrs. Hobson immediately went herself to represent the urgency of their claim, but utterly failed to procure even the use of one omnibus, to which number she at last reduced her request.

Failing to secure assistance, Mrs. Hobson returned to her charge, and at once, with the assistance of some kind friends, got the children in readiness to find safer quarters, which they hoped to do in the new building built for the Asylum on North Halsted street, near Centre, which was as yet in an unfinished condition. About ten o'clook on Monday morning the little troop of children started to find a new place of shelter, each little one able to walk, carrying some article of furniture or utensil, endeavoring with their puny strength to save something for the general good. They finally reached their destination in safety. With the aid of a cart, some ten loads of bedding, clothing, etc., were removed from the old to the new quarters, and the little ones were made as comfortable as possible, and finally put to rest. But their sleep was to be of but a short duration, for the fire-fiend threatened to pay them another visit, and again the little ones had to be removed, and again they fled from before the line of fire whose progress no human power could arrest. The streets were crowded with a multitude of people who were frantically hurrying toward the West Division, endeavoring to carry some of their household goods to a place of safety, disputing the right of way with teams of every description, loaded with every conceivable variety of household goods. Amid this thronging multitude, and under a heavy rain which had set in, the poor little children had to endeavor to pick their way along. At Clybourne avenue bridge, foot passengers and wagons had to mingle in one common road way, and nothing but an overruling Providence could have brought these little children in safety through such a hurrying and dangerous crowd. They crossed the bridge in safety and were now out of immediate danger, and by two o'clock Tuesday morning a church building was reached, the key found, and again the tired little crowd were in a place of shelter, but wet, hungry, and tired. The neighboring people kindly assisted in hunting up bread and milk for the children, and sleep once more kindly took possession of their weary frames.

But Mrs. Hobson's task was not yet done. As soon as her charge was in safety she returned to the Halsted street building, only to find that the bedding, clothing, and provisions which she and others had saved with so much labor during the day had been stolen during her absence. Disheartened, but not discouraged, she sat down on the steps, wrapped in a blanket, intending to keep guard over the building the balance of the night. And wel. it was she did so, for soon after a couple of fellows entered the inclosure and came toward the rear entrance of the building, expecting probably to have matters their own way. But Mrs. Hobson was equal to the emergency. She called out to them to stop, as they had no business there. This did not intimidate them but for a moment, and they again advanced toward the building, when Mrs. Hobson raised her arm toward them and told them if they came any further she would blow their brains out. This frightened the scoundrels, and turning about they hastily ran away. Thus was the Asylum building probably saved, and the orphans placed in security through the efforts of the noble matron, Mrs. E. L. Hobson, and a few devoted friends.

Some people had a less noble mission than this, so nobly performed. Says one:—

It was almost as ridiculous as melancholy to watch the long stream of people who poured out of the tenements on Adams street, Van Buren street, and the alleys near the river, both on the West and South Sides, and to notice what each bore. On Adams street the perambulators outnumbered every other article saved. About every third person wheeled one, and about every seventh perambulator contained a baby. One man in his shirt sleeves, and with but one boot, wheeled a child's carriage, in which was a baby, perhaps eighteen months old, astonished at its sudden awakening and the crowd, and sucking lustily at a green paper lamp-shade. These alone evidently remained of all his Lares and Penates. Another, perfectly frenzied with excitement, rushed along Harrison street, waving over his head the handle of an earthenware pitcher, and shouting at the top of his voice. The women, with hardly an exception, carried a bundle in one arm and a baby in the other, and he their shawls thrown over their heads. Perhaps a couple of older children clung, frightened and crying, to their skirts. When the hotels were menaced, out poured from each a long string of guests, each with a valise in one hand and dragging behind him a trunk. The fate of these amateur baggage-smashers is wrapped in mystery, as hardly a travelling trunk was anywhere to be seen on Tuesday.

If all our citizens, as Marshal Williams suggests, had been as fertile in expedients as the one below, much more might have been spared.

"One building on the West Side, which was saved after desperate exertions, owes its preservation to an agent, rarely if ever used before for such a purpose, and which in efficacy was a formidable rival to the Babcock. The roof was covered with wetted blankets, and when water for this purpose failed, two barrels of cider were employed with success. The flames retired, and the proprietor on the roof caroled a joyous pean, 'A little more cider, too.'"

A good story is told of Mr. Milligan's trotter, a splendid animal, worthy the industrious and successful owner, who had but recently rebuilt his magnificent store after a fire had consumed it to the ground.

Peoria sent a steam fire-engine to the relief of Chicago, and in one of the narrow streets it was so nearly surrounded by the flames that the men had given up hope of saving it, and were about being forced to seek their own safety in flight. At this juncture Mr. Milligan, of the firm of Heath & Milligan, came along with his roadster. Perceiving their peril, in a moment he had hitched the fast trotter to one side of the pole, the men caught the tongue, pole, and wheel, and with a cheery shout, out they whirled through the smoke and cinders at a four-minute gait. The Peorians saved their steamer, and vow that they will get up a subscription and purchase Milligan's sorrel if the city has to issue more bonds.

An Eastern man, who felt somewhat incredulous about the reporters' marvellous tales of the fire and its merciless devastation, thus describes

THE SCENE OF DESOLATION.

As I have said before, I had a sneaking idea, while I was yet in the suburbs, that the extent of the fire had been exaggerated in the Eastern papers, and that I would be certain to find a very different state of affairs from that which I had anticipated before I got out of the cars. But how mournfully was I disappointed! We entered the burned district by passing through State street. It was dusk as we got near where the Court-House once stood, and the feeling that came over me as I stopped my horse at this point and looked about me, was one of positive awe and dismay. As far as the eye could reach was a waste, a desert, with here and there a standing wall of some great building, through whose open windows the lurid glare of the coal fires beyond and around could be seen falling and rising with the wind as regularly as if worked by machinery. I shall never forget the scene. On, on we went, turning here and there from one street to another, picking our way carefully over the well-tried and yet perfect wooden pavement, lest by a misstep we should be plunged headlong into some cellarway or vault screened from view by a pile of brick or stone that had once been a building. After making all sorts of windings, with the same interminable view of gaunt walls and burning coal piles surrounding us whichever way we went, we reached a bridge which was solid enough to admit of our crossing to the North Side. Indeed, when I had got to the bridge I was under the impression that I had reached the full limit of the fire track; but how wonderfully mistaken did I find myself when, on getting to the other side, I saw before me a plain two or three miles ahead, as clear of anything like a house as the wild prairie itself! I noticed, as we passed along the deserted streets south of the river, which were lined with the débris of hundreds of buildings, that here and there the walls of some stanch old pile had resisted the shock of the flames and yet stood—though mere skeletons—monuments of the handiwork of the men who had put them together. But once we got to the North Side, how changed was everything! It is true that here and there a wall of some church yet reared itself above the level of the street. Yet for miles about the perspective was that of a desert waste, with nothing to break the clear view of the horizon on every side but the tall blackened telegraph poles, and the innumerable trees which still stood charred and dead, with their despoiled branches stretching out over the streets, like skeleton hands pointing to the graves of the many who were lost and buried beneath the ruins. Way out to the north, way to the south, to the east, and to the west, the view was the same—nothing but a level plain, broken slightly here and there by a pile of marble, crumbling to dust, or a great mound of brick, once red, but now white as snow, and yet so hot that not even the sentinels stationed near the safes dared to stand within a yard of them. I don't think a New Yorker can have any idea of this awful scene unless he brings it home to his own city. Let him imagine a fire to have broken out on Tenth avenue, near Twenty-eighth street, to have crossed in a straight line to Third avenue, and then to have made a clean sweep between these two lines clear down to the Battery, not leaving over a hundred walls standing, every house being levelled to the gutter, and he can then have some idea of the ravages of the awful Chicago Fire. Then let him try to do as I did, travel through the awful waste on horseback and try to find out where this and that building stood, and I guarantee he would find the task no child's play.

You would no doubt laugh if I should tell you that, if New York was ravaged as I have supposed it to have been, you could not drive down Broadway in the waste and point out where once stood the St. Nicholas. Yet I assure you my guide had been a resident of Chicago for twenty years, and, when we were about crossing to the North Side, so great was the desolation, so level the track the fire had made of wall and cellar, that he could not tell me where once stood the Sherman House. Can any better idea than this be given of what a desert the great business district of Chicago was in? But to continue my narrative. During our exploration of the North Side for an hour or so we came across —will you believe it?—a frame house amid all the ruins intact and without a singe! There it stood, with the crumbling remains of a great granite building all around it, and a few blocks off, surrounded by the blackened iron beams of a fire-proof brick building that fell a prey to the raging flames, was a neat little greenhouse, with not a pane of glass broken, not a whitened sash blackened by the smoke. What a freak of the conflagration was this! But when we rode over to the South Side again what was our surprise to find intact a frame building that stood just in front of the barn where the great fire first was started, and which it had to leap over in order to devour the city beyond. Before we had reached the North Side I was very much amused with many of the notices I caught a glimpse of as I galloped past among the ruins. There was one of a real estate man, who had been burned out, and who with wonderful enterprise had already erected a small wooden shanty as an office, upon the ruins of his former place of business. And this was his sign-board: "All lost, except my wife, my baby, and my energy." Who dare assert that that man will ever fail in the struggle of business life? Another extraordinary scene I witnessed with no small amount of interest. The safes of a safe-depository company had the day before been dug out and opened, and their contents found uninjured; and, in answer to an advertisement in the morning papers, there were right in the ruins before my very eyes, crowds of merchants hauling over their valuables to be put into the safes amid the general wreck. Just think of it—placing your treasure in a safe, surrounded by a thousand fires, and with the very stones about cracking from the yet unintensified heat. Still, the guarantee of a guard of "blue coats" appeared to make the safe investment all the safer to the merchants. What a confidence in military authority was there! But here let me pause, for just at this point myself and my guide took it into our heads to go back to the North Side, and go we did! Before we had well left the river ten blocks to the south the darkness of night was upon us. The wind at the same time began to blow at a fearful rate, and in a second a dense volume of smoke from the fires to the rear drove across the river and separated us; and thus it was that I lost my way, and had to wander out to the prairies, where I witnessed the encampment of the refugees.

I lost my way while taking a horseback ride through the ruins on Tuesday night. I was on the North Side. It was growing very late, and I knew not what to do. To turn back would have exposed me to dangers that I was unwilling to face, even with Sheridan's guards within a few hundred feet of every street, or rather roadway, one might traverse; so I chose the less of two evils, and made up my mind to keep straight ahead. I knew that straight ahead meant due north, and that by keeping on I would be certain to come into "open land" sooner or later, and not tumble headlong into cellar-ways made bristling with glass and broken iron by the falling-in of buildings that once sheltered them. So on I went. It was a long route still, although I thought that I was at the end of the city, or what I suppose most people call the North Side, when I first hesitated about my course; for I must have ridden fully twenty blocks afterwards, as I could tell by the resounding of the horse's feet on the pavement that the so-called prairie was still far out of reach. But this was not all. I could almost feel the darkness that surrounded me, rather which confronted me, for behind me were thousands of coal fires that lit up the sky for miles to the south and made the darkness ahead all the more dense by contrast—and, under the circumstances, my situation was not very pleasant. How long I continued to ride at a slow pace—for it was a funeral pace—from the moment I found the blue fires getting to the rear, I know not; but this I do know, it was an age to me. The low rumble of the wind through the ruins to the south, and the distant hum as of a bustling city, far, far off to the westward, broken now and then by what I imagined to be a piercing cry of distress, but which proved to be the sudden rushing of the wind through the yet standing open walls of the city that was, made me, it is needless to state, very anxious to get out of the wilderness of the dead city to the wilderness of the prairie itself. Suddenly the horse stumbled under me, and his hoofs made no longer an echo. At last I knew the unworked sod of the prairie had been struck. Cautiously I urged the beast to go a gentle trot, and in a few seconds came to a very abrupt halt by running plump against a sort of fence, which some worthy farmer, as I supposed, had erected to mark off his legitimate domain from outside limits. Almost at the same moment the darkness ahead of us began to clear away as the wind increased in strength (for it was the smoke from the smouldering fire to the rear that made the darkness ahead), and there right ahead of me, within a stone's throw, flashed over the plain a thousand twinkling lights. What I had not heard before I now heard plainly—the commingling of many voices, some low and some boisterous, the clinking of ware, the hallooing in the distance of men to other men nearer by, and here and there the thud of a hammer or the creaking of a cart-wheel. Dismounting, I tied the horse to the fence and jumped to the other side, and began slowly to pick my way over the field. An instant afterwards I heard a rustling in the dead grass a few feet to the right, and then came a clang as of steel against steel, followed by a loud gruff voice crying out, "Who goes there?"

I knew by this time that I was among the lots and near the park where thousands of the refugees had fled for shelter, and a feeling of relief came over me. A pass signed by the chivalrous Forsyth, and endorsed by "Little Phil," made the gruff "Who goes there?" answer himself a friend of mine, and bid me go where I listeth. Now that I had got out of the wilderness of a city, where a silence of the desert reigned supreme, to the wilderness of the open plains, where everything was bustle and confusion, I was at a loss to know how to act, glad as I was to escape from one to the other. Owing to the darkness I was at first unable to see what kind of company I had fallen in with; but as I made a slow and cautious approach to the nearest light which glimmered dimly through the chinks of what seemed to be a few planks carelessly nailed together, my ears were greeted with the cheering sounds of a woman's voice. It was a low and plaintive voice, broken by sobs that made my blood grow chill in that out-of-the-way place; but for all that it made me feel safe. From that moment my apprehensions of being attacked in the dark by the night-prowlers whose numbers the scared citizens had been for twenty-four hours increasing by hundreds, or falling into a den of encamped fire-fiends, vanished. I felt that the voice belonged to a woman who was a sufferer from the great sorrow of the great city, and the sobbings that came clearer and clearer to my ear as I felt my way nearer towards the light, were, I felt, my strongest guarantees of safety. I had got within a few feet of the light when in the dim distance I espied hundreds of forms moving about quietly and hundreds of others sitting upon the grass, and yet hundreds of others rolled up like mummies in blankets close to a fence, or half covered with such things as tables and chairs, and, in fact, every kind of household furniture which could be turned into a temporary roof. There were here and there lights, but they were not many, and as I went up to the first one that I had seen and exclaimed quite loudly: "Is there anybody here?" I felt a cold chill creep over me, so still did everything for a second seem about, although through the misty smoke, driven into the plain by the ever-changing wind, I could still see the forms of many moving, moving, moving, never seeming to stand still for a single second, yet no one saying a word. Need I say that it was quite a relief to me when a quiet, gentle voice greeted me with "What is wanted?" and the light of a candle fell upon my face. She was a little girl, not over twelve years of age, that held the candle, pale as marble, with large black eyes and a wealth of black hair, all tangled and neglected, that hung and swung in drifts over her face as the wind ruthlessly threw it about her shoulders. She had a blanket wound about her and was barefooted, and the little feet were covered with blood. This scene I took in at a glance, and for a moment I hardly knew what to say to the shivering little creature who stood before me, her teeth chattering with the cold and her pale face wearing such a pitiful, oh, such a pitiful look! Presently came from out of the "shelter"—for shelter it was—composed of a high top buggy with several planks resting against it and the ground, having to stoop low as he came, a man about thirty years of age, with a lantern in his hand. He was the very picture of despair. His eyes were swollen; dark lines under them gave to them an unnatural, haggard expression, half wild, half pleading, and wrinkles that one would expect to find traced on the face of a very old man alone furrowed the otherwise youthful face. What passed between us I need not now repeat. Sufficient it is that my horse was secured to the buggy with a halter and I was permitted to occupy a corner beneath the shelter, after the man, trying to be humorous in his sorrow, had excused himself for "not having any blankets in the house."

When I lay down the lantern was still burning, and before it was extinguished I saw that on the other side of the "house" were huddled together, under one blanket and an old dress, a woman and two small children. I thought as I was falling into a gentle sleep, with all I had seen during the day passing like a panorama before my eyes, that I heard the same sobbing that had attracted my attention a few minutes before, and every now and then a manly voice soothingly saying, "It will not last forever." But it may have been a delusion. It was bright daylight when I awoke, feeling rather stiff and cold, but, after all, considerably refreshed. The sun had not yet risen, and a keen, cutting wind swept over the lots, and what I discovered on getting up to be a large park adjoining. My kind host and his little family had risen before me, and had taken the precaution on going out to throw over me the only blanket visible in the shelter. It took me quite a while to collect my thoughts at first, and try to remember where I was and how I had got just where I was; but if I had not known anything about my whereabouts before, I certainly was not long left in ignorance once I had got outside the "shelter," where I had slept so soundly all night The sight that met my eyes fairly took my heart away. If I should live a thousand years I do not think I could ever efface it from my memory; and even now as I write, the impression it made upon me at the time comes back so strong that I seem to see staring at me from every quarter of my room the same pale, haggard, woe-begone faces, the same huddling crowds, the same weeping women and crying babes, that I beheld on emerging into the full light of the early morn. Words cannot describe the scene; and no one who did not behold it, without expecting to behold it, as was the case with me, can imagine anything that could approach the reality. For a moment I stood rooted to the ground, as it were. My good friend, who had acted so gracious a part toward me the night before, met me at the very threshold; but as I grasped him by the hand in greeting, I stood speechless before him, the scene that met my gaze beyond where we both were, striking me with an awe that was unspeakable. And how could it have been otherwise? As far as the eye could reach was a vast concourse of men, women, and children, all huddled together over the Park and in the lots, amid wagons, horses, and carts innumerable. Hundreds were still lying sound asleep; some with a sort of wooden shelter over them; some under tents; and yet others, and by far the greater number, with no shelter at all but the canopy of dark smoke that came wafting along overhead in thick rolling masses, that one could almost imagine to hear moving in the air. I fancied, even, in the midst of all the confusion that I witnessed among those who were already up and going about like ghosts from place to place, seeking apparently for some articles they had mislaid the night before, that I could tell one family from another, so distinct in the hustling, bustling crowds that moved here, there, and everywhere, was each little group from the other. There must have been on all sides fully 30,000 persons; and yet one of the most striking features about the wonderful scene was the absence of that very thing which, under almost all circumstances, is supposed to be inseparable from a large and mixed gathering of men, women, and children — boisterous noise. There was confusion, there was pushing and there was crowding in places, there was talking and there was a moving of wagons and carts from one place to another; but otherwise over the whole scene reigned a sad quietness that reminded me of the quiet crowds I have often seen at a funeral in a large church. There was not a joyous face about me. It was in vain that I tried to imagine I heard a laugh from some group which looked less disconsolate than another; but in every case the laughter I thought I heard, turned out to be a wail of anguish. My pen almost refuses to write further of this terrible evidence of what the disaster had done in one of its phases—of how it seemed to have stabbed to the heart, without actually putting out of life, each one of the hundreds that were within calling distance of my voice, yet every one of whom seemed as full of physical health as could be. I wandered about among the crowds that were walking in little groups and talking in low tones together, feeling as though I were the only person with life and thought in me, and that all who passed me, heedless how I knocked against them or got in their way, were so many automatons, with power of sorrowful speech only. It would be a futile task for me to attempt to describe the many little scenes that were so intimately interwoven with the main scene of the encampment, and which made it the mournful gathering it was. To the right and left—no matter what way I turned or how anxiously I tried to peer beyond the groups for a vacant space—my searching eyes were met with little knots of men, women, and children, some sitting, some standing, some lying on the ground, and all, even to the little prattling ones, wearing a look of such supreme sadness that my heart bled as I gazed and continued to gaze, fascinated in a strange way by the sorrow and anguish depicted upon every countenance. Did you ever look upon the face of a man who escapes from a shipwreck and gets to the shore, knowing that his wife and little ones had gone down, down under the pitiless waves, never to rise? Did you ever notice how, if he be a man of will, he says not a word; how his face, pale as the whitest marble, seems to you paler every time his eye meets the pitiful glare of his neighbor; how the lips tighten and the hands clench, and he thinks all the while he is concealing his great sorrow within his own breast? Such was the look of about every man I came across in this field of woe, for they had every one of them, it is true, escaped a danger that was past; but how many—oh, how many, as they wandered about with that agonizing look upon their faces and turned their eyes toward the black clouds that hovered over where their homes were once, were thinking of those loved and dear ones who had tried to escape when they did, and who are now—God knows where? What if a child was missing, a wife not found, a sister not heard from since the roof of the little home fell in, was it not

RUINS OF TRIBUNE BUILDING.

GENERAL VIEW OF

SILENT FOREVER.

Post-Office Cat—Four Days in the Fire.

RUINS OF RUSH

CHICAGO WILL RISE AGAIN

NEW CHICAC

:S OF THE NORTH DIVISION

RUINS OF FIELD, LEITER & CO.'S STORE.

:AL COLLEGE.

RELIC FOUND IN THE RUINS OF THE CHURCH OF THE HOLY NAME.

almost the same as actual death to them in their great desolation? They knew not where those they sought for were; and the mere thought that the smouldering ruins which lined the distant roadways for miles around had hid the missing ones forever from their sight, was of itself as harrowing as the dread certainty itself. When I now recall the low, suppressed cries of anguish that greeted my ears from one shelter after another, as I passed my mournful way along through crowd after crowd of these victims of the great disaster; when the scenes that I beheld come now vividly back to my mind as I write; of the mothers that I met, with their babes closely pressed to their bosoms, and refusing, with a half cry, half shriek, the relief of food that kind hands proffered; of the little children I found lying sleeping sweetly amid the whole confusion, as though their mothers had not been lost to them, and other mothers who had lost their own were caring for them in their stead; of the men who sat with their heads bowed on their hands, and swayed to and fro, and looked up at you when you spoke to them as though they heard you but saw you not; of the girls of a tender age who hid behind one another in their shelter as you passed, lest you would notice that they had reached the prairies with scarce enough covering to hide their nakedness; of the hundreds upon hundreds, that were about me in the park and out of it, all so sad, all so silent in their sadness, yet so many crying, strong men, too, crying in a stifling way for fear wife or child would see them so weak—when I recall all these things now, is it a wonder that I find difficulty to portray that terrible scene on the prairie, which probably, after all, had less of real agony, real suffering about it in itself than any other sorrow that befell the unfortunate people of Chicago during the fatal days of last week. Yet in it we saw reflected all the suffering, all the losses, all the heart-breakings, all the trials endured elsewhere and still to be endured; and that was what made it seem to me, and what would have made it seem to any one who beheld it, the saddest scene of all the sad scenes witnessed during the wreck of desolation. I had scarcely the heart, as I went about among the people, to say a word to any one I met. It was no place for words of pity, and expressions of sympathy would have been horrid mockery; yet occasionally I plucked up courage enough to speak to a few of the men and women, careful all the while to speak feelingly of the great misfortune that had befallen the people, without for a moment making my sympathy look like pity for their own particular desolations; for, strange as it may seem, the more I wandered among these unfortunate, brave-hearted refugees, the more did I become impressed that they felt their own sorrow so deeply, that a third person who dared to express sorrow for them in particular would have been treated to a quick rebuff. I sat down beside one young man and his wife in the park, and partook with them of the food which the young man had procured, I believe, from the place where the Relief Committee had sent food for distribution. I could not but pity the poor young wife as she sat with her head upon her husband's knees, and her face covered with her hands, while she cried, oh so bitterly! And clinging to her were three little children—one a girl of about seven and another about five, and a chubby faced boy about two years of age. They were all three beautiful children, and they seemed to know something awful had happened, without exactly knowing what, to make "mamma" cry so. And they toyed with her hair with their tiny hands, and the little boy would every now and then put his little arms round the mother's neck, place his lips against her cheek, and murmur: "Don't ki, mamma, don't ki."

Do you wonder that the man gulped down his food chokingly when he beheld this? I more than once saw him turn his head from me and wipe his eyes with his sleeve. I got into conversation with him after awhile, and he said to me, as I rose to go: "Well, sir, it can't be helped. I had a home and was worth

$10,000 a day or two ago, and now here's all I've got between me and the grave;" and he put his hands in his pockets and showed me a two-dollar bill; and then pointing to his wife and children, and smiling through his tears, he exclaimed, as he laid his hand softly on his wife's covered head: "Thank God, I have not lost these. I am better off than many."

This was but one of many instances of the same kind I came across in the camp; but now let me draw a veil over the picture. It is too sad even to think about. Thank God, most of the "campers" are now housed, and, let us hope, the time is not far distant when they will one and all have their own homes again. But what, oh what of those whose now missing ones are destined never to return?

"Sorrow never comes single spies, but in battalions," is Shakspeare's observation on human life, which many men find true to the last letter. And often a city attacked, as ours has been, becomes the field where those battalions deploy and assault successfully what remains of human joy and pride. Among several instances of the accumulation of disasters, we read the following:—

Dr. Henrotin, who lived on the North Side a little more than a fortnight ago, was among the thousands who were compelled to pull up stakes and fly before the fiery breath of the great conflagration. He succeeded in accomplishing no less than six different moves, leaving some goods at every fancied place of security, until at last he found he had nothing left him but his family and a horse and buggy. He had congratulated himself on saving the horse and buggy, for the reason that both were of a superior quality. His horse he had refused to part with for a large sum of money, and he put a high valuation on the vehicle. On Saturday he was driving along Ashland avenue, and, when about to cross the railroad track, found a locomotive almost upon him. The signalman's hat and a long line of fence had intercepted the view, nor did the signalman think proper to show himself until the locomotive was close up. The horse was frightened, leaped across the track, threw the Doctor out of his buggy, smashed the vehicle to sticks and shreds, ran like a streak to Western avenue, plunged his head against a curbstone, and broke his neck. The Doctor had, on the previous day, invested in a pair of cheap shoes, which saved him from injury, as the lines caught round his heel. He would doubtless have been dragged some distance had not the cheap heel come off.

A New York paper describes a Fire-wedding:—

CHICAGO (October 18).—Among all the pictures of "Chicago as it is" which have been photographed with the pen, I wonder whether any one has seen the chronicle of a "Fire-Wedding"—a wedding whose whole aspect and circumstances were so altered by the fire as to be inextricably connected with it forever after in the minds of the lookers-on. There was such a wedding in our poor desolated North Side the other day. The first house outside of the burned district on the north contained a most motley crowd for several days after the fire, for the owner had received everybody, high and low, till the house would have furnished excellent material for a new "Decameron" or "Canterbury Tales," if the fire had only unearthed a modern Boccaccio or Chaucer. As it was, wonderful stories flew about, rather monotonous as to tone, but evidently diversified as to incident: "Have you seen the three-days'-old baby in the barn: they picked it up with the mother in the park." "That German, covered up with greatcoats, on the corner, was found almost dead with cold and exposure." "Three men have just come in who have had but four soda-crackers between them since Sunday, and this is Wednesday morning." And so on, till one of our couriers brought in a story before which the others paled their ineffectual fires. A friend had just told him of meeting a woman, during the fire on Monday, who was struggling along under a heavy bundle wrapped in a sheet. Offering to help her, she said· "Do you know what is in here? God help me! the bodies of my two children, who were suffocated in the fire; but I could not leave them to burn." The atmosphere was full of startling and blood-curdling rumors, and every hour brought a new excitement. Incendiarism was said to be rampant; frightful and summary vengeance was reported as meted out to even supposed evil-doers. The house had twelve revolvers, loaded and capped, arranged on the parlor windows every night, and no one thought of sleeping in a house not guarded by a patrol.

So, when it was whispered about that sweet Minnie T——, a relative of our host, and who was to have been married according to the strictest sect of the fashionables, if Superior street had not been burned, would really carry out her intention and take the holy vows on the twelfth, the house was in the wildest excitement. How could they get a license? how find a clergyman? The trousseau was burned; the intended guests were burned out; the caterers and florists had neither flowers nor food. How could a fashionable young lady make up her mind to be married without these things? But she did! And what was more, being a girl of exceeding sweetness and womanliness, she did not seem to care a whit about her lost splendors. Thursday came, and with it a tremendous sweeping and dusting of the house of refuge; for, let me tell you that a running fire of running guests does not leave a house swept and garnished, but quite the contrary. A license had been obtained, and Minnie's own burnt-out clergyman had come to marry her, but what should be done for decorations? The large house-parlor had never been furnished, and there was not even a mantelpiece in the room. But it seemed as if the fire had developed as much feminine ingenuity as it had destroyed feminine property. Theodore Winthrop said that if the order had been given in the Seventh New York Regiment, "Poets to the front," a goodly company would have answered to the call; and so now a call for decorators of burned property brought a perfect rush of talent to the rescue In the unsightly chimney-hole was placed an inverted soap-box, covered with a crimson cloth. On this sat a tall slop-jar, cribbed from a bedroom, filled with lovely crimson and green autumn leaves. To be sure, a slop-jar is not *per se* a handsome ornament; but then some refugee had left a magnificent stag's head and antlers, which, set up in front of the objectionable crockery, left nothing to be desired. A white cravat, lost by some city exquisite, who probably found it was impossible to save both that and his neck, tied more autumn leaves into a true-lover's knot of colossal size, and hung high in the pier. Branches of richest color filled all sorts of niches and corners, and the room was declared magnificent. Some one, however, suggested that there was no sort of table or altar for the minister's use. But fortune favors the brave; a pair of library steps was produced from somewhere, and a sheet pinned around them. Another treasure-trove was a scarlet cloth in illuminated work, with the motto, "Cast thy care upon Him, for He careth for thee." Our little white altar, with this pinned to the front face, and surmounted by a big Bible and prayer-book, made a very canonical appearance indeed.

Then the room was ready and we had a rehearsal. But, alas! what bride and groom of the present day could kneel on a bare floor and get up again gracefully? In vain blocks of wood, boxes, and books were tried—one was too high, another too low. At last a bright thought came—carriage cushions! For it must be said that in burned Chicago now there are forty carriages and pairs of horses to one house, as these first were mostly saved; so that beggars ride where beggars never rode before. Four cushions were brought from the barn, and an Afghan converted them into a lovely hassock. As it is (or was) impossible in these days to have a wedding without showing "the presents," a vine-wreath ed table in one corner held a most elaborate display. A beautiful jewel-case contained what was set forth as the bridegroom's gift—a set of exquisite pearls, which you had to look at very nearly to discover that they were moulded from the fine white ravellings of cotton cloth. Other cases contained sets of pickle-forks, preserve-spoons, and so forth, cut out in pasteboard, with mouldings and monograms in lead-pencil. Valuable jewelry was plenty, only unfortunately the lava earrings had been dug out once too often, and the cameos looked very black and queer round the edges. A pewter table-spoon, a german-silver fork, and some valuable aids to housekeeping in the broom and dust-pan line, completed the array, which certainly was unique, and interested the spectators much more than the usual show. But when it came to dressing the bride, serious difficulties occurred. Her wedding-dress and veil had never come from Field & Leiter's. Never mind, she had a white cambric morning-dress, which, looped over a nice petticoat, would make her slender figure look lovely, and her married sister had saved her own wedding veil. Some simple white flowers from a neighbor's yard took the place of orange blossoms, and a set of pearls was borrowed from a friend who had brought them out of the fire in her hands. As for stockings, handkerchiefs, etc., the various guests provided these from among them. So the pretty bride looked, after all, as sweet as a rose, and the long-laid-away tulle veil became her soft, fair locks to a charm. The groom was dressed in borrowed clothes from head to foot, as was the first bridesmaid, while the brother who gave away the bride complained that he, being five feet nine, was obliged to borrow the dress suit of a man who stood six feet six in his stockings, and that consequently, when he stepped forward to perform his brotherly duty, he was obliged to take a reef in his habiliments to prevent falling over on his nose. At last all was ready. I wonder if just such an assemblage ever met together at a wedding before. There were about forty guests, all but one of whom had been driven from their houses by the actual presence of fire—the bride and groom had hastened their wedding so as to go away together among friends who could shelter and help them. The minister who married them had promised his congregation that he would stay among them, and work with his hands if necessary, being bereft of all he had in the world. And this was all that was left of one of the richest congregations in the richest city of the West. But never have I seen among rich or poor a sweeter and more holy-seeming wedding; and when, after the solemn words were said, the congregation, at a sign from the minister, dropped on their knees and offered a solemn thanksgiving to the Almighty for their preservation through the horrors of the last few dreadful days, broken voices and tender heartfelt tones attested the reality of the service. And so the warm biscuit and cold water that stood for wedding-cake and wine were partaken of with a plentiful seasoning of cheerful words and even jests, and all felt that to be poor in such good company robbed ruin of half its sting.

Before closing this chapter of incidents and individual experiences, we must allow the Fire Marshal, Mr. Williams, to give his vindication of himself, which was addressed to the Editor of the *Tribune* :—

SIR: You, as well as the readers of your paper perhaps, have wondered somewhat at the non-appearance of a card from me, as Fire Marshal, relative to the late conflagration. I deemed it best to wait until the excitement and bustle had quieted a little, and the community had time to settle again. I have noticed and read the suggestions and articles in regard to the "man Williams." The authors of those extracts being so modest as not to sign their names, I cannot hope to see them individually, but take this method of extending to them my heartfelt thanks for the "praise and honor" they so kindly and profusely bestowed upon me in the high and noble manner in which they have done.

They should have considered well the difficulties that the Fire

Department had to encounter on that dreadful night before trying to "comment" upon it. They had just passed through a severe fire twenty-four hours previous, and part of the companies had left the scene of the old fire but a few hours when they were called again, tired and worn out from hours of hard labor, to another still more fearful than the one they had just dealt with.

While we were working on the original fire, which was surrounded and under our control, the fearful gale which was raging at the time, carried not only sparks but brands and pieces of boards on fire, the distance of two to four squares. To our surprise we were informed that a church over two blocks to the north was on fire. We were then obliged to form a new base of operations to protect the property around the burning church. While thus engaged in staying the fiery element, which we also conquered and had under our control, we were again informed that the fire had taken another leap, and had broken out in the match factory, lumber yard, and shingle mills of W. B. Bateham.

People living in the vicinity had carried out bedding, furniture, etc., into the street for safety, which was soon all ablaze. The strong wind carried the burning material to the east side of Canal street, communicating the fire to the wooden structures on that side of the street; these buildings being elevated to the height of five to seven feet above the ground, together with the sidewalk, formed a complete tunnel, and the draught carried the flames for a whole square without meeting even the resistance of a common board partition. In the mean time the intense heat drove us, and we were compelled to remove some of our apparatus, which occupied quite a length of time; also losing considerable hose. During this time the fire made fearful progress, and while trying to rescue one of the steamers from its dangerous situation, it was discovered that the fire had crossed to the South Division, into the second street east of the river. I immediately ordered part of the Department, and went myself to the new field of action. On my arrival there, I found not only a few buildings on fire, but the largest portion of two squares. So rapid did the fire spread, that the wooden buildings on Quincy street, the Armory building, the square known as "Conley's Patch" (all composed of wooden shells), the Gas Works, and the roofing material yard of Barrett, Arnold & Powell, were one sheet of flame in a short space of time. This yard being composed of combustible material, together with the Gas Works, threw out a terrific heat, more especially after the gas was allowed to escape to prevent an explosion and the destruction of the works. Through the agency of the burning of this yard, large fire-brands, composed of tar and pitch felting some two or three feet in length, were whirled through the air for a number of blocks, and would alight on some building, and hardly a minute would elapse before the whole structure would be involved in one mass of fire, thus starting in different parts of the city what you might call different fires, and all burning at the same time.

I rallied the greater part of my force to the South Side, but it was of no avail. The wind blew so hard at this period as to cut a solid stream of water into spray before it had gone the distance of twenty feet from the pipe. The fire made such rapid headway that we were constantly compelled to move some of our steamers to save them from destruction, and by so doing lost large quantities of our hose, so much so that we were soon short of a supply.

When at work at Monroe street, I was informed that the flames were in our rear as far as Madison street. I immediately repaired to that street, and found that a large building, known as the Oriental Block, was on fire in the rear, and in a few minutes the entire block was enwrapped in a sea of burning matter. The next to be seized in the embrace of the fire-fiend was the Court-House and Chamber of Commerce buildings, which burned very wickedly.

Shortly after this I received intelligence that the Water Works were on fire. I then had no hope whatever of staying the flames.

Nearly four years ago, when I was appointed Fire Marshal, in my estimate of wants for the Fire Department I recommended to the Board of Police the necessity of having one or more "Floating Fire Engines," for the protection of the property along the river. They acquiesced in my recommendation, and asked for an appropriation to purchase the same, but without success. They have made several attempts since to obtain the necessary appropriation. This floating engine, or engines, would have done much in stopping our late great fire, as there could be two powerful pumps on board of each, throwing two or three streams of water, which would have been sufficient to keep wet the buildings on the sides of the river for a number of squares, and in protecting the elevators.

The public are probably aware of the fact that the foot of our streets are leased from time to time for dock purposes, which has always interfered with the Fire Department in obtaining a supply of water from the river. We have been deprived of the use of these docks, and also of the floating engines, which we have had occasion to regret in this as in many other instances, as either would have rendered great aid to the department.

One other circumstance that has greatly crippled our Fire Department, is the scanty supply of hose purchases from year to year; also an insufficient number of fire engines. I have always failed to obtain the amount of hose I have asked for from time to time, as in the case of the present year I requested 15,000 feet, which was small enough an amount for the number of fires we are having in Chicago (amounting to nearly 700 during the last year) Instead of allowing me the full amount, I was cut down one-third, and allowed 10,000 feet. I was also cut short of one additional steamer.

After all this, it has been stated by one of your correspondents that, had there been some engines placed upon flat-boats, canal-boats, or scows, and propelled in the river by the aid of tugs, it would have prevented the fire from crossing. Allow me to ask, where were those scows? We tried to obtain one to enable us to extinguish the coal fires; none were to be found this side of Bridgeport, and it took from four to five hours to obtain it. What time did the Fire Marshal have to hunt scows and flat-boats at that stage of the fire?

Correspondents in different papers have asked why the Fire Department did not do thus and so? Why did not the people try to protect their buildings from sparks and embers as did the watchman at the crib in the Lake? Instead of standing in the way, and finding fault with the Fire Department, the drought of the season, the state of the atmosphere, and the high gale, they should have turned their attention to the roofs of buildings. How could the firemen be battling the fire and watching buildings squares in their rear?

To a careful reader of the Marshal's letter it becomes evident that discontent exists in the public mind; and, accordingly, as after a great battle, the vanquished frequently fall into disagreements, bickerings, and mutual criticism, so also now every one is ready to justify himself, if his neighbor suffers in the result, and many seek to cast blame upon parties who did the best they could in the peculiar and trying circumstances. The Fire Department suffered a great defeat; but their enemies were peculiarly powerful, and they were wearied out by a previous day and night's struggle. Instead of ill-natured invective or attempts to shirk responsibility, there should be a calm determination to learn from experience how to avoid the recurrence of similar disasters. Such is the end the practical Chicagoans will reach.

And now must close this long chapter of personals, which have served to reveal the breadth and characteristics of that suffering which was experienced by the multitudes, who went forth homeless and impoverished on the evening of the 8th and the morning of the 9th of October, 1871.

CHAPTER XXV.

> "This kingly Wallenstein, whene'er he falls,
> Will drag a world to ruin down with him;
> And as a ship that in the midst of ocean
> Catches fire, and shivering springs into the air,
> And in a moment scatters between sea and sky
> The crew it bore, so will he hurry to destruction
> Ev'ry one whose fate was joined with his."
>
> —SCHILLER.

It is now in place to mention, more definitely and comprehensively, the losses, by this calamity, of property and life.

This city and its interests are intimately bound up with those of the whole world. The losses by the fire were not local but well-nigh universal. The representatives of all nations were here, and of all States and communities in North America—the business world were here by their money or agencies, and the fall of Chicago sent a tremor throughout the whole fabric of society. This may account, in part, for the uprising of all Christendom to assist in the terrific exigency, and roll away the burden that was crushing us into the dust.

It is scarcely possible to visit a city or enterprising village in our own country, or certain parts of Europe, without meeting persons, many or few, who say we have friends who suffered, or we ourselves have lost, by the great conflagration. Our population was not native to the soil, and our capital was largely from abroad. As a place for investment of funds none was deemed preferable; and this drew heavily upon the resources of men who had money to spare in all parts of the country. Insurance companies had sought this field in great numbers, and their losses have been very widely felt. The home companies, having their assets very considerably, or perhaps entirely, in property burned, and the stockholders themselves being comparatively helpless from their private losses, are unable to pay their policy-holders, and have lost their own capital and business.

The following tables furnish an exhibit of all the companies doing business in Chicago There is a large number which have no connection with this city, and do not require to be named here. The latest data enable us to furnish a very full and accurate statement of the capital, gross assets, and losses of such as have offices and representatives in this city :—

NEW YORK COMPANIES.

NAME.	*Capital.*	*Gross Assets, Jan.* 1, 1871.	*Losses.*
Ætna, City	$300,000	$442,709	$660,000*
Adriatic, City	200,000	246,120	8,500
Albany City Albany	200,000	397,646	800,000*
American, City	200,000	741,405	25,000
American Exchange, City	200,000	277,350	58,000
Astor, City	250,000	405,571	400,000*
Atlantic, City	300,000	556,179	600,000*
Beekman, City	200,000	261,851	350,000*
Buffalo City, Buffalo	200,000	370,934	600,000*
Buffalo Fire and Marine, Buffalo	304,222	473,577	625,000*
Buffalo German, Buffalo	200,000	270,081	5,000
Capital City, Albany	200,000	293,766	270,000*
Citizens, City	300,000	684,798	35,000
Columbia, City	300,000	451,332	3,000
Commerce, Albany	400,000	692,877	400,000
Commerce Fire, City	200,000	249,372	25,000
Commercial, City	200,000	306,002	5,000
Continental, City	500,000	2,538,038	1,000,000
Corn Exchange, City	300,000	398,986	55,000
Excelsior, City	200,000	335,724	600,000*
Exchange, City	150,000	183,959	2,500
Firemen's, City	204,000	359,961	15,000
Firemen's Fund, City	150,000	173,477	25,000

* Suspended.

NAME.	*Capital.*	*Gross Assets, Jan.* 1, 1871.	*Losses.*
Firemen's Trust, City	$150,000	$226,269	$5,000
Fulton, City	200,000	363,002	900,000*
Germania, City	500,000	1,077,849	226,500
Glens Falls, Glens Falls	200,000	571,123	13,000
Guardian, City	200,000	279,688	40,000
Hanover, City	400,000	700,335	225,000
Hoffman, City	200,000	235,242	30,000
Home, City	2,500,000	4,578,008	2,000,000
Howard, City	500,000	783,851	275,000
Humboldt, City	200,000	251,186	24,000
Importers and Traders', City	200,000	302,589	22,500
International, City	500,000	1,329,476	500,000
Irving, City	200,000	321,745	550,000*
Jefferson, City	200,010	411,155	42,500
Kings County, City	150,000	262,573	31,000
Lafayette, L. I., City	150,000	214,751	7,500
Lamar, City	300,000	551,402	450,000*
Lenox, City	150,000	240,801	32,000
Lorillard, City	1,000,000	1,715,909	1,500,000*
Manhattan, City	500,000	1,407,788	1,250,000*
Market, City	200,000	704,684	1,000,000*
Mechanics', L. I., City	150,000	218,047	22,500
Mechanics and Traders', City	200,000	460,002	41,500
Mercantile, City	200,000	273,399	100,000
Merchants', City	200,000	442,690	10,000
National, City	200,000	282,671	37,500
New Amsterdam, City	300,000	432,638	200,000
New York Fire, City	200,000	392,278	15,000
Niagara, City	1,000,000	1,304,567	225,000
North American, City	500,000	770,305	720,000*
Pacific, City	200,000	443,557	12,500
Phenix, L. I., City	1,000,000	1,890,010	350,000
Relief, City	200,000	310,908	40,000
Republic, City	300,000	683,478	225,000
Resolute, City	200,000	252,452	75,000
Security, City	1,000,000	1,880,333	1,500,000*
Sterling, City	200,000	247,027	7,500

* Suspended.

NAME.	*Capital.*	*Gross Assets, Jan. 1, 1871.*	*Losses.*
Tradesmen's, City	$150,000	$423,181	$25,000
Washington, City	400,000	774,411	900,000*
Western of Buffalo, Buffalo	300,000	582,547	750,000*
Williamsburg City, City	250,000	539,692	60,000
Yonkers and New York, City	500,000	868,933	700,000*
MISSOURI COMPANIES.			
American Central, St. Louis	$231,370	$254,875	$275,000
Anchor, St. Louis	105,225	121,974	27,000
Boatmen's, St. Louis	106,530	51,786	20,000
Citizens', St. Louis	175,000	271,373	25,000
Commercial, St. Louis	40,660	43,896	20,000
Excelsior, St. Louis	73,087	19,815	15,000
Globe Mutual, St. Louis	125,000	150,793	65,000*
Jefferson, St. Louis	101,272	121,842	10,000
Marine, St. Louis	150,000	210,925	10,000
Merchants', St. Joseph	60,636	79,682	10,000
National, Hannibal	111,201	147,738	10,000
North Missouri, Macon	134,050	154,166	21,500
Pacific, St. Louis	25,000	36,835	10,000
Phœnix, St. Louis	108,950	126,654	10,000
St. Joseph, St. Joseph	64,000	105,729	10,000
St. Louis, St. Louis	240,000	307,342	15,000
State, Hannibal	109,820	162,099	21,500
MASSACHUSETTS COMPANIES.			
Bay State, Worcester	$104,800	$196,275	$5,000
Boylston, Boston	300,000	933,256	13,000
City, Boston	200,000	399,427	15,000
Eliot, Boston	300,000	672,212	12,500
Firemen's, Boston	300,000	1,038,330	35,000
First National, Worcester	100,000	157,356	2,500
Franklin, Boston	300,000	541,908	50,000
Hide and Leather, Boston	300,000	419,211	720,000*
Howard, Boston	200,000	358,642	27,500
Independent, Boston	300,000	646,648	1,100,000*
Lawrence, Boston	250,000	262,502	10,000

* Suspended.

GENERAL VIEW OF THE RUINS ON THE SOUTH SIDE.

NAME.	*Cash Capital.*	*Gross Assets, Jan. 1, 1871.*	*Losses.*
Manufacturers', Boston	$400,000	$1,480,464	$120,000
Merchant's, Boston	500,000	958,559	10,000
National, Boston	300,000	821,840	400,000
Neptune, Boston	300,000	852,195	60,000
New England Mutual M., Boston	200,000	1,080,973	1,000,000*
North American, Boston	200.000	601,747	10,000
People's, Worcester	400,000	887,756	300,000
Shoe and Leather Dealers', Boston	200,000	549,806	25,000
Springfield, Springfield	500,000	930,101	450,000
Suffolk, Boston	150,000	283,288	23,000
Tremont, Boston	200,000	294,543	70,000
Washington, Boston	300,000	985,975	25,000
RHODE ISLAND COMPANIES.			
American, Providence	$200,000	$374,969	$600,000*
Atlantic, Providence	200,000	326,614	325,000*
City, Providence	50,000	72,150	7,500
Hope, Providence	150,000	211,673	325,000*
Merchants', Providence	200,000	372,199	15,000
Narragansett, Providence	500,000	792,947	25,000
Providence Washington, Providence	200,000	415,149	550,000*
Roger Williams, Providence	200,000	278,966	225,000*
OHIO COMPANIES.			
Alemannia, Cleveland	$250,000	$285,555	$175,000
American, Cincinnati	100,000	125,513	12,500
Andes, Cincinnati	1,000,000	1,203,425	850,000
Burnett, Cincinnati	60,000	75,369	2,500
Butler, Hamilton	14,000	22,322	......
Capital City, Columbus	60,000	78,000	......
Central, Columbus	40,000	55,541	......
Central, Dayton	20,833	29,896	......
Cincinnati, Cincinnati	150,000	209,223	......
Citizens', Cincinnati	52,500	67,690	60,000
Cleveland, Cleveland	414,400	530,208	25,000
Commercial, Cincinnati	100,000	158,987	700,000*
Commercial Mutual, Cleveland	210,210	349,624	13,000

* Suspended.

NAME.	*Cash Capital.*	*Gross Assets, Jan. 1, 1871.*	*Losses.*
Cooper, Dayton	$23,800	$32,527	......
Eclipse, Cincinnati	27,350	46,667	$2,500
Farmers', Cincinnati	23,360	24,142	10,000
Farmers', Jelloway	100,000	131,626	......
Farmers & Merchants', Dayton	32,000	55,770	......
Farmers, Mer. & Mfctrs.', Hamilton	100,000	123,366	......
Firemen's, Cincinnati	100,000	225,600	29,500
Firemen's, Dayton	100,000	126,893	......
Franklin, Cincinnati	100,000	132,465	65,000
Franklin, Columbus	70,000	88,071	......
German, Cleveland	200,000	281,260	436,657*
German, Dayton	22,500	28,347	......
Germania, Cincinnati	100,000	127,858	3,500
Germania, Toledo	45,000	54,500	7,000
Globe, Cincinnati	100,000	178,143	40,000
Hamilton, Hamilton	17,500	41,620	......
Hibernia, Cleveland	200,000	225,000	360,000*
Home, Columbus	500,000	637,947	300,000
Home, Toledo	69,000	76,335	......
Jefferson, Steubenville	43,392	60,632	......
Merchants & Manufrs.', Cincinnati	150,000	266,780	14,500
Miami Valley, Cincinnati	100,000	141,094	15,000
Miami Valley, Dayton	26,100	51,133	......
Mutual, Toledo	90,000	90,249	3,000
National, Cincinnati	100,000	120,514	3,000
Ohio, Chillicothe	40,000	49,092	......
Ohio, Dayton	35,282	54,818	......
Ohio Valley, Cincinnati	50,760	79,921	32,000
People's, Cincinnati	20,000	48,928	5,000
Sun, Cleveland	200,000	301,340	175,000
Teutonia, Cleveland	200,000	237,016	1,000,000*
Teutonia, Dayton	25,000	46,572	......
Toledo, Toledo	75,000	105,837	......
Union, Cincinnati	100,000	130,845	27,500
Washington, Cincinnati	129,100	148,747	21,000
Western, Cincinnati	100,000	178 550	31,000

* Suspended.

NAME.	*Cash Capital.*	*Gross Assets, Jan. 1, 1871.*	*Losses.*
CALIFORNIA COMPANIES.			
Firemen's Fund, San Francisco......	$500,000	$799,627	$300,000
Occidental, San Francisco...........	300,000	474,095	300,000
Pacific, San Francisco...............	1,000,000	1,777,267	1,500,000
People's, San Francisco.............	300,000	500,000	400,000
Union, San Francisco...............	750,000	1,115,574	450,000
MICHIGAN COMPANIES.			
Detroit Fire & Marine, Detroit......	$150,000	$273,063	$175,000
ILLINOIS COMPANIES.			
American, Chicago..................	$150,000	$548,875	$1,000
Aurora, Aurora.....................	200,000	220,471	......*
Chicago Fire, Chicago..............	101,800	131,566	3,000,000*
Chicago Firemen's, Chicago.........	200,000	372,544	3,000,000*
Commercial, Chicago................	180,000	266,535	3,000,000*
Equitable, Chicago.................	100,000	120,191	3,000,000*
Farmers', Freeport.................	100,000	191,303	......
Garden City, Chicago...............	150,000	181,489	2,000,000*
German, Freeport...................	101,000	119,824	......
German Ins. & Sav's Co., Quincy....	132,900	158,951	......
Germania, Chicago..................	200,000	257,821	1,500,000*
Great Western, Chicago.............	222,831	274,125	227,000
Home, Chicago......................	200,000	245,338	2,000,000*
Illinois Mutual, Alton.............	113,000	350,016	1,100,000*
Knickerbocker, Chicago.............	160,000	170,129	750,000*
Merchants', Chicago................	500,000	878,252	6,000,000*
Mutual Security, Chicago...........	118,325	145,584	1,800,000*
Republic, Chicago..................	998,200	1,132,812	3,500,000*
Rockford, Rockford.................	100,000	235,442	......
State, Chicago.....................	425,000	460,000	3,000,000*
MARYLAND COMPANIES.			
Maryland, Baltimore................	$200,000	$251,157	$12,000
Merchants & Mechanics', Baltimore..	250,000	324,208	290,000*
National, Baltimore................	100,000	224,000	33,165

* Suspended.

Name.	Cash Capital.	Gross Assets, Jan. 1, 1871.	Losses.
Peabody, Baltimore	$125,000	$190,388	$10,000
People's, Baltimore	100,000	113,094	17,000
Potomac, Baltimore	75,651	97,209	10,000
Union, Baltimore	100,000	164,986	25,000
Washington, Baltimore	100,000	121,804	......
CONNECTICUT COMPANIES.			
Ætna, Hartford	$3,000,000	$5,782,635	$2,500,000
City, Hartford	250,000	554,287	650,000*
Charter Oak, Hartford	150,000	251,951	400,000*
Connecticut, Hartford	200,000	405,069	600,000*
Fairfield County, Norwalk	200,000	216,358	25,000
Hartford, Hartford	1,000,000	2,737,519	1,200,000
Merchants', Hartford	200,000	540,096	1,000,000*
North American, Hartford	300,000	456,503	800,000*
Norwich, Norwich	300,000	381,736	350,000*
Phœnix, Hartford	600,000	1,717,947	800,000
Putnam, Hartford	500,000	785,783	1,000,000*
MAINE COMPANIES.			
Eastern, Bangor	$150,000	$237,648	$7,500
National, Bangor	200,000	241,308	17,500
Union, Bangor	200,000	421,205	5,000
PENNSYLVANIA COMPANIES.			
Alleghany, Pittsburg	$50,000	....	...
Allemania, Pittsburg	50,000	....	....
Alps, Erie	250,000	$265,524	$185,000
Artisan's, Pittsburg	64,000	....	....
Ben. Franklin, Allegheny	2,000	....	....
Boatmen's, Pittsburg	125,000	....	....
Cash, Pittsburg	100,000	....	....
Citizens', Pittsburg	100,000	....	....
Enterprise, Philadelphia	200,000	611,654	325,000*
Enterprise, Pittsburg	25,000	....	....
Eureka, Pittsburg	175,000	....	....

* Suspended.

NAME.	*Cash Capital.*	*Gross Assets, Jan. 1, 1871.*	*Losses.*
Federal, Alleghany	$20,000	....	....
Franklin, Philadelphia	400,000	$3,087,452	$500,000
German, Pittsburg	50,000	....	....
Girard, Philadelphia	200,000	403,062	13,000
Ins. Co. of N. America, Philadelphia	500,000	3,050,536	500,000
Ins. Co. State of Pa., Philadelphia	200,000	542,908	25,000
Lancaster, Lancaster	200,000	250,349	34,000
Lycoming, Muncy	Mutual.	516,896	500,000
Manufac'rs & Merchants', Pittsburg	125,000	....	....
Monongahela, Pittsburg	140,000	....	....
National, Alleghany	50,000	....	....
Pennsylvania, Pittsburg	115,800	....	....
People's, Pittsburg	76,000	....	....
Pittsburg, Pittsburg	100,000	....	....
Western, Pittsburg	98,000	....	....
WISCONSIN COMPANIES.			
Brewers' Protective, Milwaukee	$164,175	$183,681	$200,000
Northwestern National, Milwaukee	150,000	191,202	90,000
MINNESOTA COMPANIES.			
St. Paul Fire and Marine, St. Paul	$120,000	$280,593	$100,000
FOREIGN COMPANIES.			
Commercial Union	$1,250,000	$4,000,000	$65,000
Imperial	3,500,000	5,438,665	150,000
Liverpool & London and Globe	1,958,760	20,136,420	3,500,000
North British & Mercantile	1,350,000	4,104,598	2,000,000
Royal	1,444,475	9,274,776	98,000

It will be seen that the United States companies have lost $82,821,122; the foreign, $5,813,000; and grand total of losses by all companies is $88,634,122. Such a sum is almost incomprehensible, or altogether beyond the adequate grasp of any human mind. In this connection it is curious to observe how the stable institutions at once received an increase of business almost incredible, and hope to place themselves upon a more solid foundation, which shall be practically immovable. Here is another illustration of the old truth—"To him that hath shall be given, and he shall have abundance; but from him that hath not shall be taken even that which he seemeth to have." This was also verified in the case of the elevators that survived. Their grain belonged to merchants who held checks for it. In many, perhaps most instances, these checks were destroyed, and the grain belongs to the elevator companies. Thus they must become immensely rich. In some instances, where the checks were not wholly burned up, Professor Wheeler, of the University of Chicago, took these charred and obliterated checks, and, by a chemical process, restored the numbers to a momentary existence, and enabled the owner to recover the property. Laying the thin black slip of paper on a plate, he passed a liquid over it, which burned all but the writing and printing, and gave the eye a flying glimpse sufficient to indicate the contents. This fact reminds us of the almost utter worthlessness of safes not protected by vaults. It was sad to see these vaunted "safes" turn out so generally unsafe. "If my safe is all right," said one man, "I am worth ten thousand dollars." He opened it, and instantly, so intense was the heat, all the contents flamed up under the draft of fresh air, and consumed before his eyes, and under his grasp. Probably there was no difference in the powerlessness of safes to resist that burning. My attorney's safe, containing my papers, fell into the coal bin, and lay there roasting like a chestnut, till everything was just in a condition to be blown away by a breath. Some things from some safes in favorable positions were preserved. The little fire box within, in a few instances, appeared to attract heat, and left the contents in a worse condition than those that were outside of it. Men will hereafter trust to nothing less substantial than brick vaults, built underground, or upon foundations that rest on the solid earth. Iron columns twisted and fell, and ruined the vaults they supported, and their contents.

A curious paragraph is worth preserving, to show how the government deals with the charred currency.

"I wandered into the Treasury Department a day or two ago to ask General Spinner to let me see what was being done with the charred money from Chicago. I was shown to the room occupied by the ladies employed on the burned money; shall I call it *cinder*-cates? This room is very large and pleasant, and was selected because, having a southern exposure with nothing near to intercept the light, it has special advantages for this kind of work, which needs the strongest light possible. The charred packages of money which have been almost reduced to cinders, and which crumble at the slightest touch, are brought to the ladies skilled in dealing with such cases. The contents of a safe which was in Adams Express Company's building, in Chicago, were being counted when I went in. There were National Bank notes, United States Treasury bonds, nickels, railroad bonds, and postage-stamps upon the tables. All these must be sorted and arranged, counted, and the value estimated. Such work as this, as may easily be believed, is no light task. The notes are baked to a crisp, and are perfectly black, and the idea of separating them, and deciphering the engraving on their faces, seems at first utterly absurd. Some of the packages are in tolerable order, in other cases three or four hundred notes which have been carelessly thrown into a box, are so melted together that it seems impossible to separate them; in others, bonds have been tied up in a roll for convenience' sake, and are in the worst condition possible to be separated. And here I would give a word of warning. Anybody is liable to be burnt out; any fire-proof safe is subject to being brought under the test of extreme heat, and its contents roasted, so that all persons having notes, bonds, or postage-stamps put away for safe keeping, should take the precaution to keep them spread out their full size, one placed neatly over the other, and, in case of an accident or calamity such as that at Chicago, very little will be lost in the process of redemption. All notes, whose value can be made out, are redeemed at full value. There is no discount on burnt money. The safes or the boxes containing the money are sent at once from the Treasurer's office to the ladies, whom long experience has proved qualified for the delicate and difficult task of handling it and deciphering its value. They take it carefully from its receptacles, and proceed to separate the notes with the utmost skill. Those notes which are so far gone that they crumble at the slightest touch, have their cinders carefully pasted together on sheets of tissue-paper. Great care is taken to prevent the loss of a single note. The ladies are supplied with various aids in their work. Each has a magnifying-glass and several small, thin, sharp, steel instruments with flat blades, which last are indispensable in separating the notes. With National Bank notes the name of the State, the bank, and the denomination of the note must be deciphered, that the money may be returned to the banks which issued it for redemption. The counter certifies to the number of packages, of pieces, denomination, and the total amount. In the case of the Treasury notes, the counter furnishes a schedule for the office of the Secretary of the Treasury, another for the Treasurer, and a third for the Register. These schedules are carefully looked over in these bureaus, signed, and afterward the notes are burned in the presence of representatives of the three officers above named. This work is not only complicated, but imposes great responsibility upon those having it to do; nevertheless, it is proper to state that the ladies receive but $900 per annum for their labor.

The postage stamps found in the express safes are arranged, counted, and returned to the General Post Office; the railroad bonds are returned to the railroad companies who issued them.

The National Bank notes, after being returned to the banks which issued them, are sent to the agents of the banks in Washington, and by them sent to the office of the Comptroller of the Currency, in the Treasury Department. In the redemption division of this office they are counted by ladies, one a representative of the Secretary, a second a representative of the Comptroller, and a third a representative of the Treasurer, and lastly by the agent of the bank, making four countings in all. Accuracy is thus secured, and each counter is a check upon the rest. Afterward the certificates are signed by gentlemen representing the officers above named, and the money is taken in strong boxes, securely locked, to be burned. There is a considerable degree of ceremony attending upon the burning of the notes, although they have already been cancelled and reduced to the value of waste paper. The representatives of the offices named and the agent of the bank whose notes are to be burned go down into the cellar of the Treasury building into a small room resembling a prison cell more than anything else. The furnace resembles an oven, and is set in the wall. It has an iron door which is fastened with three padlocks. Each lock will open only to its own key. The gentlemen acting as representatives of the three officers before mentioned have each a key, and each in turn unlocks the padlock which his key fits. The boxes containing the money are opened by the Secretary's representative; the messenger in attendance sweeps back the ashes of yesterday's burning, piles shavings in the furnaces, throws in a package of notes as a first offering, closes the furnace door, and the fire begins to roar. The door is opened again, and package after package of notes is thrown in; mutilated notes, defaced and time-worn notes, and the charred relics of the Chicago disaster are tossed in. There is a species of excitement in throwing money into the fire. There is a dash of recklessness in it which is fascinating to sober-minded persons accustomed to economy. I know of no better way to ease one's mind after being forced to look twice at every penny before spending it, than to be allowed to participate in the incremation of that money-god which has been tormenting you. You have your revenge on yourself for your enforced niggardliness, and on the dire necessity which has caused your straits. It is almost exhilarating to toss $20,000 that doesn't belong to you into the flames. Nothing equals it, except perhaps Artemus Ward's self-sacrifice in sending his wife's relatives into the army.

After all the money is thrown in, the door of the furnace is locked with the same ceremony with which it was unlocked, and the money is left to burn alone.

Once upon a time this draught in the furnaces used to burn Treasury notes and fractional currency, and was so strong as to carry notes up the chimney, whence they would fly a short distance in the air and fall in the court-yard. It was discovered that these were picked up and used again as money, so measures have been taken to prevent any such occurrence in the future.

This writer has not told us how the gold and silver coin that was melted into masses was separated by the experts, but it was ingeniously counted, and as far as possible restored without reminting. Vast sums of money perished of which no record can be made.

There were in the vault of the Sub-Treasury, at the time of the fire, $1,500,000 in greenbacks, $300,000 in National Bank notes, $225,000 in gold, and $5,000 in silver; making a total of $2,030,000, of which $230,000 was in specie.

In an old iron safe which was left outside the vault was deposited $35,000 consisting of mutilated bills and fractional currency. When the building caught fire, and blazed with fervent heat, the immense vault, with its fabulous treasures, fell to the basement, burying the insignificant safe and its mutilated contents. The contents of the latter were saved, while $1,800,000 in currency was burned to ashes and hopelessly lost.

The specie was scattered over the basement floor and fused with the heat. There were lumps of fused eagles valued at from $500 to $1,000, blackened and burned, but nevertheless good as refined gold. The employés raked the ruins of the whole building and recovered altogether about five-sixths of the whole amount.

It was a fortunate circumstance that only a week before $500,000 in gold, and $25,000 in silver, had been shipped from the city.

Not a single indictment was left on record by the fire against any rogue in Chicago; not a paper to show that there is a suit pending in any of the six courts of the county; not a judgment, not a petition in bankruptcy in the federal courts. And worse yet, so far as is known, all the records of deeds and mortgages are destroyed. The loss of deeds must entail immense trouble upon the owners of lands, and require special legislative enactments to secure proprietors of real estate in their titles. The wisest men are maturing plans to provide against losses and litigations, and make investments in real estate as safe as ever.

INDIRECT LOSSES.

A naked estimate of the value of property actually destroyed cannot contain any adequate conception of the immense damage sustained by the city in its industries and in near and remote business prospects. If we say that 1,100 squares, or more than 2,200 acres, were swept by the remorseless flames in the space of twenty-four hours; that from 20,000 to 26,000 buildings were utterly devoured or left in heaps of unsightly ruins; that the value of the buildings alone was fully $75,000,000, and of their contents at least as much more, we are oppressed by the magnitude of our statements and really comprehend nothing. Regarding the $150,000,000 of property consumed as productive capital —and most of it was that or its equivalent—the income therefrom, reckoned at the moderate rate of six per cent., was no less than

$9,000,000 a year; a sum sufficient to pay perpetually the wages of 7,000 workmen at two dollars a day each, and 3,000 salaried men with salaries of $1,500 a year each; in other words, a sum sufficient for the comfortable support of no less than 40,000 souls.

In saying that the direct losses, regarded as capital, represented the wages fund of 10,000 men, and that the arrest of business represents for the time being a wages fund even greater, it is not by any means meant that more than 20,000 men are thrown out of employment, and 100,000 human beings deprived of the means of support. Thanks to the modern system of insurance, to the modern spirit of enterprise, and to the energy and large-heartedness of our own people, but very few willing hands will long remain idle. Common laborers and such mechanics as are willing to rough it for a season, will find plenty to do in clearing away the rubbish and erecting either permanent or temporary structures.

The Chicago Law Institute reports that, on the 8th day of October, 1871, it had acquired about 7,000 volumes of law books, valued at about $30,000. In October, 1867, it owned, by actual count, 4,681 volumes, which number has since rapidly and steadily increased, and embraced a nearly perfect series of American reports; all the reports of the English courts, and many of the most valuable Irish and Scotch reports; all the law journals of the United States and England; most, if not all, the modern text-books published in this country and England, and also the old English digests, together with a large collection of rare and valuable works on the civil law. While the institute had been aided by many generous gifts from personal and professional friends, yet most of the library had been procured with its own funds, derived from the sale of its stock and assessments upon its members. The library was the property of the shareholders, and freely used by them and all subscribers to the stock who were not in default in the payment of their dues, and was free to all judges and lawyers living outside of Cook county, either in this or any other State. It had always been kept in rooms in the court-house furnished by the county of Cook; convenient to all the State courts, and freely used by all the judges holding courts in this city; and was in charge of a librarian and assistant, one of whom was always in attendance, and was insured for $20,000, divided among different companies, as follows:—Five thousand each in the Lumberman's, Merchants', Firemen's, and Equitable Insurance Companies, all established in the city of Chicago, and organized under charters granted by the State of Illinois.

Besides its library, the Institute had, on the 8th day of October, 1871, $1,318.58 in the hands of its treasurer, and owed only about $350 for all purposes.

On the night of the 8th of October, 1871, a memorable fire destroyed all the books, records, vouchers, and papers of the Institute, with every record of deeds and wills, and all the files and records of all the State and Federal courts established and held in the city of Chicago. The Law Institute thus lost everything it possessed, except its name and legal organization, the balance of $1,318.58 in the hands of its treasurer, and what may be realized upon its insurance, which will not, in the present judgment of this committee, exceed $1,500. In addition to their loss as members of the Institute, the lawyers of this city, with, so far as now known, but one or two exceptions, lost at the same time all their private libraries and papers, although some of them saved a very few sets of reports—mostly those of the State of Illinois and the Supreme Court of the United States.

The following is a complete list of the breweries destroyed by the late fire. The insurance on the property was generally light, and much of it uncertain pay:—

Lill's Brewing Company....................	$500,000
J. A. Huck....................................	400,000
Sand's Brewing Company........................	335,000

Bush & Brand	$250,000
Buffalo Brewery	150,000
Schmidt, Katz & Co	60,000
Metz & Stage	80,000
Doyle Bros. & Co	45,000
Mloeler Bros	20,000
K. G. Schmidt	90,000
George Hiller	35,000
Schmidt & Bender	25,000
Mitinet & Puopfel	12,000
John Behringer	15,000
J. Miller	8,000
William Bowman	5,000
George Wagner	5,000
	$2,025,000

The loss of Mr. Lill has been estimated at $240,000. Mr. Lill's residence alone was filled with furniture valued at over $10,000, much of the furniture being made in imitation of old English furniture, and constructed of the finest materials.

There were eighty-nine newspaper establishments burned, embracing dailies and monthlies.

Thus eloquently has Townsend, "Gath," described the resistance of the fire-proof *Tribune* building, which long withstood but finally succumbed with everything around it.

> Oh! thou, my master, champion of the people,
> TRIBUNE august, who e'er kept righteous court,
> Long after fire had toppled church and steeple,
> Thou stoodst amidst the ruins like a fort.
>
> High and serene thy cornices extended,
> Though scorched by smoke, and of the flame the prey,
> Above the vault where, grim, and calm, and splendid
> The sleeping lions of thy presses lay;

Till looking round on the wondrous pity,
  Thyself alone erect, intact, upreared,
Disdaining to outlive the glorious city,
  With innate heat transfigured, disappeared.

The following estimate of losses of city property under the jurisdiction of the Board of Public Works, is given by Commissioner Redmond Prindiville, who has devoted considerable attention to the subject. This estimate does not include the school-houses, engine-houses and apparatus, police stations, sidewalks, etc. The item of sidewalks only refers to those in front of city property, together with all street and alley crossings, which are constructed by the Board of Public Works. The item of the City Hall embraces only the west half of the Court-House, the remainder being owned by the County. The list is as follows:—

City Hall, including furniture..............	$470,000
Water Works, engines....................	15,000
Water Works, buildings and tools...........	20,000
Rush street bridge.......................	15,000
State street bridge.......................	15,000
Clark street bridge.......................	13,800
Wells street bridge.......................	15,000
Chicago avenue bridge....................	26,700
Adams street bridge......................	37,860
Van Buren street bridge..................	13,470
Polk street bridge........................	29,450
Washington street tunnel..................	2,000
La Salle street tunnel....................	1,800
Lamp-posts.............................	25,000
Fire hydrants...........................	15,000
Street pavements.........................	250,000
Sidewalks and crossings..................	70,000
Reservoirs.............................	15,000
Docks................................	10,000

Sewers	$10,000
Water service	15,000
Total	$1,085,080

It is safe to estimate that the aggregate losses of the several religious denominations by the Great Fire, by the destruction of churches, schools, and other property, approximate $4,000,000. The Roman Catholics alone lose $1,500,000, and the Methodists $600,000. The Presbyterians lose probably about $250,000. The other denominations do not lose so heavily, but the Congregationalists, Baptists, Unitarians, Swedenborgians, Universalists, and Israelites all lost valuable church buildings. In several instances not only the churches were destroyed, but all their members lost their homes.

The Rev. Dr. Bolles writes from Chicago, under date of October 11, as follows:—"Our church (Episcopal), on the North Side, with its 70,000 or 80,000 inhabitants, is completely burned out of existence. Not only has every church edifice been destroyed, but there is not a single parishioner whose private dwelling has not been annihilated in the great conflagration, except that of Mr. Ogden. Among the sufferers are the clergy, and especially the Rev. Mr. Street, the Rev. Mr. Dorset, and the Rev. Mr. Bredburg the Danish missionary; all of whom are reduced to the lowest depths of poverty and destitution."

The *Interior* thus summarizes the Presbyterian losses by the fire:—"Our three oldest, largest, and wealthiest churches are utterly destroyed; a number of our mission schools are burned up; the homes of nearly 1,500 members of our congregations are in ashes; almost every prominent business man in any one of the churches—whether of those destroyed or of those saved—is crippled if not ruined, by losses sustained by the fire; our Seminary is placed in straitened circumstances because of the failure of its invested funds to yield a revenue sufficient to meet its expenses."

HON. ISAAC N. ARNOLD.

"Out of two hundred and fifty families on my list," said Rev. Mr. Parkhurst, of the Grace M. E. Church, "not one has a roof left. Church, parsonage, homes, all were gone, literally annihilated."

Some idea of the proportion of losses may be gained from the following estimates:—

Dry Goods	$6,045,000
Groceries	2,452,500
Clothing houses	1,911,000
Stationers, blank books, etc.	1,110,000
Jewellers, watches and clocks	1,335,000
Hardware	1,280,000
Millinery	1,100,000
Hotels	1,210,000
Church societies and corporations	4,240,000
City property	1,005,000
Railroads	2,000,000
Boots and shoes	975,000
Drugs, paints, and oils	621,000
Books	864,000
Hides and leather	428,000
Restaurants, saloons, etc	528,000
Furniture	510,000
Music dealers	775,000
Hats, caps, and furs	423,000
Glassware, crockery, etc	133,000
Auctioneers	306,000
Tailors and outfitters	178,000
Commissions, etc	128,000

Nothing gives a clearer idea of the magnitude of the great fire than the fact that no one man or body of men can, of themselves, give any idea of the damage done. Every man doing business in the city has it in his power to contribute a valuable chapter to the history which will some time be written of the conflagration.

An untold amount of literary and art treasures have gone down into ashes. In addition to the hundreds of private libraries and collections of works of art, all our public libraries, and an immense number of law libraries, are among the lost.

The collection of the Historical Society, which was among the largest in the country, cannot be replaced. It was the work of years to get it together, but a few hours served to destroy it. The Young Men's Association Library, and the Farwell Hall Library, and several other lesser ones, have passed away. Many gentlemen had extensive private collections of rare and valuable works. A large portion of the members of the bar were sufferers, in this respect, to the aggregate extent of tens of thousands of volumes of costly law books. And then many of the more wealthy of the citizens were liberal patrons of the fine arts, and had brought from Europe valuable paintings with which to adorn their residences. These were left behind in their flight for life before the great sea of flame, which, with the irresistible tread of fate, was sweeping towards them. And last, there are the Sunday-school Libraries of at least half a hundred churches—all gone. And libraries of clergymen, and the great and the small bookstores, scores of them. There is scarcely any end to the loss of the literary and art treasures of the city. It will be long years before our people will be in a condition to restore these adjuncts of our civilization.

Says a correspondent:—

"I walked down the avenue with Robert T. Lincoln, son of the late President. He entered his law-office about daylight on Monday morning, after the flames had attacked the building, opened the vault, and piled upon a table-cloth the most valuable papers, then slung the pack over his shoulder, and escaped amid a shower of falling firebrands. He walked up Michigan avenue, with this load on his back, and stopped at the mansion of John Young Scammon, where they breakfasted with a feeling of perfect security. Lincoln went home with his papers, and before noon the house of Scammon was in ruins, the last which was sacrificed by the Lake side."

Mr. Benjamin F. Taylor, the poet and lecturer, who was formerly the literary editor of the *Journal*, and one of the pioneer citizens of Chicago, was in Buffalo when he learned of the great calamity, and thus, in a note to a Buffalo editer, tells in a few words the feelings and the spirit of Chicago :—

"What time but this," he says, "ever showed the world an extinguished city and an extinguished press? It seems to me as terrible as the day of doom. I cannot realize it. To me it is as if a best-known, best-loved part of the planet had been stricken off with a hammer and lost in space. To speak of small things, but things very near home: here am I, a roll of paper I can carry in my pocket is all I have to show for twenty-one years of daily writing, such as it was. Not a paragraph but these few left to prove I ever penned a line. I feel as if somebody had set me adrift in a boat with a biscuit, and nobody in all the world to make a signal to. But for those who have lost their all—whose magnificent, tangible monuments of wealth, money, and skill, have perished like a wisp of smoke—those who are homeless on the footstool, there is no rhetoric to meet their case. They stand literally *disastered* in the world. And how much grander than eloquent words was the action of the city of Buffalo that sent aid and hope and good cheer to them that stand desolate on the shore of Lake Michigan. God bless Buffalo! And yet I cannot think that the *soul* of the West is scorched at all—that there is so much as the "smell of fire" on the garments of the enterprise that found Chicago like little Moses in the bulrushes, and reared it into a mighty leader, and set the star of the West upon its brow; for I believe the spirit of the little Scot, when 'whelmed beneath the falling house in the Canongate of Edinburgh, is not extinct—the lad who cried out from his living tomb, and so lent muscle and heart to the rescuers, 'Heave away, chaps; I'm not dead yet!' I can hear that voice this morning from away there on Michigan, and even here rises the shout of a rescuer where she sits on the shore of Lake Erie."

The following paragraph is deeply interesting and painfully true:—

No calculation can begin to tell the tale of ruin. Banks, hotels, wholesale houses, all gone—this is indeed fearfully significant; but of the amount and multiplicity of losses no estimate can be brought home to the mind. A volume needs to be written to portray all that Burnt Chicago means. I sat at a railway station the other day by a gentleman who told me his story. He had had nothing burned up. But nearly all that he possessed was in the keeping of the banks. Twenty years ago he went into business with a dentist who has been of late among the first in the city. He prospered greatly, but worked too hard, and for three years he has been wandering into all lands and all climes where possibly he might find relief from intense pain. He looks a young man still, but is broken down, and *Will the Banks Pay?* is his problem of existence. His case is a typical one in this respect, the illustration which he is of prosperity gained here by too great strain upon body and mind. There is no more terrible feature of this calamity than the condition, from excessive overwork, of many of the minds on which it falls most heavily. Can they look into the gulf of madness which this ruin opens at their very feet, sobered instead of crazed, or will they plunge over the brink, either into instant insanity or into utter madness of new excess of exertion? The rage of speculation which has run such a course here vastly complicates all the perplexities of our present situation. All these speculative values—boulevards, suburbs, South Side, etc., etc.—are gone for the present. What might have been available resources in the hands of active business men have been risked and lost. One move of fate has blocked the whole game. Here is my neighbor who was considered worth a million and a half over his debts, yet was under so many mortgages that he must be penniless now; and he is away looking for the power to sleep. I might enumerate many typical instances of enterprise overwhelmed by the descent of this storm while carrying too much sail. The men that had great liabilities on account of real estate speculations, and those who had been taxed in brain and nerve already to the breaking point, were far too many in our city, even compared with the average downward tendency of civilization, in this respect, at the present time. Then there were very many, including many widows and heirs, who had obtained very comfortable means by the rise in value of cheap houses and lots, and whose property had behind it no habit or capacity of self-help. A great deal of the property of this class was in small loans on property. Now all is gone. Time may give some value to the titles or the claims, but all income is cut off. Even the metes and bounds are blotted out.

A description like this below, which some friend of the family has written, enables us to look into the inner circle of losses, and apprehend their exceeding greatness.

Among the many beautiful homes destroyed by the great fire which laid Chicago in ruins, few, if any, were more attractive and home-like than that of Hon. I. N. Arnold. The house was a large, plain, double house, situated nearly in the centre of the block bounded by Erie, Huron, Pine, and Rush streets. The grounds were filled with the most beautiful shrubbery and trees, and entirely secluded by a very luxuriant lilac hedge. Perhaps the most noticeable feature was the vines of wild grapes, Virginia creeper, and bitter-sweet, which hung in graceful festoons from every tree, and covered with a mass of foliage piazzas and summer-houses. There was a simple but quaint fountain playing in front—beneath a perfect bower of overhanging vines—a great rock, upon whose front had been rudely carved the features of an Indian chief, which had been pierced, and a way made for water, and through the head of the old chief the water of Lake Michigan was always throwing its spray. On one side of the entrance was a little green-house, always gay with flowers. Two vineries of choice varieties of foreign grapes, a large green-house and barn, constituted the out-houses. On the lawn was a sun-dial, with the inscription,

"*Horas non numero nisi serenas.*"
("I number none but sunny hours.")

Alas, its tablet was broken with the destruction of the house it seemed to guard—but a brighter day may come, and "sunny hours" be again numbered.

But pleasant as was the outside, it was the interior where its great attractions lay—and chief of these was the library.

Here were the collections of a life—a law library, and a miscellaneous library of about seven or eight thousand volumes. Many of the books were specialties and the objects of pride and affection. The speeches of Burke, Sheridan, Fox, Pitt, Erskine, Curran, Brougham, Webster, Wirt, Seward, Sumner, etc., all superbly bound; a pretty full collection of English literature, poetry and history. Among the notable books were the Abbotsford edition of Scott's novels in full russia binding, Pickering's Bacon in tree calf, six copies of Shakespeare, Knight's illustrated edition, a full set of the British poets, all of Bohn's Libraries, Milton, Bolingbroke, Hume, etc., etc. In American literature and history the library was rich. Beautiful editions of the works of Irving, Cooper, Paulding, Willis, Bryant, Longfellow, Prescott, Holmes, the writings of Washington, Madison, Jefferson, Hamilton, Marshall, Story, Bancroft, etc.

The pictures were not numerous, but of very decided merit. Landscapes by Kensett, Brown, and Mignot, family portraits by Healy, the original study of Webster's reply to Hayne, now in Faneuil Hall, Boston, in which were some forty portraits of distinguished Americans, many of them from life a portrait of Webster by Chester Harding, etc. Mr. A. had a very complete collection of the proceedings of Congress, and the debates, from the organization of the Federal government down. In this library was perhaps as full a collection of the books and pamphlets in relation to slavery, the rebellion, the war, and President Lincoln, as existed in any private hands.

There were ten large volumes of manuscript letters written by distinguished military and civil characters, during and since the war of the rebellion, including many from Lincoln, McClellan, Grant, Farragut, Sherman, Halleck, Seward, Sumner, Chase, Colfax, and others, of great personal and historic interest.

For the last ten years Mr. Arnold has been collecting the speeches, writings, and letters of Lincoln, for publication, and had many volumes of manuscripts and letters, the material for a strictly biographical work upon Mr. Lincoln, several chapters of which were ready for publication. These, with many rare and curious relics, prints, and engravings, have all perished.

The failure of Mr. Arnold to save anything was the result of a most determined effort to save his house, and a confident belief that he could succeed. This confidence did not seem to be unreasonable. The house standing in the centre of an open block, with a wide street, and Newberry block with only one house in front, the Ogden block with only one house directly in the pathway of the flames, it is not surprising that he believed he could save his home. Besides he had connections by hose with the hydrants both in the front and rear. Mrs. Arnold had a better appreciation of the danger, and calling up the family and dressing little Alice, a child of eight years old, she left the house, and went to her daughter's, Mrs. Scudder's, leaving Mr. A. and the remainder of the family, consisting of a daughter, a lad of thirteen, a school-girl of fifteen, and the servants, to fight the battle with the flames. There was a sea of fire to the south and southwest; the wind blew a gale, carrying smoke and sparks, shingles, pieces of lumber and roof directly over the house. Everything was parched, and dry as tinder. The leaves from the trees and shrubbery covered the ground. The first thing was to turn on the water to the fountains in front and on the east side of the house to wet the ground and grass, and attach the hose. He stationed the servants on each side of the house, and others on the piazzas, and for an hour and a half, perhaps two hours, was able by the utmost vigilance and exertion to extinguish the flames as often as they caught. During all this time the fire fell in torrents; there was literally a rain of fire. It caught in the dry leaves; it caught in the grass, in the barn, in the piazzas, and as often as it caught it was put out, before it got any headway. When the barn first caught, the horses and cow were removed to the lawn. The fight was continued, and with success, until three o'clock in the morning. Every moment flakes of fire falling, touching dry wood, with the high wind, would kindle into a blaze, and the next instant would be extinguished. The contest after three o'clock grew warmer and more fierce, and those who fought the devouring element were becoming exhausted. The contest had been going on from half-past one until after three, when young Arthur Arnold, a lad of thirteen, called to his father, "The barn and hay are on fire." "The leaves are on fire on the east side," said the gardener. "The front piazza is in a blaze," cried another. "The front green-house is in flames, and the roof on fire." "*The water has stopped!*" was the last apalling announcement. "Now, for the first time," said Mr. A., "I gave up hope of saving my *home*, and considered whether we could save any of the contents. My pictures, papers, and books, can I save any of them?" An effort was made to cut down some portraits, a landscape of Kensett, Otsego Lake, by Mignot—it was too late! Seizing a bundle of papers, gathering the children and servants together, and leading forth the animals, they started. But where to go? They were surrounded by fire on three sides; to the south, west, and north raged the flames, making a wall of fire and smoke from the ground to the sky; their only escape was east to the Lake shore. Leading the horses and cow, they went to the beach. Here were thousands of fugitives hemmed in and imprisoned by the raging element. The sands, from the Government pier north to Lill's pier, a distance of three-quarters of a mile, were covered with men, women, and children, some half-clad, in every variety of dress, with the motley collection of things which they sought to save. Some had silver, some valuable papers, some pictures, some old carpets, beds, etc. One little child had her doll tenderly pressed in her arms, an old woman a grunting pig, a fat woman had two large pillows, as portly as herself, which she had apparently snatched from her bed when she left. There was a singular mingling of the awful, the ludicrous, and the pathetic.

Reaching the water's edge Mr. A. says he paused to examine the situation, and determine where was the least danger. Southwest toward the river were millions of feet of lumber, and many shanties and wooden structures yet unburned, but which must be consumed before there could be any abatement of the fires. The air was full of cinders and smoke, the gale blew the heated sand worse than any sirocco. Where was a place of refuge? W. B. Ogden had lately constructed a long pier north of and parallel to the United States pier, and it had been filled with stone, but had not been planked over, and it would not readily burn. It was "a hard road to travel," but it seemed the safest place, and Mr. Arnold and his three children worked their way far out on this pier, but it became so uncomfortable that he at length determined to cross the Ogden slip to the light-house, situated well out on the United States pier. With much difficulty the party crossed the Ogden slip in a small row-boat, and entered the lighthouse, and here they and all others met the kindest reception and hospitality.

The party remained prisoners in the light-house and on the pier in which it stood, for several hours. The shipping above in the river was burning; the immense grain elevators of the Illinois Central and the Galena railroads were a mass of flames, and the pier itself some distance up the river was slowly burning towards the light-house. A large propeller fastened to the dock a short distance up the river caught fire, and the danger was that as soon as the ropes by which it was fastened burned off it would float down stream and set fire to the dock in the immediate vicinity of the light-house. Several propellers moved down near the mouth of the river and took on board several hundred fugitives, and steamed out into the Lake. If the burning propeller came down it would set fire to the pier, the light-house and vast piles of lumber, which had as yet escaped in consequence of being directly on the shore and detached from the burning mass. A fire company was organized of those on the pier, and with water dipped in pails from the river the fire kept at bay, but all felt relieved when the propeller went to the bottom. The party were still prisoners on an angle of sand, and the fire running along the north shore of the river. The river and the fire prevented an escape to the south, west and north. The fire was still raging with unabated fury. The party waited for hours, hoping the fire would subside. The day wore on, noon passed and one and two o'clock, and still it seemed difficult if not dangerous to escape to the north. Mr. Arnold, leaving his children in the light-house, went north towards Lill's, and thought it was practicable to get through, but was not willing to expose the females to the great discomfort and possible danger of the experiment. On this occasion Mr. A. saw his gardener with the horses and cow, which could not approach the light-house on account of Ogden's slip. The faithful fellow had ridden the horses far out in the Lake, and he sat on the horse's back several rods from the shore, holding the pony by the halter and the cow by the horn. He saved the animals.

THE GAUNTLET OF FIRE.

Between three and four in the afternoon the tugboat Clifford came down the river and tied up near the light-house. Could she return—taking the party up the river—through and beyond the fire to the West Side, or was it better and safer to spend the night at the light-house? If it and the pier, the lumber and shanties around should burn during the night, as seemed not unlikely, the position would not be tenable, and might be extremely perilous; besides Mr. A. was very anxious to *know* that Mrs. A. and little Alice were safe. The officer of the tug said the return passage was practicable. Rush, Clark, State, and Wells street bridges had all burned and their fragments had fallen into the river. The great warehouses, elevators, storehouses, docks on the banks of the river, were still burning, but the fury of the fire had exhausted itself. The party resolved to go through this narrow canal or river to the south bank, outside the burned district. This was the most dangerous experience of the day. The tug might take fire itself, the woodwork of which had been blistered with heat as she came down; the engine might get out of order and the boat become unmanageable after she got inside the line of fire, or she might get entangled in the floating timber and *débris* of the fallen bridges. However, the party determined to go. A full head of steam was gotten up, the hose was attached to the engine so that if the boat or clothes caught it could be put out. The children and ladies were placed in the pilot house, and the windows shut and the boat started. The men crouched clear to the deck behind the butt works, and with a full head of steam the tug darted past the abutments of Rush street bridge, and as they passed State street bridge the pilot had to pick his way carefully among fallen and floating timber. The extent of the danger now became obvious, but it was too late to retreat. As the boat passed State street the pump supplying cold water ceased to work, and the exposed wood in some parts was blistering. "Snatching a handkerchief," says Mr. Arnold, "I dipped it in water, and covering the face and head of Arthur, whose hat the wind had blown away, I made him lie flat on the deck, as we plunged forward through the fiery furnace. On we sped past Clark and Wells streets. "Is not the worst over?" he asked of the Captain, as the boat dashed on and on. "We are through, sir," answered the Captain. "We are safe." "Thank God!" came from hearts and lips as the boat emerged from the smoke into the clear, cool air outside the fire lines.

Going ashore near West Lake street, Mr. A. obtained a hack at the depot of the Northwestern Railroad, and drove to Mr. Geo. Davis', and leaving his children there he borrowed a horse and rode north on the West Side of the North Branch to get around and above the fire, which was still raging, to try and find Mrs. Arnold and Alice. He crossed at North avenue and went to Lincoln Park, but could get no intelligence of them until he reached General Stockton's, at the end of the Lake shore drive, whose house was filled with North Side fugitives. Here, on the Lake shore, a mile north of the park, he was relieved to learn of Mrs. A.'s safety, and he was advised that she had gone with some friends and neighbors to Lake View or Evanston. It was now dark, and Mr. A. returned to the house of Mr. Davis for the night.

Early Tuesday morning he started to renew the search. Passing through Lincoln Park and the Lake shore drive, he went north, inquiring at every house until reaching Mrs. Snow's, where he learned that Mrs. A. had gone to the West Side. Returning, on his way he met friends who gave him the cheering words that Mrs. A. and Alice, with many neighbors and friends, had on the evening before taken the cars for Winfield, and were all well at the house of Judge Drummond, and there, Tuesday evening, the family all met, and returned thanks to God for each other's safety.

The *Tribune's* account of Allan Pinkerton's loss reveals a curious feature of our modern city life :—

Thousands of thieves, and hundreds of thousands of respectable people, were as fully acquainted with the name of Allan Pinkerton as of Thomas Jefferson, or Jack the Giant-Killer, and his detective agency was as famous an institution as Boston Common. The system over which he presided was the result of years of patient toil and persevering energy, and the reputation enjoyed by him was the fruit of that toil and energy and perseverance. With a huge central office in Chicago, and branches in New York and Philadelphia, the champion thief-catcher had his prey so uncomfortably situated as to be all the time in the toils, only they didn't know it.

The system was not destroyed, but a portion of its foundation has given way, and the savings of twenty years—not in dollars and cents, but in records which dollars and cents can never replace—vanished in about half an hour. Mr. Pinkerton started his famous detective agency in Chicago in 1852, and two years later, when it began to assume large proportions, the records were commenced. The most minute details of every case were all faithfully recorded; the statement of every applicant for assistance in receiving property; the detectives to whom the "job" was intrusted; his orders; his report of the operations; the disposition by the thief of the property stolen; the amount recovered, and indeed every detail of the case. Then, when the thief was brought to trial, the whole of the testimony in the case was taken down, and the final disposition of the prisoners duly recorded, so that from the time a complaint was made at Pinkerton's headquarters that money or property had been missing, a complete history of the thief and his pursuers until the disappearance of the former in the Penitentiary or his acquittal, was recorded. The amount of matter thus created was astonishing. For the mere clerical work upon it more than $50,000 had been paid. Of such curious records there were no less than 400 gigantic volumes of great value. These were nearly all stowed away in six of Harris' largest safes, while the remainder were placed in wooden cases. It is needless to state that every one of them was destroyed. That in itself would have been a public calamity.

Mr. Pinkerton also possessed complete records of the secret service of the Army of the Potomac. They were of immense value, being not only the complete, but the only set in existence. Mr. Pinkerton, whose facilities for obtaining correct information during those days were, of course, very much greater than those of any one else, valued them at $50,000. The government had already offered $30,000 for them—59 volumes altogether—and negotiations were still going on. The whole set perished. That was also a public calamity.

Pinkerton had in his employ a large number of preventive policemen, whose occupation was "to watch while all the city's sleeping, to chase the rogues that prowl by night," as the two *gendarmes* were wont to sing. These men had orders every night to make out a report when they came in. They had to give an account of their proceedings on their beat, the condition of the weather, what unusual circumstances they witnessed, who they saw, and what they said or did to him or her. These reports were all copied into the records. There were forty of these ponderous volumes, which were obtained at a cost of $40 each. Their value may be imagined. They were frequently consulted in court proceedings for the purpose of gaining information as regards the weather, the condition of the streets, the presence or absence of the moon, and other policemen. There were in all forty-eight patrolmen, who gave each an account of these particulars, and it is presumable that their accounts generally coincided so far as atmospheric conditions were concerned. Of course the records were all destroyed. In the first-made rush, when the men seized everything on which they could lay hands, they carried two of these volumes down-stairs and threw them on a wagon, into which much other miscellaneous matter was thrown also. The remainder shared the fiery fate common to everything in the burnt district.

In a small room adjoining Mr. Pinkerton's private office were a number of plain wooden cases in which were stored the files of the daily papers since 1854. There was not a copy of a daily or weekly paper issued since 1854 of which Mr. Pinkerton had not a duplicate. Many were bound together, and 105 volumes covered them all. There were printed instructions posted all round the wall, giving directions to the men, in case of fire, to move these perishable goods first, trusting the safes to protect the records. It was supposed that the Harris safes, which cost $370 each, and a Herring safe which cost $600, would be worth something as a protection against fire, but the result proved them valueless. Only one safe preserved its contents uninjured, and that one by a fortunate accident. Owing to some misconstruction in the building one safe, containing some of the account books and receipts from express companies of money restored, fell into the street, and escaped being melted. The contents were valuable in their way, but to Mr. Pinkerton only. The other records were completely wiped out, and with them was wiped out the foundation for a complete and exhaustive history of the Northwest, besides matter enough for thrilling stories without end; groundwork for sensational stories innumerable. It was Mr. Pinkerton's intention to have some of them published in due time, when the parties were dead or forgotten. Indeed, many of them were already written, and were waiting but a favorable opportunity for introduction to the world. The recent heated term has interfered materially with his designs, and crumbled plans and papers into one common ruin. When the fire first broke out a rush was immediately made to save the most valuable property. All that could be moved in Mr. Pinkerton's room was transferred to a wagon, and the newspaper files were made ready for removing. But before the lowering tackle could be put into satisfactory operation the flames were darting through the hatchway, and the wagon containing the trifling proportion of salvage drove quickly off to Mr. Pinkerton's home on West Monroe street. Besides these losses there were others. The storeroom where the disguises and other paraphernalia of a detective were stored; the dormitory and the extensive household arrangements necessary for the accommodation of the small army of men in constant employ in the building also were burned. The greedy fire here did all the damage it could. In half an hour the best regulated office in the country, and the most accurate and probably minutely detailed records, lay in the basement—a red-hot, indescribable heap of rubbish, the only recognizable article whereof, a fortnight later, was a heap of bricks, surmounted by the remains of a pen-holder.

The damage to the telegraph system was such that every wire in the city was disarranged, all the instruments misplaced, dam aged and removed, and, to crown all, a half-dozen wires between Chicago and New York were completely broken down. To re-establish connection, the whole post of operators moved before the fire three or four times, and the bridgeless stream has been crossed by the reconstructed wire.

Mr. Mullett, supervising architect of government buildings, after inspecting the Post-Office and Custom-House of Chicago, says he is satisfied, from the appearance of the building, that if it had been provided with fire-proof shutters and a safe roof, its contents would have been preserved. He also says that if there had been a whole street of such buildings with fire-proof shutters, it would have stopped the fire and saved the rest of the city. He says the city should be divided into fire districts, so that the firemen should have some rallying point, and that there should be a law requiring every building on certain streets to be built of materials to resist flames, and thus prevent the annihilation of the city at a single conflagration. The government will not entertain any

LIVING A

G THE RUINS.

proposition for the removal of the public buildings in Chicago, but will probably purchase the entire block if it can be obtained at a reasonable price. One of the owners agreed to sell his lot at the same price asked for it before the fire, while the owner of a small shop, learning that the government wanted to purchase, has raised his price two or three times higher than it was before the fire. The Secretary of the Treasury will probably ask Congress to condemn the property, when it will be taken and the regular price paid for it. Mr. Mullett will at once begin the plans for a new building, which will be submitted, with the estimates, to Congress. It is thought the buildings will cost from $2,000,000 to $3,000,-000, and the land $1,000,000.

The Chicago Library possessed many costly works, among which were the records of the English Patent Office, in 3,000 volumes. The destruction of the files of the *Tribune* is an immense loss to Chicago, and an irreparable one to the *Tribune*. There was a duplicate copy presented to the Historical Society. They contained a complete and exhaustive history of Chicago from its first settlement.

The following is a list of the principal edifices destroyed:

Great Central Depot,	St. James Hotel,
Palmer House,	Matteson House,
Tribune Building,	Sherman House,
Post-Office,	*Republican* Building,
Bigelow House,	Chamber of Commerce,
Evening Post Building,	Nevada Hotel,
Tremont House,	Gas Works,
Court-House,	Briggs House,
Lombard Block,	Crosby's Opera House,
Times Building,	*Staats-Zeitung* Building,
Terrace Block,	McVicker's Theatre,
Armour Block,	Wood's Museum,
Journal Building,	Dearborn Theatre,

Adams House,
Massasoit House,
City Hotel,
Metropolitan Hotel,
Union Building,
Post-Office Block,
McCormick's Block,
Western News Co.'s Block,
Manufacturers' National Bank,
S. C. Griggs & Co.'s Book House,
German National Bank,
Mechanics' National Bank,
Commercial National Bank,
Metropolitan Hall,
Arcade Building,
Merchants' National Bank,
Loan & Trust Co.'s Building,
W. U. Telegraph Co. Building,
Oriental Block,
St. Mary's Church (Catholic),
Palmer House,
First National Bank,
Trinity Church,
Third National Bank,
Jewish Synagogue,
Mayo Block,
Fifth National Bank,
Burch Block,
Lake Shore Depot,
Second Presbyterian Church,
Merchants' Ins. Building,
Academy of Design,
Water Works,
Hooley's Opera House,
Mail Building,
Shepard Block,
Honoré Block,
Reynolds' Block,
National Bank of Commerce,
Illinois National Bank,
Cook County National Bank,
Ætna Building,
Armory,
Brunswick's Billiard Factory,
Farwell Hall,
Union National Bank,
Mer. and Farm.'s Savings Bank,
Badger's Bank,
Illinois Saving Institution,
City National Bank,
Adams Express Co.,
W. Fire and Marine Building,
First M. E. Church,
Sturges Block,
Second National Bank,
Phœnix Club,
Morrison Block,
Fourth National Bank,
Catholic Cathedral,
McCormick's Factory,
Galena Elevator,
Galena Depot,
German Theatre,
Unity, N. E., and Westminster Churches,
Sisters of Mercy Convent.

Clarendon Hotel,
Diversey Block,
Lill's Brewery,
First Presbyterian Church,
Hubbard Block,
Chittenden Building,
Bryant's Commercial College,
Otis Block,
St. Paul's Church,
Academy of Music,
Drake-Farwell Block,
Stone's Block,
North Baptist Church,
Hiram Wheeler's Elevator,
Elm st., Catholic Hospital, and the Dearborn, Franklin, Mosely, Lincoln, Pierson street, Elm street, and other Schools,
Sand's Brewery,
Church of the Holy Name,
Alexian Hospital,
Armour & Dole's Elevator,
Hatch House,
Illinois street Church,
Jewish Hospital,
North Star Mission,
Historical Society Building.

The best authorities concur in estimating the total loss at from \$198,000,000 to \$215,000,000, taking the total insurance to represent one-third of the total loss. This may be divided as follows, on a rude approximate:

Loss on buildings and property..........	\$106,590,000
Loss on stock and plant................	74,560,000
Loss on furniture and personal property..	24,850,000
Total..........................	\$206,000,000

Such statements can only be relative and proximate. Actual losses have in some instances been made up by unexpected gains. Interruption to business cannot be valued and may extend over years of the future. By such a view as is here given persons may obtain impressions of great value as to the immense destruction wrought. If first reports were exaggerated in some respects they have never fully comprehended the situation, and only as we travel over the desolated region of nearly three thousand acres, can we fitly conceive what awful damage was consummated.

Before speaking of the dead who lost their lives by the fire, we give room to an appropriate paragraph from the *N. Y. Tribune's* correspondence :—

There was more spared of the remote Northwest of this North Side of Chicago than the reports had any of them admitted—an explanation of which fact I shall presently mention. In 1868, the city limits were at Fullerton avenue, the length of which, from the Lake to the north branch of the river, is two miles. North avenue, a mile back in the city, is but a mile and a half in length from lake to river. As far as North avenue there was little left, and clear up to Fullerton avenue, the more thickly occupied part was all swept away, but the limit of this part ran diagonally from near the west end of North avenue, to near the east end of Fullerton avenue. On the left or west of this limit is a large district mostly unoccupied, and yet sprinkled in various directions with residences of city people, as well as with the cottages of gardeners. Unpaved streets, deep with sand or with earth which is like ashes, are opened, and to a considerable extent sidewalks of plank are laid; and there are two or three small churches within the district. Thus in fact a territory, in shape an isosceles triangle, having the base nearly two miles long on Fullerton avenue to the north, and the sides (1) the river on the west, running there northwest and southeast, and (2) the limit of closer building on the east, running northeast and southwest, was not swept by the fire, and is now the equivalent of a small, very sparsely settled village. Oak openings covered with a young and low growth of trees, squares bare even of fences and thickly covered with thistles, gardens occupying four to eight acres, make up a very large proportion of the district. The city limits were not long ago removed half a mile north of Fullerton avenue, adding a district of more than a square mile, the whole of which is as much "country" as if no city had ever been thought of in the vicinity. The fire actually crossed Fullerton avenue into this district, and ran across its southeast corner, near the Lake on the east, and above Lincoln Park on the north. But it was the least possible snip of ground which was burned over here, and only one small building which was reached. The residence and grounds of Mr. Huck, one of the great North-side brewers, who lost $500,000 lower down on the Lake shore by the destruction of his brewery, occupies the Lake shore front on the north side of Fullerton avenue, his barn standing nearest the southeast corner of the premises, and just beyond it to the southeast is the small house which the fire reached. By great efforts, and aided by the police, whom Mr. Huck stimulated by the promise of $1,000 reward, the barn was saved, and the fire checked at that point. On the site of this one small house, therefore, just over Fullerton avenue, and right at the edge of the wide sands beyond which is the Lake, one stands at the finishing point of the conflagration. And here I may correct the common accounts even of persons resident at the extreme north end, in regard to the distance run by the fire. From Fullerton avenue south to Kinzie street is two and one-half miles by the survey. Kinzie street is the second street north of the main channel of the river. From Kinzie street south, across the river, and as far as Harrison street, is exactly one mile. Nearly all of one block was saved north of Harrison street, the last block to the east, directly on the Lake. Excepting this block, the distance due north from one limit of the fire to the other, or from Harrison street to Fullerton avenue, is precisely three-and-one-half miles. This, therefore, is the length of the broad sweep of conflagration. The average breadth on the south side is three-fourths of a mile, until one reaches Randolph street, going north, which is the third street south of the river. Here the great Central Depot grounds, at the foot of Lake and Water streets, push the line of breadth out to exactly one mile. Thus the conflagration crossed the main trunk of Chicago River with one mile of front. Over the river the line of breadth pushes still more into the Lake, enough to give the fire a front of a mile and one-sixth, and this front is fully kept for the first half-mile north, and nearly or quite kept for the second half-mile; it did not lose much of it for the third half-mile. But for the last mile not more than half of the square mile was run over, the burnt half being a triangle, of which the base was about a mile in length, and the upper point was the finishing point of the fire. This whole region was not burned by a direct northward progress of the fire, but in vast swaths from the river on the west, diagonally across to the Lake. First one vast sweep was made of the triangle the base of which is the main channel of the river, and the upper point of which is the Water Works. After this there struck in a dozen other sweeping scythes of flame, the fire first creeping a block or two along the bank of the north branch, and then tearing madly across in a northeast direction to the Lake. The swinging terrors did not sweep evenly forward, but sometimes one behind outran one which had the start, and they made horrible dashes into each other. As each new start was made higher up on the river bank, and the course was diagonally across, the effect was to maintain a general line of advance directly north, until the last start on the river was taken, when the front commenced steadily narrowing until the fire ended in a point as I have described. The effect of thus moving corps after corps of fire-terrors, their racing side by side, and their fierce mutual interferences, was one of compounded horrors and of amazing sublimity. It seemed as if the earth shook with the awful breathing of the fire monsters, while their voices roared in horrid unison or more horrid discord, as if earth and sky were rushing to ruin. The trampling of the fire-chased throngs, vast whirls of smoke and sparks constantly sweeping over them, frantic men dragging bundles or trunks, women hurrying forward little children, teams dashing recklessly or choked by their own mad rush, women's clothes constantly taking fire, and combustible bundles bursting into flame, while sighs and groans and shrieks made an undertone to the fire-tempest—such was the scene at the moment when the fullest and fiercest course of the manifold conflagration was reached, after successive starts of the fire had been made along the river bank, and when the full number of the reapers of destruction were in mad career across the doomed plain.

A morgue, or dead-house, was early established, where all corpses were gathered for recognition, previous to interment.

Here were enacted scenes of pathetic interest, as friends came to seek their lost ones, and were disappointed; or, discovering the objects of their affection, were overwhelmed with grief. Two girls, looking for their father, recognized him as he lay upon his face, by the hair and shape of his head. They were motherless before the fire, and this robbed them of their chief earthly protector. It was a sad funeral, when we buried him, amid all the excitement and tumult of the day succeeding the conflagration. But, away in the green recesses of the cemetery, there was sweet rest. Let us hope that the repose of Heaven is more sure and satisfying, after the excitement and agony of life.

In the presence of death and woe will men forget the better part? How insignificant seemed man as we stood by the dead in the morgue! Mere pailfuls of charred bones and flesh indicated the existence of those who but the day before were full of lusty life. Oh! helpless man, call upon God, the living God. Here lay the body of a beautiful young girl, of perhaps two-and-twenty.

This poor victim has a wealth of rich brown hair, and brown eyes; she is four feet in height, and posesses a handsome figure. She must in life have been exceedingly lovely. Not being burnt at all, she suffocated in the smoke, as did many of the other victims whose remains were afterwards consumed by the flames.

The fire, whose intensity melted all things, was able to so destroy human bodies that not a trace of them should remain. This fact serves to account for the utter loss of many persons known to have been in the vicinity where the fire appeared and wrought most suddenly and rapidly. It will, therefore, never be known who perished, and how many, until God finally reveals all secret things. Besides the actually burned, many were so shocked as to sink down into death. A lovely aged woman, Mrs. Wright, had long been ill, and was convalescing finally, when her son came home, and said, "Mother, everything is gone." The old lady answered with a smile, "James, then you won't have enough to bury me;" and immediately she began to decline, and soon dropped away into that blessed sleep, "from which none ever wakes to weep." A little girl, dying, said to her mother, "I knew it would rain, because I asked Jesus to send it;" and amidst the falling drops, so grateful to a whole cityful, the trusting child went to her Saviour. Many infants saw the light only to close their eyes upon it forever. And while hundreds were gathered up out of the ruins others have not been discovered, others survived the wreck for a few hours or days, and others linger, who will owe their decease to the terrors, and anxieties, and sorrows of this signal calamity.

Fair she rose,
Lifting high her stately head,
Victor-crowned,
Stretching strong and helpful hands
Far around;
Full of lusty, throbbing life,
In the strife
Dealing quick and sturdy blows.

Sudden swept
Through her streets a sea of fire;
Roaring came
Seething waves, cinders, brands,
All aflame;
Blood-red glowed the brazen sky;
Far and nigh
Smoke in wreaths and eddies crept.

Oh! the cries
Shrill, heart-rending! Oh! the hands

Frantic wrung!
Oh! the swaying buildings vast!
Pen or tongue
Ne'er the awful tale can tell,
How they fell
Underneath the dizzy skies.

Low she lies,
Bowed in dust her stately head,
Desolate;
Yet, by all her glory past,
Let us wait,
Stand beside her firm and true;
Built anew,
Watch her, help her upward rise.

V.—MINISTERED TO BY CHRISTIAN CHARITY.

CHAPTER XXVI.

The greatest human ills have their compensations. Every picture, however dark, has its bright side. Pain and sorrow save from evils deeper and more enduring. Misery and sin develop pity, compassion, patience, and enterprises of recovery and salvation, which bring out the grandest heroes of history, and call forth the most beautiful and sublime qualities of our nature. The war of Revolution and the war against Secession, alike, had their compensations, so vast and real as to cover all the woe and loss they occasioned and entailed.

As great exigencies develop great men, and peculiar sorrows call forth the best elements of human nature, thus compensating for labors and loss in some measure, glorifying mankind, and bringing down God's richest blessings, so on the bosom of this mighty sea of trouble rose a light that brightened into perfect day, and the people of this and other countries put forth their energies to relieve distress and provide for the army of sufferers.

Severe and terrible though our sufferings were, and immense our losses, and the world's losses, yet the spontaneous and magnificent uprising of our countrymen and of people across the ocean, to aid the poor, to help the fallen, to relieve suffering, and prevent despair, was a spectacle unprecedented in history, and may be productive of results that shall be an abundant recompense for so painful a catastrophe. Persons abroad seemed to comprehend our case more perfectly even than we who were almost paralyzed by the shock. The telegraph made our situation known at once to all parts of the world; and while the grounds were red with the embers of the conflagration, men and women began to take measures for the relief of the one hundred thousand sufferers in Chicago.

Nor did they prepare a moment too soon. It will be seen that such a destruction, so sudden, speedy, and complete, must have left a great army utterly destitute of the commonest necessities of existence. Those who were able and accustomed to provide for the needy were many of them as poor as the poorest they had ever assisted. This was our extremity. All were alike in a condition of partial demoralization, and the rush of needy ones from the flames was now turned towards the immediate supplying of their pressing wants. And they were destitute of everything but life and the little they carried away in their hands and saved from plunder. On Monday and Monday night the farmers and inhabitants of the towns close at hand began to gather up clothing, to cook provisions, to empty their cellars, and pour their bounty upon us by means of the railroads.

"An old man from Iowa no sooner heard of the conflagration than he took instant passage for the city to succor his son's family. It was his first visit to Chicago, and it is to be presumed he was ignorant of our geographical position. Still he meant well, so well indeed that on being informed at a way-station that the people were suffering from a scarcity of water, he alighted from the train, purchased a cask, filled it with water, and brought it to the city in triumph. It did not transpire, but is likely to have been the case, that a philanthropic expressman charged him $100 to convey it from the railroad station."

"A clergyman in Athol (Mass.), whose home, we are sorry to say, is not given, was so enthusiastic in packing clothing for the Chicago sufferers that he put his own hat by mistake into the box, and it has gone on with the rest of the donations. This was a truly charitable gift, for it is evident that the left hand of the reverend gentleman didn't know what his right hand was doing; and can there be a more unconditional kind of self-surrender than that which is implied in the formula, 'Take my hat'?"

The papers told a good story of Mr. Ed. Hudson, Superintendent of the P., P. & J. Railroad, and a gentleman well known to railroad men. Upon hearing of the burning of Chicago, his first act was to telegraph to all agents to transport free all provisions for Chicago, and to receive such articles to the exclusion of freight. He then purchased a number of good hams and sent them home with a request to his wife to cook them as soon as possible, so that they might be sent to Chicago. He then ordered the baker to put up fifty loaves of bread. He was kept busy during the day until five o'clock. Just as he was starting for home the baker informed him the hundred loaves of bread were ready.

"But I only ordered fifty," said Ed.

"Mrs. Hudson also ordered fifty," said the baker.

"All right," said Ed., and he inwardly blessed his wife for the generous deed.

Arriving at home he found his little boy, dressed in a fine cloth suit, carrying in wood. He told him that would not do; he must change his clothes.

"But mother sent all my clothes to Chicago," replied the boy.

Entering the house he found his wife, clad in a fine silk dress, superintending the cooking. A remark in regard to the matter elicited the information that she had sent her other dresses to Chicago.

The matter was getting serious. He sat down to a supper without butter, because all that could be purchased had been sent to Chicago. There were no pickles—the poor souls in Chicago would relish them so much.

A little "put out," but not a bit angry or disgusted, Ed. went to the wardrobe to get his overcoat, but it was not there. An interrogatory revealed the fact that it fitted in the box real well, and he needed a new overcoat anyway, although he had paid $50 for the one in question only a few days before. An examination revealed the fact that all the rest of his clothes fitted the box real nicely, for not a "dud" did he possess except those he had on.

While he admitted the generosity of his wife, he thought the matter was getting entirely too personal, and turned to her with the characteristic inquiry:

"Do you think we can stand an *encore* on that Chicago fire?"

Rival cities forgot all the hard words uttered by Chicago, and rallied to our aid with a magnanimity unparalleled, and never to be forgotten. Milwaukee, Cincinnati, and St. Louis were princely in their liberality, which has been eloquently celebrated in these ringing lines:—

I saw the city's terror,
I heard the city's cry,
As a flame leaped out of her bosom
Up, up to the brazen sky!
And wilder rose the tumult,
And thicker the tidings came—
Chicago, queen of the cities,
Was a rolling sea of flame!

Yet higher rose the fury,
And louder the surges raved
(Thousands were saved but to suffer,
And hundreds never were saved),
Till out of the awful burning
A flash of lightning went,
As across to brave St. Louis
The prayer for succor was sent.

God bless thee, O true St. Louis!
So worthy thy royal name—
Back, back on the wing of the lightning
Thy answer of rescue came.
But alas! it could not enter
Through the horrible flame and heat,
For the fire had conquered the lightning
And sat in the Thunderer's seat!

God bless thee again, St. Louis!
For resting never then,
Thou calledst to all the cities
By lightning and steam and pen.
"Ho, ho, ye hundred sisters,
Stand forth in your bravest might!
Our sister in flame is falling.
Her children are dying to-night!"

And through the mighty republic
Thy summons went rolling on,
Till it rippled the seas of the Tropics
And ruffled the Oregon.
The distant Golden City
Called through her golden gates,
And quickly rung the answer
From the City of the Straits.

And the cities that sit in splendor
Along the Atlantic Sea,
Replying, called to the dwellers
Where the proud magnolias be.
From slumber the army started
At the far-resounding call,
"Food for a hundred thousand,"
They shouted, "and tents for all."

I heard through next night's darkness
The trains go thundering by,
Till they stood where the fated city
Shone red in the brazen sky.
The rich gave their abundance,
The poor their willing hands;
There was wine from all the vineyards,
There was corn from all the lands.

At daybreak over the prairies
Re-echoed the gladsome cry—
"Ho, look unto us, ye thousands,
Ye shall not hunger nor die!"

Their weeping was all the answer
That the famishing throng could give
To the million voices calling
"Look unto us, and live!"

Destruction wasted the city,
But the burning curse that came
Enkindled in all the people
Sweet Charity's holy flame.
Then still to our God be glory!
I bless Him, through my tears,
That I live in the grandest nation
That hath stood in all the years.

New York crowned her record of benevolence by gifts that were positively enormous. The Old World, thrilled to the heart, by the flash of the telegraph that showed our city burning and our people roofless, responded with promptness and munificence. Indeed, from one end of the land to the other there was a generosity, such as declared that He who "went about doing good" had not lived in vain. Even the "Heathen Chinee" has a heart in his bosom to feel for others' woes.

The San Francisco *Alta* says that when the Committee in that city to solicit contributions from the Chinese merchants for the relief of the Chicago sufferers made known the object of their visit, the response was a credit to the representatives of that race who have been treated with indignity on so many occasions, and are liable at any time to be assaulted when passing through the streets. In one case an intelligent merchant said to the collectors: "Me leadee in *Alta*, Melican man town all same hap gone—burnee up. Melican man wantee dollas; some time poor Melican man strikee Chinaman with blicks; Chinaman no care. Allee people Chicago losee everything—wifee and childlen burn out. Chinaman say allee same my countree peoplee—wantee help. How muchee dollas you wantee? Hundled dollas? Alee light: you nct find enough monee comee me again, give another hundled." The contributions thus given by the merchants reached $1,290.

From the South responses were slow and feeble. Yet Baltimore, Louisville, and some other towns gave nobly, and their representatives labored personally with efficient energy and wisdom in the distribution of relief. From Falkland, North Carolina, Annie Jones wrote this letter—

Dr. E. J. Goodspeed, Chicago, Ill.:

Having just read in the *Religious Herald* of the great suffering of the Baptists of Chicago, by the late fire there, and wishing to give a *little* aid, you will please accept one dollar from a poor Baptist. Give it to some poor sufferer, and may the Lord open the hearts of *many* others to aid them, is the sincere prayer of

Annie Jones.

This may offset some of those bitter words written upon our fallen city, and printed in Southern papers, to

"Show how its sins invoked the Sovereign's frown."

This seemed to have been a time for sympathy, and the cementing of ties, and not for malediction and savage triumph. So dire a misfortune gave men opportunities to wipe out a dark past; for charity hides a multitude of sins.

And who, hence looking backward o'er his years,
Feels not his eyelids wet with grateful tears,
If he hath been
Permitted, weak and sinful as he was,
To cheer and aid in some ennobling cause,
His fellow-man?

If he hath hidden the outcast, or let in
A ray of sunshine to the cell of sin—
If he hath lent
Strength to the weak, and in an hour of need,
Over the suffering, mindless of his creed
Or home, hath bent—

THE NATIONAL HAND OF FELLOWSHIP.

THE RELIEF COMMITTEE IN SESSION.

He has not lived in vain. And while he gives
The praise to Him, in whom he moves and lives,
With thankful heart
He gazes backward, and with hope before,
Knowing that from his works he nevermore
Can henceforth part.

Among cities east of us, Cleveland was first to arrive with bread and raiment; among cities south, Springfield perhaps took the lead on that memorable morning; among cities north, Milwaukee; and, indeed, from every point of compass, and grade of life, help came, and the one aim seemed to be, to do the utmost, in the speediest possible way, for the miserable sufferers. Philadelphia, city of brotherly love, showed its fraternal spirit in ample gifts. Pittsburg, city of iron, rained gold upon us. Boston, seat of all noble charities and beautiful accomplishments, lavished her thousands, and gave her heartiest toil. Montreal, the American city of Canada, was glorious in her liberality. And so, all around the galaxy, every star seemed to excel in brilliancy to light our darkness; and when we begin to enumerate each bright particular star, they multiply, till we are dazzled and confounded.

CHAPTER XXVII.

The following official communication sheds clear light upon the first steps of the citizens' course, and the initiatory acts of relief, which heralded the incoming of the river-like beneficence of mankind. It was addressed by the president, Hon. Charles C. P. Holden, one of our best citizens:

To the Mayor and Aldermen of the City of Chicago, in Common Council assembled:

Gentlemen: On the 8th and 9th of October last past the heart of our city was destroyed by fire. The territory covered by this terrible conflagration, and the municipal, commercial, and private losses sustained by this fire are all familiar; many of you, having been embraced in its territory, know full well the effect of this great calamity by sad experience.

The undersigned desires to call your attention to the manner n which preliminary measures were taken and arrangements made for succor and relief.

On Monday morning I tried to get the city government together, or portions of it, for the purpose of taking some action to meet the great emergency; in this we failed. The North Division was at the time being burned to ruins; its officers were busily engaged in trying to save their families and the lives of its inhabitants. The Mayor was in the South Division, using every available means to stay the further spread of the fire; indeed, at this particular time, all seemed to be in a state of chaos, and all who had thus far escaped the terrible calamity, expected hourly to be numbered among its victims. At noon of that ever-memorable day, the undersigned called upon Orrin E. Moore, and after a few moments' conference with this gentleman, a general plan of action was fixed upon. In company with Mr. Moore, we at once drove to the Police Station, corner of Union and Madison streets, and after leaving word with Capt. Miller to have certain parties sent for, and to meet us at the Congregational Church, corner of Ann and Washington streets, at the earliest possible moment, we repaired to that church, and at a quarter to one o'clock, in the name of the City of Chicago, we took possession of the same. Capt. S. M. Miller, Deputy Superintendent Wells Sherman, of the Police Department, reported at once for duty. The Mayor was sent for, and before three o'clock the Mayor, Police Commissioner Brown, Hon. S. S. Hayes, Ald. Wilce, Ald. Witbeck, Ald. Bateham, H. Z. Culver, Dr. Goodwin, and very many other citizens had assembled.

Mr. Hayes drew up a proclamation for general distribution,

pledging the credit of the City of Chicago for the necessary expenses for the relief of the sufferers; calling upon the entire police force, the Fire Department, and the Health Department to maintain the peace and good order of the city; establishing the head-quarters of the city government at the Congregational Church, corner Ann and Washington streets. This proclamation was signed by the Mayor, Comptroller, the President of the Common Council, and President of the Board of Police. An organization was immediately effected for the great work in hand, and consisting of the following gentlemen: Orrin E. Moore, Ald. Buehler, Ald. Devine, John Herting, Ald. McAvoy, and N. K. Fairbanks. Orrin E. Moore was chosen President, C. T. Hotchkiss was made Secretary, and C. C. P. Holden, Treasurer.

All the churches and school-houses were thrown open to the distressed. Delegations were sent out to relieve such as they could. Scouts were sent to all parts of the city to watch for incendiarism, and also to watch and report the progress of the fire, where it was then raging, and before midnight of Monday many thousands of special patrolmen had been sworn into the service, and were doing patrol duty. Major Phelps had been detailed to get together a corps to aid him in looking after the sufferers in the South Division. As daylight came on Tuesday, also came E. B. Harlan, the Private Secretary of Gov. Palmer, tendering money, troops, and arms; in fact, John M. Palmer saw at once our situation, and took immediate steps to meet the trying emergency. Committees from the nation commenced arriving—at the head of them was the St. Louis delegation—headed by the Hon. H. T. Blow. Vast quantities of supplies commenced arriving. Ald. Gill, Ald. McCotter, and Supervisor Pierce took charge of the work to receive and distribute supplies from that point; Gen. Mann and Col. Ray took charge of receiving supplies from the railroads in the West Division, and Gen. Hardin had charge of all supplies arriving on the railroads in the South Division.

Various parties were placed in charge of the various churches school-houses, depots for supplies, etc., etc., to the end that all the sufferers by the fire should be cared for at the earliest possible moment. Ald. Wilce was requested to cause to be erected at once from 100 to 2,000 houses, to be occupied by families then homeless. He was to take possession of any land suitable for this purpose. Most energetically did he perform his duty, in company with Ald. Bateham. The Water Works had been destroyed, and not only was there great suffering by those who had been burned out for this most important commodity, but the suffering was being felt by all classes. Water carts in various numbers, trucks, drays, express wagons, carriages, buggies, in fact every vehicle which would not volunteer to aid in the noble work was pressed into the service—water from the parks and artesian wells was distributed throughout the city.

The sufferers were brought from the streets and other places to those where shelter was provided, and before eight P.M. of Tuesday it was reported by a well-known city officer that every homeless soul had shelter, food, and water, and when we recollected that 100,000 or more of our citizens had been rendered homeless by the fire, the result of this day's work must be satisfactory to you. At a meeting of the Committee early Wednesday morning, the Treasurer made a statement to the effect that all moneys should be paid into the City Treasury, where the safety guard of our municipal government would be thrown around it; and further, that this would meet the approval of the country at large, whose moneys were then *en route* here for our succor. David A. Gage was therefore appointed Treasurer.

Mr. Moore and his association had now the work well in hand, considering that the undertaking was less than forty-eight hours old. An arrangement had been made with the railroads, and a Bureau established for the issuing of passes to all sufferers by the fire, another for the lost and found, another for medical purposes, and so on, till there were some eight or ten heads of departments working for the common good in that church, corner of Ann and Washington streets. During this day (Wednesday) numerous quantities of supplies were arriving by every train and on every road—committees from every principal city in the Union and Canadas kept pouring in, bringing words of cheer as they came. Governors of States, too, came—particularly do we remember the deep interest for the sufferers manifested by Gov. Hayes, of Ohio. The committee from the nation held their meetings in the church, and gave us such advice and information as was calculated to inspire us with courage.

The Cincinnati committee commenced at once the erection of a mammoth soup-house, indeed it seemed that these committees from abroad comprehended the situation even better than ourselves. Everything that could be done in that hour of great distress, by them was done. At their meeting held in the evening of Wednesday, they had more than one hundred present; the result of the meeting was the issuing of an address to the nation. The effect of this address has had a wide-spread influence in making known to the country our real wants and needs. Thursday, the 12th, the Chicago Relief and Aid Society took charge of the great work then fairly commenced. On Friday evening the committees from abroad held their final meeting. At this meeting they issued an address to the citizens of Chicago. In it they said: "We are perfectly satisfied to recommend to the country that all moneys intended for your relief be sent to the City Treasurer, because we believe they will not only be safe, but will be expended in accordance with the wishes of the contributors. It was signed: H. T. Blow, Chairman Western Committee; A. J. Goshoon, Chairman Cincinnati Committee; W. M. Morris, Chairman Louisville Committee.

The undersigned remained at the head-quarters first established until Thursday evening, Oct. 24, doing all that he could do in behalf of the city to carry aid and relief to all the sufferers. In this great work there had been voluntarily engaged during the first week an army of our citizens, both male and female, and very many of them are still in the traces and at work. During this time great expenses were incurred in the procuring of lumber, nails, etc., for the building of temporary houses; the providing of all classes of vehicles for the moving of families and their supplies; during the same time the undersigned received numerous advices of the sending forward for the relief of the sufferers vast sums of money; he also received in person the sum of $42.50 in cash, to wit: Committee from Valparaiso, Indiana, the sum of $40, and from two ladies $2.50, all of which was immediately turned over to D. A. Gage, treasurer. Before closing this report I desire to call the attention of the Council to the great good performed by the Board of Health, who were at the head-quarters night and day till the 24th, doing all that could be done in the line of their profession to relieve the distressed. To all the members of your honorable Board I bear witness to the aid and efficiency rendered by you. Many of you lost your homes and places of business by the fire; even this did not deter or keep you from rendering aid and assistance to others, as well became those occupying the positions you do. To the ladies, who rendered great assistance on this most trying occasion, no words can express the encomiums they have earned—their names are legion.

Gentlemen, I have deemed it my duty to make this statement to you of matters pertaining to the great fire and subsequent thereto, and would ask your kind consideration of the same.

In connection with this important contribution to the history of relief, we publish the following address to the citizens of Chicago, written October 13th, which was referred to above:

The undersigned respectfully call your attention to the following facts: The committees from the principal cities of the West, with food and supplies of all kinds, have been in your city since last Monday night; they assembled at the head-quarters of the Mayor and City Council, corner of Ann and Washington streets, and have since co-operated with Alderman Holden and other members of the Council. Mr. Moore and his associates being the only organization known to them in the city for the relief of the sufferers by the great fire, the St. Louis supplies, with large quantities intrusted to the delegation from Indiana and Illinois, were distributed by General Hardie, who in person, under orders of General Sheridan, placed them, as we believe, most judiciously. We attest most heartily to the unselfish and arduous services rendered by Alderman Holden, Mr. Moore and his associate members, the Mayor, and many of the Common Council, Mr. Preston, of the Board of Trade, and especially General Sheridan and his aids, and yet deem it a duty to say to you that it is now absolutely essential that the work be systematically and economically extended, that ample arrangements *should at once* be made for the reception and careful distribution of coming supplies, by an organization which will satisfy yourselves and encourage your friends to continued action. We are perfectly satisfied to recommend to the country that all moneys intended for your relief be sent to the City Treasurer, because we believe that they will not only be safe, but will be expended in accordance with the wish of the contributors; but from the facts presented we trust you will see the actual necessity for the systematic arrangement alluded to; and now that your best men can calmly survey the condition without fear of the future, we again most earnestly beg that you will take immediate steps for a thorough and permanent organization, that will be entirely equal to the great work before them.

Henry P. Blow,
Ch. of the Western Committees
A. T. Goshorn,
Ch. of Cincinnati Committee.
Wm. M. Morris,
Ch. of Louisville Committee.

And herewith is presented the Mayor's order, which gave universal satisfaction:

"I have deemed it best for the interests of this city to turn over to the Chicago Relief and Aid Society all contributions for suffering people in this city. This Society is an incorporated, old-established organization, and has possessed for many years the entire confidence of our community, and is familiar with the work to be done. The regular force of this Society is inadequate to this immense work, but they will rapidly enlarge and extend the same by adding prominent citizens to the respective committees; and I call upon all citizens to aid this organization in every possible way. I also confer upon them the power, heretofore exercised by the Citizens' Committee, to impress teams and labor, and to procure quarters so far as may be necessary for the transportation, distribution, and care of the sick and disabled.

"General Sheridan desires this arrangement, and has promised to co-operate with this association. It will be seen from the plan of work detailed below, that every precaution has been taken in regard to the distribution of the contributions."

Up to the time of this step towards a more thorough and judicious management of supplies for relief, there had been various points selected in the unburnt district, especially churches, where the houseless found shelter and food.

The rush to these depositories of food, and places of rest for the outcast multitude, was in many cases overwhelming and fearful. In my own church, every lower room was occupied by the sick as a hospital, by mothers as a nursery, by the committee on distribution, and for storage of goods and provisions. Orders from our committee were honored at the Rink, where the supplies were gathered for general distribution, and immense loads would melt away like snow in the summer sun. There was no lack of helpers to succor the unfortunate. We could not find work enough for those who were anxious to assist in caring for their more unfortunate fellow-citizens. Hundreds were comfortably lodged on the benches, which were cushioned. There was a record of missing, lost and found, kept in the church, and hundreds daily searched it; and in several instances the long-separated met together in the sanctuary. A colored girl saved a charming white baby, and the exigencies of flight drove her here, and the mother found her beautiful child safely cared for by its nurse. Death came also to some who were hospitably entertained, and they gave up their lives in peace within the walls which often echoed to the message of eternal life.

When shelter, tents, and barracks had been provided, one by one the lodgers left the church, every one being presented with a cushion and a blanket. The same scenes were enacted on every hand in the churches, which were homes, where the beautiful hand of charity gave cheer and aid, with kind words and tender acts. One learned to love the Chicagoans more, when we saw their self-sacrificing devotion to the welfare of their neighbors, amidst their own desolation, losses, and forebodings of coming want, or fears of present peril. Yet there were instances of despicable thieving, pilfering, and hypocritical pretence, which outrivalled anything we ever read of in history. Some parties made raids upon the public bounty, and supplied themselves with a winter's stock. There was a woman in one of the churches who got upon her person and in her bundle twenty-seven dresses. Wherever these instances were found they were speedily punished, and imposition was checked. But in the first hurry and pressure of want there was too little opportunity for discrimination; and people said, we must not let any one suffer, even though impostors share with the actually destitute. It was soon seen that there must be careful, faithful discrimination, or the supply would be gone and the want unrelieved. At this juncture the entire matter was committed to the organization called the Chicage Relief and Aid Society.

CHAPTER XXVIII.

While the boundless charity of the great-hearted American public made it possible to feed, clothe, and comfort one hundred thousand persons in an incredibly short period, so that the very poor fared better than it was their wont to do, and all classes were blest in some measure, the necessity of an efficient association for permanent and deeper work was instantly apparent, and grew more urgent every hour. This was the crisis, too, for the machinery of our Aid Society to be applied to the greatest problem of the century; and nobly has it met the emergency. Under the superintendency of a warm-hearted and large-minded Christian gentleman, Mr. O. C. Gibbs, it had been for years efficient in providing for the large number of poor people always crowding around its doors, and so investigating their claims that imposition was well-nigh impossible. It was found all ready for indefinite expansion, and assumed the control of all contributions of every kind, except those sent to individuals. Its visitors were sent through districts to every house, and all applications were investigated thoroughly, and when worthy sufferers applied, they were at once provided with what they needed for the time, and arrangements made to issue them rations till they could become self-supporting. The accompanying directions and information were furnished by printed circulars:—

To all Superintendents, Assistants and Visitors in the Service of the Chicago Relief and Aid Society:

In the distribution of supplies give uncooked instead of cooked food to all families provided with stoves; flour instead of bread, etc.

The Shelter Committee furnish all families for whom they provide houses and barracks, with stove, bedstead, and mattrass, and no issue of those articles to such families will be necessary on your part.

Superintendents of Districts and Sub-Districts will so keep an account of their disbursements as to give a correct report to me at the end of each week, the number of families aided during the week, and the amount, in gross, of supplies distributed.

Superintendents will also ascertain and report, as early as possible, the amount of furniture, number of stoves, amount of common crockery, etc., which will be needed in their respective districts.

Superintendents will also organize their working force as early as possible, retaining upon their force those who have proved themselves the most efficient and capable in the discharge of their duties, reducing the number of paid employés to the smallest number consistent with the efficient performance of the work of their districts.

A special organization charged with the relief of special cases is being effected, to which all that class of persons whose previous condition and circumstances in life were such as to make it unsuitable that they should be relieved through the ordinary channels of relief, can be referred.

No person in the employ of the Society will be allowed to receive for his own use any supplies of any kind whatever, except it be through the ordinary channels of relief, and recorded on the books of the office in which he is employed.

In all cases of applicants moving into your district from another, you will, before giving any relief, ascertain, by inquiry at the office of the district from which they came, if they had been aided in that district, and to what extent.

In the issue of supplies you will discriminate according to the health and condition of the family, furnishing to the aged, infirm, and delicate, supplies not ordinarily furnished to those in robust health.

The following has been adopted by the Society as the standard daily ration for a family of five persons; you will vary from the amount according to the income of the family from labor or other sources:

Bacon or pork	2 pounds.
or beef,	3 "
Beans	1 pint.
Potatoes	2 quarts.
Bread	3 pounds.
or flour	2 "
Tea	1 ounce.
or coffee	2½ "
Sugar	4 "
Rice	4 "
Soap	4 "
Soft coal	¾ ton per month.

The Department of Sick and Hospitals have adopted the system of Districts and Sub-Districts established by this department, and appointed a medical officer for each District. Visitors will report all cases coming to their knowledge requiring medical attendance, and the person in charge of each office will have such reports at all times in readiness for the medical officer of the District, when he calls. All possible aid must be given the medical officer of the District, and he is to be allowed free access to the office and books of the Society at all times.

The bread now being furnished is contracted for by the pound. You will be furnished with platform scales, and required to weigh and receipt for all bread delivered to you.

Superintendents and Visitors in those districts in which the Shelter Committee have furnished houses to men who were burned out, will inquire carefully into the condition and circumstances of all persons who have been furnished houses by the Shelter Committee, and report to Mr. Avery, Chairman, all cases in which parties have obtained lumber or building material by fraudulent representations.

The Chicago Relief and Aid Society will, for the coming winter, have to provide for all of the poor of the city, as there will be no distribution of the out-door relief by the County Agent as heretofore. While your first care should be for those who have lost all by the fire, those that are not direct sufferers by it must be aided according to their necessities. The loudest complaints will come from those least deserving, who are always on hand for their share when any distribution is to be made or relief given.

You will have to refuse the application of many worthy people, who, having lost heavily by the fire, will think themselves entitled to a share of the relief fund, although still possessed of the means or ability to meet their present wants. You will explain to such as kindly as possible that the relief fund is not intended to make good losses by the fire; that it can be used only to prevent and relieve actual suffering.

We are not yet in a condition to be even liberal in disbursements. Three months hence we will be in better condition to decide how far we can be liberal than now.

In the matter of fuel, soft coal only will be furnished to those whose stoves will burn it; hard coal only to those who cannot burn soft. No wood will be furnished, except for hospital use, and in case of sickness in families where it is necessary.

Those having wood stoves will be furnished with grates to enable them to burn soft coal. The Chicago & Wilmington Coal Company, and the Chicago Relief Society's yard can furnish only soft coal; until further orders hard coal will be furnished by Ames & Co. and B. Holbrooke & Co.

As fast as your stores will permit, give out a week's supply of food to those families whose cases have been thoroughly investigated—this can soon be increased to two weeks, which by so largely diminishing the number of daily applicants will enable you to dispense with a large part of the working force in your offices and stores, and relieve the applicants of the humiliation of daily attendance upon your office to obtain their supplies.

You will instruct those families who have been visited and found worthy, and who will require aid during the winter, to make their applications to you hereafter in writing, either through the mail, or by the hand of a child, or some other messenger. It is a terrible trial to a sensitive woman or honorable-minded man to be compelled to make a personal application at a relief office, and we must so arrange our work to relieve such as far as possible of this necessity. On receiving such application the necessary orders for supplies can be made, and the supplies sent directly to the family. To fill these orders you will require the services of an experienced retail grocery clerk, and one or more express or grocery wagons for delivery.

I am informed that large numbers of servant girls are unemployed in the city, who refuse to go to employment at good wages in the country or other cities. Be sure that none such are fed by the Chicago Relief and Aid Society. If there was ever a time when every person capable of earning his or her own support should be made to do it, it is now. Help must even be withheld from families who harbor persons able to work, but who are unemployed. In all cases where help is discontinued or refused to families, your books must show the reason for such discontinuance or refusal.

There are several thousand men and boys working this week whose families we are feeding, who will be paid for their work on Saturday night, sufficient to meet all the wants of the family for food or fuel next week. Be sure that every such family is known in your district, and reported at the office, so that no more supplies be given to it. Our supplies are going at a *fearful* rate. If any men, boys, or women are not working, apply St. Paul's rule: "If any man among you will not *work*, neither let him *eat*."

I think it will be conceded that the generous confidence be

stowed upon us in the following paragraph from a New York editorial, was justified by the manner in which these funds were distributed, and the supplies continue to be dispensed: "To feed, shelter, and clothe these suffering thousands, without waste or misapplication, will require all that executive capacity which the Chicago people eminently possess. But no one need fear that the relief so generously poured out will not be judiciously distributed. Difficult as must be the organization of a force to superintend and move the machinery to be called into operation, we know enough of keen, practical Chicago, to confide to the hands of its business men all the gifts which they are to receive in trust for the whole suffering people."

There were cases where men, dressed in a little brief authority, or impatient under the accumulation of petty annoyances from the vast stream of applicants at the depots for distribution, gave just cause of offence on the part of sufferers. There were insults given and hardships endured. Wild rumors of extensive peculations ran over the city. Fault-finding was as prevalent then and there as might have been expected; but the gentlemen connected with the Society labored zealously and with extraordinary judgment and patience to satisfy the clamors of the eager thousands who thronged them. In their instructions to employés the Society said:—

In all your intercourse with applicants for relief, your manners to and treatment of them should be kind and considerate. You will have to render aid to many families whose condition is one of chronic pauperism, resulting from their vices or improvidence. This class you can never satisfy; like the daughters of the horse-leech, their constant cry is "give," but the great majority of your applicants will be people who have suddenly been reduced from a condition of self-support, and in many cases of affluence, to one of partial or entire dependence. Their case is a sufficiently painful one without anything in your intercourse with them to remind them that they are now dependent upon charity. You will give such persons the preference over the class first named, so far as it is possible for you to do so, in receiving their application and supplying their wants, and let your intercourse with them be such that they will ever after look upon you as a friend in their time of need. While you may not be able to supply all their wants, convince them, by the kindness of your manners and your interest in their behalf, that you are doing all that is in your power to do for them.

In the press of business at your office, you will not be able to give much personal attention to a statement of their wants and necessities, but the visitors at their homes can do so; hence it is of the utmost importance that your visitors be persons fitted by character and experience for these delicate duties.

The Superintendents will be required to dismiss from their further employ any person whose manner has been uncourteous or unkind to applicants for relief.

The Bureau of Special Assistance is now in active operation, with head-quarters at the Church of the Messiah, Wabash avenue, near Hubbard court. Applications to this Bureau can be made either in person or by letter, addressed to its head-quarters, or through any pastor of the churches of the city, as may best suit the inclination or convenience of the applicant.

Superintendents of districts and sub-districts will fill from their stores all orders addressed to them by this Bureau without question, the necessary investigation having in all cases been made by the Special Bureau.

This allusion to the Bureau of Special Assistance requires a few words of explanation, since it grew out of the exigencies of the situation and supplemented the regular society's work. There was a vast number of cases where families or individuals had suffered the loss of all things, whose circumstances in life had been above all need, and whose delicacy of feeling would not per-

THE WEST SIDE RINK, CHICAGO—GENERAL DEPO

OF SUPPLIES FOR THE SUFFERERS BY THE FIRE.

mit them to stand in line with hundreds of the very poor, degraded, and foreign applicants who unblushingly pushed themselves into the front ranks. Many had been educated to abhor dependence as something worse than death. I recollect one boy of seventeen, who said one morning after sermon, "I can't stand it any longer, pastor; we six are eating from a wash-stand, and sleeping on the floor; I must tell you about it. I thought to work and get along, but we can barely get enough to eat. We were burnt out and lost everything except what we had on, and I have my three younger brothers to look after." "Of course," I replied, "you shall be attended to at once," and before forty-eight hours things were changed in that house, and no application was made to the Society. This boy has no father or mother, and I wrote of the case, after we relieved it, to a friend, who writes "To the boy who takes care of his helpless brothers":—

"East Orange, N. J., Nov. 27, 1871.

"My Young Friend:—I do not know your name, but Rev. Mr. Goodspeed, in a letter to me, spoke of you as one of those worthy ones who had suffered by the fire. He spoke of your courage and brotherly care over some younger brothers.

"That letter I read to some of my friends, and one of them, some days after, handed me these same bills, "for the boy who took care of his younger brothers." Fidelity, my young friend, will always be rewarded.

"Very truly yours,

"Wm. D. Hedden,

"*Pastor of E. O. B. Ch.*"

To meet the multiplying cases of this kind that were known and suspected, a meeting of pastors and representatives of benevolent organizations took place, at which a committee was chosen, by consent of the Relief and Aid Society, to constitute a

Bureau cf Special Relief. This splendid measure of assistance has proved of incalculable benefit to thousands, who otherwise must have suffered alone, and unknown to any but God who seeth all. At first there was a delay in securing men who could give their time to this important service. When the Committee had been filled, another meeting was held, at which these resolutions were passed, and the New Bureau was fully launched :—

Whereas, The great exigency of public relief demands immediate, large, and constant service in special council and assistance, and

Whereas, We learn that a portion of the Executive Committee originally appointed by this Bureau has been unable to meet the demands of this great work, on account of their inability to serve at all, and of others to give any considerable service; therefore,

Resolved, That the Bureau cordially endorse the action of the Executive Committee in filling the vacancies in that body by adding to its members gentlemen widely known as wise, efficient, and eminently fitted to carry on the work.

Resolved, That the Church of the Messiah, the depot of special supplies, be also the place of special meeting of the Committee, and that said Committee be hereby instructed to make arrangements among themselves so as to have at least three of their members, two gentlemen and one lady, in attendance at the office during all the hours in which the depot is open for distribution.

Resolved, That all churches and beneficial societies should regard themselves as special bureaus for council and relief, and feel responsible, not only for looking up and bringing to the relief supply through the appointed channels those who have been overlooked and are deserving, but also especially to guard the munificent bounty of the nation from plundering waste and ravaging imposture. And they are hereby earnestly exhorted to use their easy approach to the masses, through necessary meetings and supervision, to prevent the Executive Committee of this Bureau and the General Board of Relief from being overwhelmed by a countless multitude of unworthy or doubtful applications.

This gave pastors and others great opportunities, and imposed grave responsibilities. Their hands were soon full, and the revelations of need yet unprovided for, after three weeks had elapsed, were truly startling. We had not realized the appalling magnitude of the calamity, though in its very midst.

And we may ask, who will ever apprehend it, in all its gigantic proportions?

The "Special Relief" committee, charged with aiding cases of peculiar delicacy, from the former respectability of the sufferers, learned of a gentleman who, before that terrible night of the fire, was worth between $150,000 and $200,000. He boarded with his family at one of our splendid marble palaces known as hotels, where his elegantly furnished apartments and luxurious table, indicated his wealth and ministered to his ease. The next that was heard of him was some days after the fire, when he applied to the committee, saying that the fire had literally burned up every dollar of his fortune, and he had no money, no home, no clothes, no furniture, and no food! His family were living in a stable, sleeping on the hay, and eating the cold potatoes and bread which the children begged from the neighbors!

The duties of applicants were thus set forth in a notice by this Bureau, which shows the public what care was exercised with their bounties.

All applications to this Committee for Special Relief, must be certified by the pastor of a church, or proper officer of some organized benevolent society, or by a member of the Executive Committee of the Relief and Aid Society, or of this Committee, who shall state in such certificate that the condition and needs of the applicant have been duly investigated, to the satisfaction of the persons so certifying, and stating what amount and kind of relief should be afforded to such applicant.

In every application, the name, residence, and relief district in which such applicant lives should be plainly written.

Such application should state whether the applicant is married or single, the number of persons in the applicant's family, the age and sex of each member, and should set forth in detail the articles which are wanted, and the number, amount, or quantity of such articles. In applications for clothing, the kind of clothing, and number of pieces needed of each kind, should be distinctly stated, the proper sizes, where necessary (as of boots, shoes, and other articles), given.

Applications for groceries should state specifically the articles wanted, and amounts of each article, and where crockery, or furniture, or bedding is needed, the specific articles wanted, and the number of such articles should be stated.

The committee desire to call the attention of the applicant specially to the following points, in regard to which information will be desired, and which should be stated :—

The present and former occupation of the applicant, whether burned out and what loss they suffered, amount of insurance and in what company, what property applicant has, and what aid they have received from any source, or expect to receive.

Careful attention to these requirements will save the applicant delay and trouble, and insure prompt action in the case.

The railroads gave free transportation to seven or eight thousand persons, who left the city for refuge under friendly roofs elsewhere, or to obtain employment, and brought in the stores that were contributed without charge, thus conferring benefits of immense value upon our people. On the eleventh of October, two days after the fire, the Erie Railroad had its relief cars on the way at ten in the morning. The train consisted of seven cars heavily laden with provisions. Mr. George Crouch went with it as supercargo, and delivered the freight to the Mayor of Chicago. The train averaged about fifty miles an hour to Port

Jervis. It reached Susquehanna at 3.05 P.M., and was last reported at Elmira, making unprecedented time to that point. Dense crowds of enthusiastic people were assembled at the depots in the principal towns, and many attempted to throw bundles on the train as it flew past.

On the evening of that day, Col. James Fisk, Jr., writes:—

We received to-day, since the departure of the lightning relief train at 10 o'clock this morning, over 10,000 consignments for the sufferers at Chicago, which were forwarded by the express train at 7 o'clock this evening. It would be almost impossible to enumerate the contents of the packages or their value; but as far as we can judge, taking the entire shipment, nothing could be more appropriate had a month been occupied in the selection. I find that in a single consignment there were shipped 100 coats, 100 pairs of trousers, 100 vests, while another consignment included 400 barrels of sugar and coffee, and still another consisted of 100 barrels of flour. A person competent to judge, who inspected the goods forwarded to Chicago by this single train, estimated their cash value at over $100,000.

We have, from appearance, as much, if not more, to receive to-morrow, which we shall forward by our express trains only at 9 A.M., 12 M., 5½ P.M., and 7 P.M.

It were idle to attempt an enumeration of the kindly offices of the railways, which made Chicago, which have ministered to it in distress, and must recreate and secure the future.

AMOUNT OF MONEY RECEIVED.

From the appended circular it will be seen what had been received in contributions from every source, down to November 7, 1871.

The Executive Committee of the Chicago Relief and Aid Society are aware that the public desire to know the amount of the subscriptions to the Relief Fund. It is impossible at present to give a detailed account of the amounts, for the reason that purchases made in some cities, invoices of which have not yet reached us, are to be deducted from the gross amounts of the subscriptions. The previous report of our Treasurer stated the amount actually received at that date. We are now able to give the amount received to this date, November 7th, and the probable amount of the entire subscriptions, with approximate accuracy. We have actually received two million fifty-one thousand twenty-three dollars and fifty-five cents ($2,051,023.55). Arrangements have been made by which the Society draws five per cent. on all balances in bank. So far as our present information goes—and we think we have advices of all sums subscribed—the entire fund will vary but little from three million five hundred thousand dollars ($3,500,000). This includes the funds in the hands of the New York Chamber of Commerce, amounting to about $600,000, and the balance of the Boston Fund, about $240,000, both amounting to $840,000, not yet placed to the credit of this Society, but which may undoubtedly be relied upon to meet the needs of the future. As to our disbursements, we can only say that we are at present aiding 60,000 people at our regular distributing points. Some of this vast number we relieve in part only, but the greater portion to the extent of their entire support. This is in addition to the work of the Special Relief Committee for people who ought not to be sent to the general distributing points, and which is largely increasing upon our hands. It is also in addition to the expenditures of the Committee on Existing Charitable Institutions.

The great matter pressing upon the Committee is shelter for the coming winter. We may feed people during the mild weather, but where and how they are to be housed—permanently housed—we regard as the serious question. To this end we have been aiding those burned out to replace small but comfortable houses upon their own, or upon leased lots, where they can live, not only this winter, but next summer, and be ready to work in rebuilding the city. Of these houses—which are really very comfortable, being 16 by 20 feet, with two rooms, one 12 by 16 feet, and one 8 by 16 feet, with a planed and matched floor, panel doors, and good windows—we have already furnished over 4,000, making permanent homes, allowing five for a family, for 20,000 people, and with the 7,000 houses which we expect to build, shall have homes for 35,000 people. These houses and some barracks, in both of which there is a moderate outfit of furniture, such as stoves, mattresses, and a little crockery, will consume, say $1,250-000, leaving $2,250,000 with which to meet all the demands for food, fuel, clothing, and general expenses, from the 13th of October last—when we took the work—until the completion of the same, which cannot possibly end with the present winter. We may say that particular attention has been paid to sanitary regulations. The entire work in this respect, as in others, is district ed. Medical visitors, dispensaries, and hospitals are provided.

The Committee need hardly say, that if the demand should continue as great as at present, the fund would be exhausted by midwinter; but we hope to cut this down very largely as soon as we can get people into houses, so that they can leave their families and find work. Indeed this is being done already. Within a few days we shall arrive at the exact daily expense of food and fuel rations; but the demand, as might be expected, is a fluctuating one. If the weather is good and men can work, it falls off; if cold and stormy, it at once increases at a fearful rate.

The work has so pressed upon us, night and day, that we cannot present a detailed report to the public, but furnish this statement for the purpose of affording a general idea of what we have done and are trying to do, with an organization necessarily composed largely of unskilled forces, but the only one at hand for the emergency. Within the next ten days we shall be able to give a detailed report of the work as well as all sums contributed.

December 1st, this sum had been swelled to $2,508,000, with the current still flowing steadily into the treasury.

Of other gifts, the value is known to One, who sees the widow's mite as well as the millionaire's mightier help. But we cannot estimate the worth of all that vast store which was made up as rivers are—by ten thousand rivulets, brooks, and streams incessantly emptying in their precious contents.

CHAPTER XXIX.

In addition to these organizations, there were movements among the citizens, and the Young Men's Christian Association early entered the field, with their forces generalled by Rev. Robert Patterson, D.D., who labored during the war in the Christian Commission. Their head-quarters were at the Seventh Presbyterian Church, and their charities were immense. Parties from Boston came on and superintended the distribution of the supplies from that city. Their work was largely among those who were not well served at the general relief depots, or who were looked up and searched out, on account of pride, or sickness, or some inability, moral or physical, to make application in person.

There was also the Woman's Christian Union, a society particularly concerned with the employment of women, who now supplemented this service with a relief duty, always aiming at securing means by which the poor women could become self-sustaining. They rendered most valuable aid to the suffering.

Societies sprang into being on all sides, and made their appeals to the churches and benevolent societies, and their acquaintances throughout the land. They obtained clothing, made up garments from new cloth, nursed the sick, and gave a helpful hand to any whom they found neglected in the crowd of miserable beings that overflowed into every street, alley, garret, and cellar of the unburnt city.

Private citizens donated from their own houses all that they could spare, in many instances, and vied with the outside public in liberality. It was pitiful to see the burned-out parties, the morning of Monday and of Tuesday, begrimed, soiled, scorched, bearing a little truck, a trifling remnant of their possessions, to the homes of their friends, and begging for temporary shelter. But it was grand to observe the nobleness of the many, and their perfect sympathy with the distressed. And rising to the height of their obligation, our people are preparing for a campaign against poverty and misery, by leaguing in societies for service to the poor, that shall relax no effort during the long winter now marching down upon us. Rich and prosperous communities become greedy, selfish, covetous, and money-worshippers. It remains for us to prove the benefit of adversity by opening our hearts and hands to give and not to save. "Alms the salt of riches," is an old proverb. I know a man ruined by this fire, who was very rich, and refused to lend a poor woman fifty dollars, in an extremity, to save her house and furniture from the sheriff. He had known her for many years, and she had claims upon him; but no, he loved his money, and turned the poor woman off with stern denial. The week before the fire, she came to the gentleman who lent her the money—a poor man, and a minister—and paid it with interest, thanks, and tears. This gave the clergyman spending-money after the fire. And how must the miser feel—yea, and many others like him—who have been close-fisted, hard-hearted, niggardly, and avaricious? The worldling saves his money, yea, and the Christian too, for his children. But how often does its possession curse them. An eccentric D.D., in the course of a sermon in behalf of some charitable object, once said, "There are twenty men in this congregation who can give $20,000 each to this charity, and then have money enough left to ruin their children." Now there is an opportunity for the young men and women to show the quality of their characters. We shall be a better people for this trial, if we give with full-handed generosity, and learn that

"To give is to live,
To deny is to die."

Among the impossibilities is any just account of the aid received by the sufferers; because large amounts of moneys and supplies have been sent to private individuals, for personal use or disbursement; and thousands have gone home to their friends, who have proffered shelter and food for the winter. In time a book will be written, acknowledging these grand charities in a befitting manner.

From a New York paper we clip the story of the reception of certain refugees from the fire.

A few evenings ago eight newsboys of Chicago arrived in this city and sought shelter at the Newsboys' Lodging-House in Park Place. On Saturday six more arrived and went to the same home, and on Sunday four more. The ages of these youths were from sixteen to nineteen years. One of them, a lad of eighteen, had his face very much scorched by the fire, and some of the others were disfigured to some extent from the same cause. Those who arrived on Friday night left the next morning to seek friends in this city or in Brooklyn, the six who came on Saturday, with the four who arrived on Sunday, remained in the Lodging-House until Monday afternoon, when they, too, left in pursuit of friends. The New York boys gave their brothers of the West a very cordial reception, and as far as their little means allowed, lavished upon them a generous welcome. The New York boys, all so much younger than the Chicagoans, were profuse in their expressions of sympathy, albeit uttered in the vernacular of the profession, and poured out volleys of inquiries as to the state of trade in the ill-fated city. The usual sports of the evening were stopped immediately on the arrival of the immigrants, and each visitor, after a hearty meal, formed a centre of attraction for a score of boys, each of whom had something to learn of the great fire. Sunday was a great opportunity for the exchanging of notes, it being comparatively a dull day, and the new arrivals of the evening previous were escorted to favorite haunts and lionized to an extraordinary degree. Those of the boys who had belonged to the boot-blacking profession very warmly discussed the depression of prices in that line, and though it was unanimously agreed upon that the profession should be on all occasions retained by a "choker" fee, yet, all things considered, it was deemed best just now not to enter upon that dangerous experiment, a "strike." The Chicagoans were loud in their admiration of Peter B. Sweeny, who, they said, deserved the presentation of a set of complimentary resolutions on account of his great services to the shine-em-up boys in beautifying the City Hall Park. Regrets were expressed that all the fountains were not in working order, as they are very inviting to customers. The Chicagoans urged upon their New York brothers to establish their headquarters around the fountain in front of the City Hall when completed, and not under any circumstances to yield their right on this point. It was also suggested that as many portable chairs as possible be provided, with a view to placing business on a footing more conformable to ordinary mercantile pursuits. It was said that the experience of a series of years has demonstrated that "chairs are good." The newsboys learned with great satisfaction that the people of Chicago were a newspaper reading community, and did not stick at trifles. All that was necessary to do in case of small change was to delay a few minutes in procuring it, and the "gent" was sure to "get." Harmony among members of the profession was also an admirable feature in Chicago, everybody "working his own route" on the square and with no nonsense. The Chicago delegation was enlightened as to how trade stood in New York, and a comparative estimate given as to the daily receipts afforded by every evening paper in the city. Discussing these topics and similar ones the boys passed the day, and after a hearty supper at the Lodging-House in the evening, again resumed the entertainment. Cordial invitations were extended to the Chicagoans to join the honorable brotherhood of New York, and assurances extended that a most friendly reception awaited them in the arena of competition. No decisive answer was given to the New Yorkers' offer; but evidently the Chicago boys were deeply impressed with the tone and boyish bearing of their new acquaintances, and promised that if ever they should again return to "that line of business" New York City should be the theatre in which their ambition should have a chance. When it was approaching bedtime it was felt by New York that it was necessary to do something grand on an occasion like that then being celebrated, and why should not the newsboys have a say of sympathy as well as every other branch of business? This idea became so impressed upon the mind of one of the boys—a sort of leader of a set—that he summed up courage, and, rising, said:—

GENTLEMEN,—You know about the Chicago fire, and that these gentlemen (pointing to the ten Chicagoans) are sufferers. I now want to tell 'em that we're sorry for 'em. Our subscription list is making up, and I heard Mr. O'Connor say 'twill amount to $8.25, which they will get, though it's small and not as much as we'd like to. That's all I have to say, except that if these gentlemen stay here we'll post 'em.

ANOTHER BOY.—Billy, propose a resolution.

BILLY.—I move that we're awful sorry for the sufferings of the newsboys and black-a-boots of Chicago, and that if they stay we post 'em, and that anything we can do we'll do to help 'em, and that we're sorry it ain't more than $8.25.

Great applause followed from all the other boys.

One of the Chicago youths then rose, after some hesitation, and said:—

"Thankee, gents, for what ye've done, an if it weren't that we had to go and see some friends we'd like to stay. Maybe, though, we'll come back."

At this moment the Superintendent appeared on the scene, and this was the signal for the adjournment of the meeting *sine die.*

The boys then went to their "little beds" and to sound sleep, New Yorkers to dream of Chicago, and the Chicagoans of the great fire and their recent hardships.

A little Irish boy, Tim, employed in a bake-shop, sent five dollars from East Orange, N. J.

And here speaks a voice from Old England:

"CONYNGHAM-ROAD, VICTORIA PARK,
"MANCHESTER, Oct. 16, 1871.

"MY DEAR MR. MAYOR: As you have convened a meeting, to be held to-morrow, of the inhabitants of Sheffield, to consider what measure should be taken to relieve the sufferers at Chicago, in the United States of America, under the calamity which has so suddenly befallen them, I beg leave, as a native of the borough over which you worthily preside as Chief Magistrate, to offer a contribution of two hundred guineas to the fund intended to be raised, for which sum I inclose a check payable to your order. I am gratified to learn that the people of America will accept the expression of sympathy and sorrow in this country, in the kindly sense in which it will be offered to them; and I consider it to be a privilege to have the opportunity of uniting in this undoubted sentiment of affection and regard of the inhabitants of Great Britain and Ireland towards the people of America.

"May we and they ever be one people under our respective

governments; and be bound together as lovers of freedom to the end of time. I have the honor to be your very obedient servant,

"GEORGE HADFIELD.

"To the Mayor of Sheffield."

The good Queen has thought of us and given for our relief. She reads every word of the tidings from our city with intense interest. Her subjects have also responded, in a most creditable manner, to the silent appeal of our distress. There is a very profound regard for our country in the old world, and the ties that bind us together are strengthened by these expressions of active charity. Scarcely a hamlet in the British Isles can be found which has not its representative here, either among the humble or the influential. We are essentially cosmopolitan, and the world have taken us up, to nurse and cherish in our fall. It is true as ever, that one touch of nature makes the whole world kin. The natural feelings of all Christendom have been touched by the unvoiced woe of Chicago. The magnitude of this generous work, and the spontaneity of the timely giving, fitly symbolize the community of interest and feeling which now bind the human family together.

CHAPTER XXX.

BESIDES the magnificent gifts for the body and for immediate comfort, there have been systematic and general efforts in aid of the churches and educational institutions, which must result in placing them once more in a position of usefulness and stability. It is impossible to chronicle these donations, as the tide is still flowing in upon us. Orders and societies are rising in their might, all over the land, to rebuild and re-establish the institutions lost in Chicago. Christians, surely, will prove their profound interest in their cherished cause, by responses that shall make the future of our city worthy of the Lord Jehovah, and a centre of evangelical power.

Capitalists came forward instantly, with offers of money and credit to any extent, for the reconstruction of what was their pride as well as our own. Merchants and business men received the heartiest assurances of sympathy from those to whom they were indebted, from their creditors and customers; and every leniency was afforded and extended, compatible with safety and creditable to the heart.

Governors of States took up our cause, and commended us to the philanthropy of their citizens.

BY THE GOVERNOR OF WISCONSIN.

To the People of Wisconsin:

Throughout the northern part of this State fires have been raging in the woods for many days, spreading desolation on every side. It is reported that hundreds of families have been rendered homeless by this devouring element, and reduced to utter destitution, their entire crops having been consumed. Their stock has been destroyed, and their farms are but a blackened desert. Unless they receive instant aid from portions not visited by this dreadful calamity, they must perish.

The telegraph also brings the terrible news that a large portion of the city of Chicago is destroyed by a conflagration, which is still raging. Many thousands of people are thus reduced to penury, stripped of their all, and are now destitute of shelter and food. Their sufferings will be intense, and many may perish unless provisions are at once sent to them from the surrounding country. They must be assisted now.

In the awful presence of such calamities the people of Wisconsin will not be backward in giving assistance to their afflicted fellow-men.

I therefore recommend that immediate organized effort be made in every locality to forward provisions and money to the sufferers by this visitation, and suggest to mayors of cities, presidents of villages, town supervisors, pastors of churches, and to the various benevolent societies, that they devote themselves immediately to the work of organizing effort, collecting contributions, and sending forward supplies for distribution.

And I entreat all to give of their abundance to help those in such sore distress.

Given under my hand, at the Capitol, at Madison, this 9th day of October, A.D. 1871.

LUCIUS FAIRCHILD.

BY THE GOVERNOR OF MICHIGAN.

STATE OF MICHIGAN, EXECUTIVE OFFICE,
LANSING, *October* 9.

The city of Chicago, in the neighboring State of Illinois, has been visited, in the providence of Almighty God, with a calamity almost unequalled in the annals of history. A large portion of that beautiful and most prosperous city has been reduced to ashes and is now in ruins. Many millions of dollars in property, the accumulation of years of industry and toil, have been swept away in a moment. The rich have been reduced to penury, the poor have lost the little they possessed, and many thousands of people rendered homeless and houseless, and are now without the absolute necessaries of life. I therefore earnestly call upon the citizens of every portion of Michigan to take immediate measures for alleviating the pressing wants of that fearfully afflicted city, by collecting and forwarding to the Mayor or proper authorities of Chicago supplies of food, as well as liberal collections of money.

YOUNG LADIES MINI:

G TO THE HOMELESS.

Let this sore calamity of our neighbors remind us of the uncertainty of earthly possessions, and that when one member suffers all the members should suffer with it. I cannot doubt that the whole people of the State will most gladly and most promptly and most liberally respond to this urgent demand upon their sympathy; but no words of mine can plead so strongly as the calamity itself.

HENRY P. BALDWIN,
Governor of Michigan.

BY THE GOVERNOR OF IOWA.

To the People of Iowa:

An appalling calamity has befallen our sister State. Her metropolis—the great city of Chicago—is in ruins. Over 100,000 people are without shelter or food, except as supplied by others. A helping hand let us now promptly give. Let the liberality of our people, so lavishly displayed during the long period of national peril, come again to the front, to lend succor in this hour of distress. I would urge the appointment at once of relief committees in every city, town, and township, and I respectfully ask the local authorities to call meetings of the citizens to devise ways and means to render efficient aid. I would also ask the pastors of the various churches throughout the State to take up collections on Sunday morning next, or at such other time as they may deem proper, for the relief of the sufferers. Let us not be satisfied with any spasmodic effort. There will be need of relief of a substantial character to aid the many thousands to prepare for the rigors of the coming winter. The magnificent public charities of that city, now paralyzed, can do little to this end. Those who live in homes of comfort and plenty must furnish this help, or misery and suffering will be the fate of many thousands of our neighbors.

SAMUEL MERRILL,
Governor.

DES MOINES, *Oct.* 10, 1871.

BY THE GOVERNOR OF OHIO.

CHICAGO, *Oct.* 12.

To the People of Ohio:

It is believed by the best informed citizens here that many thousands of the sufferers must be provided with the necessaries of life during the cold winter. Let the efforts to raise contributions be energetically pushed. Money, fuel, flour, pork, clothing, and other articles not perishable should be collected as rapidly as possible—especially money, fuel, and flour. Mr. Joseph Medill, of *The Tribune*, estimates the number of those who will need assistance at about 70,000.

R. B. HAYES,
Governor of Ohio.

BY THE GOVERNOR OF ILLINOIS.

STATE OF ILLINOIS,
EXECUTIVE DEPARTMENT.

John M. Palmer, Governor of Illinois, to the People of the State of Illinois:

A fire of unexampled magnitude has devastated the city of Chicago, depriving thousands of our citizens of shelter and food and clothing.

Under these painful circumstances, I call upon you to open

your hearts for the relief of the suffering. Contribute of your abundance everything that you can—food, clothing, money; organize committees and systematize your efforts.

Remember those, our fellow-citizens who have always responded so nobly to every call.

In testimony whereof, I have hereunto set my hand and caused the great seal of State to be affixed.
[SEAL.] Done at the city of Springfield, this 10th day of October, A.D. 1871.

JOHN M. PALMER.

By the Governor,
EDWARD RUMMELL, *Secretary of State.*

STATE OF ILLINOIS,
EXECUTIVE DEPARTMENT.

John M. Palmer, Governor of Illinois, To all to whom these presents shall come, greeting:

Whereas, in my judgment, the great calamity that has overtaken Chicago, the largest city of the State; that has deprived many thousands of our citizens of homes and rendered them destitute; that has destroyed many millions in value of property, and thereby disturbing the business of the people and deranging the finances of the State, and interrupting the execution of the laws, is and constitutes "an extraordinary occasion" within the true intent and meaning of the eighth section of the fifth article of the Constitution.

Now, therefore, I, John M. Palmer, Governor of the State of Illinois, do by this, my proclamation, convene and invite the two Houses of the General Assembly in session in the city of Springfield, on Friday, the 13th day of the month of October, in the year of our Lord 1871, at 12 o'clock noon of said day, to take into consideration the following subjects:—

1. To appropriate such sum or sums of money, or adopt such other legislative measures as may be thought judicious, necessary, or proper, for the relief of the people of the city of Chicago.

2. To make provision, by amending the revenue laws or otherwise, for the proper and just assessment and collection of taxes within the city of Chicago.

3. To enact such other laws and to adopt such other measures as may be necessary for the relief of the city of Chicago and the people of said city, and for the execution and enforcement of the laws of the State.

4. To make appropriations for the expenses of the General Assembly, and such other appropriations as may be necessary to carry on the State Government.

In testimony whereof I have hereunto set my hand and caused the great seal of State to be affixed.
[SEAL.] Done at the city of Springfield, this 10th day of October, A.D. 1871.

JOHN M. PALMER.

By the Governor,
EDWARD RUMMELL, *Secretary of State.*

In response to the call of the Executive, the Legislature assembled, and received this further message from the Governor, whose contents met the warmest approval of all our citizens:—

STATE OF ILLINOIS, EXECUTIVE DEPARTMENT,
SPRINGFIELD, *October* 16, 1871.

Gentlemen of the Senate and House of Representatives: On the 8th day of the present month a fire broke out in the city of Chicago, which, in a few hours, destroyed a large portion of that city.

It is useless to attempt to describe the awful and saddening spectacle of the destruction of the most wealthy and populous parts of our great city. The destroyer came suddenly, and under circumstances well calculated to impress us with a sense of our littleness.

Chicago is situated on the shore of a great lake; it is intersected by rivers; it was provided with all the means for protection against fire that are the product of the united efforts of the advanced science and skill of modern civilization; yet in the presence of the destructive element men were powerless, and it pursued its course until nothing was left for it to destroy.

In the course of this remarkable conflagration, which has already taken its place in history with the greatest calamities that have afflicted mankind, the flames, with unexampled fury, swept over the eastern half of the devoted city, destroying many lives, consumed churches, hospitals, schools, dwellings, warehouses, stores, bridges, and structures of every kind. Everything perished at their touch, and whole wards of the city were left without a house or an inhabitant. No reliable estimate of the number of lives lost can be made, but the amount of property destroyed is estimated at three hundred millions of dollars.

In view of the circumstances, I felt it to be my duty to convene a session of the General Assembly, and, accordingly, on the 10th day of October, 1871, issued the proclamation which I have the honor to lay before you.

At the time of the meeting of the General Assembly, all were still so far under the control of the feelings excited by this extraordinary calamity, that no scheme had been formed for the employment of the powers and resources of the State to meet the duties that are imposed upon it by this unexpected condition of affairs.

But before proceeding to invite your attention to the details of the business of the session, I must be permitted, in the name of the people of the State, to express their grateful thankfulness for the exhibition of outpouring sympathy and benevolence that this great and sudden calamity has excited in all civilized lands. Not only have our own people and the people of our sister States distinguished themselves by an active liberality that is without a parallel, but in foreign countries the hearts of men and women have throbbed with pity for Chicago, and their hands, filled with contributions, have opened to supply the wants of its suffering people. Where all have aided, and all have done so much, it is impossible to give even the names of our benefactors. Their example, so honorable to them and to human nature, is worthy perpetual remembrance, and I trust that the General Assembly will provide for the preparation and publication of a memorial volume, in which their names shall be preserved. The people of the State should be permitted to know the names of those who, when their brethren were hungry, fed them, and when they were naked, clothed them.

The first question to be decided by the General Assembly, after a careful review of the situation, is, what can be done for the relief of the people, and for the discharge of the duties of the State? In finding an answer to this question, there are some difficulties and causes of embarrassment that are yet to be stated; and these are, the court-house, jail, and public offices, and records of Cook county are destroyed. The tax-books are consumed, so that the collection of unpaid taxes cannot, without great difficulty, be enforced. The courts are powerless. The utmost confusion, as to the titles of lands, must soon prevail. All the offices and most of the records of the city of Chicago are lost. Still the question, What can be done by the State? presses for an answer—and all the wisdom, experience, and patience of the General Assembly is invoked to furnish a full, complete, and satisfactory response.

The general political proposition, that that government is to be regarded as the best that interferes with the people the least, will remain forever true; and experience has conclusively shown that intelligent men and women are, under all ordinary circum stances, more capable of providing for their own wants, managing their own affairs, and regulating their own conduct, than any government can be, however organized or administered. It seems to me, then, that the people of Chicago and Cook county, who have suffered losses, require nothing from the State but to be left free to employ their unexampled and unbroken energies in the great work of rebuilding their homes.

They need no loans or gifts from the United States or the State of Illinois; and, unless I greatly mistake them, they will ask no more than that the State shall assume the discharge of its own proper duties, and relieve them from burdens—that, from their peculiar situation, were always heavy, but have been cheerfully borne—so that they may be left to apply all their resources to their own great task. It is primarily the duty of the State to provide for the poor, the blind, the insane, and all other helpless classes, and for the enforcement of its laws everywhere within its limits. It is also its duty to provide for the construction of its highways, building bridges, and the support of schools. The State of Illinois has always recognized the obligation of these duties, and for the more convenient performance of many of them, counties, townships, cities, towns, and other organizations have been established by law. They are but parts of machinery employed in carrying on the affairs of the State, and the authority and the duties of each are confined to certain well-defined territorial as well as legal boundaries, that may be modified or destroyed, as the exigencies of the public may demand. And whenever, from any cause, any of these agencies become unequal to the discharge of the duties assigned them, or the public duties imposed upon them become too burdensome or oppressive to the people embraced within their limits, it is the duty of the State to provide other means for their performance. It is a fact that requires no proof, that the county of Cook and the city of Chicago, two of the most important of the classes of public agencies to which they respectively belong, are, from causes that are well understood, unable to continue the full discharge of all the duties that were imposed upon them. From an inevitable accident, their resources are diminished and their local burdens vastly increased, so that they are no longer available to the State as governmental agencies for all the purposes for which they were created, and it follows from that fact that to the extent that the requirements of such duties are in excess of the legal resources of the county and city—such duties must be resumed by the State, and the General Assembly must devise other methods for their performance.

It is a most remarkable illustration of the difficulty of providing for every possible contingency by constitutional regulations, that certain provisions of the constitution of 1870, that were intended to restrict the powers of municipal corporations, and were resisted upon that ground, will be found to operate to relieve the county of Cook and city of Chicago of what would otherwise be intolerable burdens. Every part of the constitution abounds with proof that its framers regarded the municipal organizations of the State as mere administrative agencies, and that they intended to deprive them of all emergent or discretionary authority, except within very narrow limits.

By the twelfth section of the ninth article of the Constitution it is provided that "No county, city, township, school district, or other municipal corporation shall be allowed to become indebted, *in any manner or for any purpose*, to an amount, including existing indebtedness, in the aggregate exceeding five per centum of the value of the taxable property therein—to be ascertained by the last assessment for State and county taxes." . . . . And by the eighth section of the same article, county authorities are prohibited from assessing taxes, the aggregate of which shall exceed seventy-five cents on the hundred dollars valuation. Then, whatever power to raise money for necessary public purposes the State has denied its local or municipal organizations it has reserved to itself, to be exercised by the General Assembly. The financial resources of municipal and local organizations are necessarily limited to their powers to contract debts and to impose taxes. When these powers have been exerted to the utmost legal or possible limit, and are inadequate to the complete performance of their duties to the State, they must be relieved of such duties altogether; for the accepted construction of the constitution forbids the General Assembly to pay, assume to pay, or to become responsible for the debts or liabilities of, or in any manner give, loan, or extend its credit to or in aid of any public or other corporation or individual—(Sec. 20, Article 10, State Constitution). This provision of the Constitution was adopted for reasons well understood, and but few will doubt its policy or wisdom, and no one will, I apprehend, be willing to relax its stringency, or narrow its interpretation by constructions however ingenious or plausible.

It has been proposed to give immediate aid to the city of Chicago, by discharging the lien of the city upon the Illinois and Michigan canal, authorized to be created by the act approved February 16, 1865; and it is claimed that if the State should now refund to the city the amount of money secured upon the revenues of the canal, with the interest thereon (which would be, in round numbers, about three millions of dollars), the city would be enabled to rebuild its bridges and public structures, remove the obstructions from and repair its streets, pay the expenses of its government, and other expenses pertaining to its own organization, and discharge its general duties to the State.

I am not prepared to express an opinion upon the question: whether even that sum of money would be sufficient to supply all the essential wants of the city; but my impressions incline me to admit that it would; and I am prepared to say that while, under ordinary circumstances, influenced alone by my views of the proper policy to be pursued by the State, I would not advise the acceptance of the option secured to the State in the fifth section of the act of 1865, to refund to the city the sum of two millions and a half dollars, with interest thereon. Under present circumstances, if the money can be raised by any satisfactory means for the purpose, it seems to me that it should be done. The county of Cook, alone, has heretofore contained nearly one-sixth of the taxable property of the State, and a proportion of this, which falls very little short of the whole, was situated in the city of Chicago. Now, nearly one-half of the productive property of the city is detroyed, and its present resources are crippled; but the day is not distant when its walls will be rebuilt, its wealth and population not only restored but increased, and instead of requiring aid from the treasury of the State, it will be again its chief resource, and money now appropriated to meet its necessities, will be bread cast upon the waters, to be gathered again after not many days. But while policy as well as duty concur in support of the propriety of an appropriation from the State treasury, either to discharge the duties heretofore imposed upon the city, and which unaided it can no longer perform, and for that reason they now devolve directly upon the State—or to refund to the city the sum of money used by it in deepening the canal, and for which it has a lien upon the property of the State—it remains to be considered how the money is to be raised to meet such appropriation.

Two methods have been suggested for the accomplishment of this object. I am informed that the amount of the taxable property, as reported to the Auditor, for 1871 is about five hundred millions of dollars, which is probably less than one-tenth of the actual cash value of all the property in the State. From that sum will probably be deducted fifty millions, on account of the destruction of property in the county of Cook. Calculating, then, upon the basis of an actual assessment of four hundred and fifty millions, the rate of taxation required to raise three millions of dollars is sixty-six and two-thirds cents upon the hundred dollars; and when to this is added the probable rate of fifty-five cents, that may be required for revenue and school purposes, the rate of taxation for the year 1871 will be one dollar and twenty-one and two-thirds cents upon the hundred dollars. And I confess to a preference for this mode of raising all money required for public purposes. It is simple, direct, and, of all modes of raising money, it is the cheapest. It proposes that each generation shall discharge its own duties, and it conforms to the golden rule of business morality: "Pay as you go."

But the demands of the city of Chicago, for whatever sum may be appropriated for its use, are urgent and immediate, and months may elapse before the proceeds of taxation can be realized, and it may be the judgment of the representatives of the people, that the rate of taxation that it will be necessary to impose is, under present circumstances, too heavy to be conveniently borne; and for some or all of these reasons, some other method of raising the requisite sum may be preferred.

The only other mode of raising money that has occurred to me is that of borrowing the amount required. But it has been asked, with some degree of anxiety, under what clause of the present Constitution is the exercise of the power to contract a greater debt by the State than $250,000 to be justified? and to find a satisfactory answer to the question, is thought by some to be a task not altogether free from difficulty. The provision of the Constitution relied on by those who question the power of the General Assembly to borrow money (and thereby contract a debt) to a greater amount than two hundred and fifty thousand dollars, is found in the proviso to the eighteenth section of the fourth article. The language of this proviso is: "The State may, to meet casual deficits or failure in revenue, contract debts never to exceed in the aggregate two hundred and fifty thousand dollars; and moneys thus borrowed shall be applied to the purpose for which they were obtained, or to pay the debt thus created, and no other purpose; and no other debt, except for the purpose of repelling invasion, suppressing insurrection, or defending the State in war, . . . shall be contracted, unless the law authorizing the same shall have been submitted to the people at a general election." Those who deny the power to contract a debt to raise money to discharge the lien on the canal insist that the amount of money expended by the city of Chicago to deepen the canal does not, when tested by the proviso of the thirty-seventh section of the third article of the Constitution of 1848, constitute a debt against the State, and that now to borrow money to discharge the lien of the city would be to create a debt in violation of the eighteenth section of the fourth article of the Constitution of 1870; and they contend that the words employed in the section last referred to, that prohibit the General Assembly from contracting debts, "except for the purpose of repelling invasions, suppressing insurrection, or defending the State in war," are to be construed literally and strictly, and that their effect is to absolutely prohibit the State from contracting debts except for the very purposes and under the precise circumstances specified.

It must be confessed that if those who thus reason are correct, the only mode that can be adopted to afford either direct or indirect aid to the city of Chicago, is that of direct taxation; and it is an argument in favor of the last-mentioned mode of raising money, that we thereby avoid the necessity of giving any other than the precise and literal construction to the words of the proviso that is insisted upon. But, as has often been suggested, with reference to other instruments, "the true construction is the only one that is admissible;" and a literal construction is not necessarily true, for the object of construction is to ascertain the sense and purpose for which the words in question were introduced into the instrument, and that sense, when discovered, is to be accepted; and in that sense the instrument, if a Constitution, is to be obeyed and enforced.

I do not believe that those who insist upon confining the power of the General Assembly to contract debts to the precise occasions of invasion, insurrection, or war, do justice to the purposes of the framers of the Constitution. They did intend, beyond all doubt, to deny to the General Assembly the power to contract debts beyond the sum of two hundred and fifty thousand dollars (which they have authorized it to do, substantially, at its own discretion), except under circumstances of extreme peril to the State. In defining the degree of peril that they intended should warrant the exercise by the General Assembly of the power that had been so much abused, they employed the strongest language; but it cannot be inferred that they intended that the State should be defended from invasion—that it might employ its resources to suppress an insurrection, or to prosecute a war—but should be powerless to resist the greatest evils, or prevent the most threatening dangers that might arise from any other possible cause. It seems to me that they intend to define the degree of urgency, rather than to express the particular occasions when the power in question might be employed. The framers of the Constitution were statesmen familiar with the practice as well as the science of government, and well understood, from the examples in which history abounds, that occasions might arise in the future of the State, when money would be required to be raised before the people could be consulted at a general election, to meet other exigencies than those of actual invasion, insurrection, or war. They knew that the safety of a State is often imperilled by the feebleness of its Government—by its inability to respond to the requirements of extraordinary duties, and that dangers sometimes impend over States, and evils overtake them (of which the dangers and evils produced by invasions, insurrec tions, and wars are but types and examples), that might require that all its resources should be employed at once to prevent or remove them; and with that knowledge, it cannot be presumed that they intended that the State, abounding in wealth, should submit to an unhappy fate, or invite an invasion, excite its people to insurrection, or engage in a war, to find a pretext for employing its own resources to avert it.

It was not the purpose of the framers of the Constitution to deprive the State of the power to discharge its vital and essential functions, as the narrow interpretation of the Constitution I am disputing undoubtedly does; and the circumstances of the case of the city of Chicago, now under consideration, serves all the purposes of the most complete and satisfactory illustration. In that city, within a few hours, many millions of property was suddenly destroyed; nearly or quite one hundred thousand of its inhabitants deprived of food and shelter; the ordinary agencies created by the State were, by the same overwhelming calamity, deprived of their power and resources, and were helpless to feed or shelter them. The Legislature of the State was convened by the Governor; they find the moneys in the treasury inadequate to meet the demands upon the State, but its credit is practically limitless, and the means to feed and give protection to the hungry multitude abound on every hand.

The General Assembly cannot, as is claimed, draw upon the resources of the State, or anticipate its revenue beyond an amount limited—not by the urgency of its duties, but by certain technical words contained in the Constitution. If this is the proper conclusion, and the people were not otherwise relieved, one of the conditions upon which the power to contract debts is said to depend, would be soon supplied, for the cravings of hunger will madden any population on earth to the point of insurrection.

It is to be borne in mind that the State of Illinois is so far independent of all other governments that it must at all times be equal to the perfect discharge of its own obligations. It cannot rely upon the voluntary charities of the benevolent to feed or give shelter to its destitute population without at the same time ceasing to exist.

It cannot and has not abdicated the most essential function of its existence, of raising moneys required for the discharge of its most important duties, by regular modes, for the safety of all the interests of the people forbid it. To claim that the people of the State have locked up their property so it cannot be reached by constitutional methods, to be used for the most urgent purposes of government and discharge the highest social obligations, is not only to do injustice to their character for humanity, but to their intelligence and discernment; for the power to raise money to meet the great and sudden emergencies in the affairs of States is essential to their existence.

Entertaining these views of the proper construction of the language of the proviso of the 18th section of the 4th Article of the Constitution, I feel no hesitation in recommending that if that course is deemed by the General Assembly most judicious, the amount necessary to meet the urgent demands upon the resources of the State be borrowed, and at the same time provision be made for its early and prompt repayment.

It is proper that I should also invite the attention of the General Assembly to the necessity of providing by law for the reassessment of property in Cook County for State and county purposes, and it is probably true that some legislation will be necessary to enable the authorities of the city of Chicago, and of the school and other minor districts of the county, to enforce the collection of taxes.

I am not prepared to express an opinion as to what legislation is necessary, but feel that my duty is discharged, though imperfectly, by commending the matter to your attention.

There is too much reason to apprehend that the destruction of the public buildings and records that pertain to the county of Cook and the city of Chicago have resulted in producing much mischief. How far such anticipated mischief, losses, and inconveniences can be remedied by legislation, must remain a matter of uncertainty and doubt.

Invoking your sympathies for that portion of our people who have suffered such unexampled losses, I can only express my most earnest desire to co-operate with you in any proper plan that may be devised for their relief. JOHN M. PALMER.

The members of both Houses adjourned to visit Chicago, and there saw what was needed, and returned to pass, with great unanimity, the following Act:—

"A BILL for an 'Act to relieve the lien of the City of Chicago upon the Illinois and Michigan Canal and revenues, by refunding to said city the amount expended by it in making the improvement contemplated by an Act to provide for the completion of the Illinois and Michigan Canal, upon the plan adopted by the State in 1836, approved February 16, 1865, together with the interest thereon, as authorized by section five of said Act, and to provide for issuing bonds therefor.'

"WHEREAS the city of Chicago has expended a large amount of money, to wit: the sum of two and a half millions of dollars, to secure the completion of the Summit division of the Illinois and Michigan Canal, under and pursuant to the provisions of said Act, so approved February 16, A.D. 1865, and Act supplementary thereto; and whereas the said city has a vested lien upon the said canal, with its revenues, subject to any canal debt existing at the time of the passage of said Acts; and whereas said then existing debt due by the State has been fully paid and cancelled; and whereas the canal trustees have delivered to the State of Illinois possession and control of said canal; and whereas it is provided by section five of said Act, as follows: 'The State of Illinois may,

OPENING THE VAULTS OF THE MERCHANTS' SAVINGS, LOAN

TRUST COMPANY, CORNER LAKE AND DEARBORN STREETS.

at any time, relieve this lien upon the canal and revenues, by refunding to the City of Chicago the amount expended in making the contemplated improvement and the interest thereon.' Now, therefore,

"SEC. 1. *Be it enacted by the People of the State of Illinois, represented in the General Assembly,* That the sum of two million nine hundred and fifty-five thousand three hundred and forty dollars ($2,955,340) with interest thereon, until paid, be and the same is hereby appropriated, for the purpose of relieving the lien as aforesaid, being the principal expended and the interest thereon; which said sum is hereby refunded to said city, and when paid, said city shall execute and deliver to the State of Illinois a proper release of said lien to the satisfaction of the Governor; and the auditor of State, under the direction of the Governor, is hereby directed to draw his warrants for said sum of money and interest, payable only out of any moneys in the Treasury belonging to the fund hereafter provided, to be known as the 'Canal Redemption Fund.'

"That for the purpose of providing said fund, any funds that are now or may be hereafter in the State treasury, paid in on the settlement of the canal commissioners with the trustees of the Illinois and Michigan canal, as well as from the revenue of the canal, also all funds that are now or may hereafter be paid into the State treasury, known as the "Illinois Central Railroad fund," shall be transferred by the State treasurer, upon the auditor's warrant drawn for that purpose, to said redemption fund; that a tax of one and a half mills on each dollar of the assessed value of all the taxable property of the State be levied as a special tax for the years 1871 and 1872, and to meet any deficit in said revenues to meet said appropriation, the governor, auditor, and treasurer are hereby authorized to issue bonds of the State of Illinois, to the amount of two hundred and fifty thousand dollars; said bonds to bear interest at the rate of six per cent. per annum, payable semi-annually, in the city of New York, and shall be paid at pleasure of the State, at any time after three years after the date thereof, and shall be of such denominations as the governor may deem advisable, and be known as the 'Revenue Deficit Bonds,' and shall be delivered to the city authorities of the city of Chicago, at par, as a part payment on above appropriations: *Provided, however*, that not less than one-fifth, nor to exceed one-third of said sum so appropriated, shall be received by said city, and be applied in reconstructing the bridges, and the public buildings and structures destroyed by fire, upon the original sites thereof, as already provided by the Common Council; and the remainder thereof to be applied to the payment of the interest on the bonded debt of such city, and the maintenance of the fire and police department thereof.

"WHEREAS, by reason of a great conflagration in the city of Chicago, the public buildings, bridges, and other public improvements have been totally destroyed, and the business of the courts is suspended, whereby an emergency exists as a reason why this act shall take effect before the first day of July next; therefore,

"*Be it further enacted*, That this act shall take effect and be in force from and after its passage."

This bill having received the Governor's approval became a law, and will work out its measure of relief.

In addition to material aid, there fell upon our ears grand, cheering utterances from the pulpits, platforms, and presses of the world, which stirred again the pulses of charity, and gave strength and courage to a staggering people enshrouded in the smoke and gloom of battle and defeat. The muse of poetry thrilled to the tale of woe, and sent her sweet voice through the pall of grief, and woke the pride and hope of our people by her glowing and tender strains. Rebuking those who would attribute our disaster to God's anger against our special sinfulness, the poet proceeds;

Bright, Christian capital of lakes and prairie,
  Heaven had no interest in thy scourge and scath;
Thou wert the newest shrine of our religion,
  The youngest witness of our hope and faith.

Not in thy embers do we rake for folly,
  But like a martyr's ashes gather thee,
With chastened pride and tender melancholy,—
  The miracle thou wast, and yet wilt be!

Not merely in the homages of churches,
  Or bells of praise tolled o'er the inland seas,—
Thou glorifiedst our God and human nature.
  With meeter works and grander melodies,

Of cheerful toil and willing enterprises,
  Of hearty faith in freedom and in man;
The hoar old capitals looked on in wonder
  To see the swift strong race this stripling ran.

How like the sun he rose above the marshes,
  And built the world beneath his airy feet,
And changed the course of immemorial rivers,
  And tapped the lakes for water cool and sweet.

How skilfully the golden grain transmuted
  To birds of sail and meteors of spark,
And, like another Noah, bade creation
  March in the teeming mazes of his ark.

Yet in his power, most frank and democratic,
  He roused no envious witness of his joy.
And in the stature of the Prince and hero
  We saw the laughing dimples of a boy.

Still wise and apt among the oldest merchants,
  His young example steered the wary mart,
And amplest credit poured its gold around him,
  And trade imperial gave scope for art.

His architectures passed all heathen splendor,
  The immigrating Goth drew wondering near;
To see his shafts and arches tall and slender
  Branch o'er the new homes of this pioneer.

The Greek and Roman there might see rebuilded
  In vastness equal and in style as pure,
The merchants' markets like a palace gilded,
  With marble walls and deep entablature.

His twoscore bridges swinging on their pivots,
  The long and laden line of vessels sped,
While he, impatient, marched beneath the sluices
  His hosts, like Cyrus, in the river's bed.

Then, when all weak predictions proved but scandal,
  And the wild marshes grew a sovereign's home,
A dozing cow o'erset an urchin's candle,—
  Once more a fool fired the Ephesian dome.

The artless winds that blew o'er plains of cattle,
  And cooled the corn through all the summer days,
Plunged like wild steeds in pastime or in battle,
  Straight in the blinding brightness of the breeze.

And down fell bridge, and parapet, and lintel,
  The blazing barks went drifting one by one,
The mighty city wrapped its head in splendor,
  And sank into the waters like a sun!

CHAPTER XXXI.

How did our people accept this widespread sympathy, and its godlike manifestation? It was a surprise as great as the conflagration. We scarcely believed it possible that our calamity could take such hold upon the universal heart of the race. And as the stream kept swelling till millions had been provided, and all immediate wants were supplied, and something was left for the stern winter's trials, our wonder grew. We were humbled by the spectacle. We knew not our losses, but we felt buoyant with the consciousness that the whole world felt our loss to be its own, and was rallying to succor and save from crushing overthrow. The primitive fraternity seemed to be revived, which is described in the Acts of the Apostles, when "no man said that aught that he possessed was his own, but they had all things common." Wrecked by a surging ocean of flame, with peril overhung every hour, we heard a cheering voice sounding through the gloom, and our hearts bounded like the hearts of mariners ready to perish, when a sail is discerned upon the waters bearing down towards them.

"There are men among us who have lost their all, who have seen the labors, the plans, the hopes of a lifetime annihilated in a moment, who have stood unmoved amidst universal desolation, and who have witnessed all with tearless composure, and yet whose eyes have been often splashed with the spray of tears as they read of the unanimity, the cordiality, the lavish generosity with which people everywhere have contributed to our relief. Oftener from among these ghastly walls and smoking desolation has there been heard a fervent "God bless our sympathizers!" than a "God pity our sufferings!"

Men who had not shed a tear till then, shook with uncontrollable emotion and wept for joy. The gratitude was equal to the charity, if such an equalization were possible.

We began to realize how intimately the interests of Chicago were bound up with those of the whole country and the world. We were brothers in distress. The feelings of her citizens were well expressed in the *Tribune*, which said:—

"Amid the general gloom, the public distress, and the wide spread wreck of private property, the heart of the most impoverished man is warmed and lightened by the universal sympathy and aid of his fellow-countrymen. There were cities that looked upon Chicago as a rival. Her unexampled success had provoked hostility,—amounting at times to bitterness. In the ranks of municipalities Chicago stood pre-eminent, and that eminence had drawn upon her the prejudices, and often the ill-natured jealousies, of her supposed rivals. But the fire ended all this. Hardly had the news reached those cities before our sorrows were made theirs. The noble-hearted people did not wait for details; they suspended all other business, each man giving of his money and his property to be sent to Chicago. Before the fire had ceased its ravages, trains laden with supplies of food and clothing had actually reached the city. St. Louis and Cincinnati, Milwaukee, Detroit, Pittsburgh, and Louisville were active, even while the fire was burning, in providing for the relief of devastated Chicago. Every semblance of rivalry had disappeared. Not an ungenerous or selfish thought was uttered—everywhere the great brotherhood of man was vindicated, and our loss was made the loss of the nation.

"In the light of this experience, how absurd are the criminations and controversies of men. The hospitality and humanity of those in our city who have retained their homes, toward their less fortunate neighbors, though marked by every feature of unselfish charity, has failed even to equal the zealous efforts and generous actions of the people of the country, who have laid aside all other business to feed the hungry, clothe the naked, and give shelter to the roofless of Chicago.

"The national sympathy for us in our distress has shown that in the presence of human suffering there are no geographical lines, no sectional boundaries, no distinction of politics or creeds. The Samaritans have outlived the Levites, and there has been no such thing as passing by on the other side. The wine and oil have been distributed with a lavish hand, and the moneys have been deposited to pay for the lodging of the bruised and homeless.

"Words fail to express the grateful feelings of our people. Men who braved the perils of the dreadful Monday, who witnessed the destruction of all their wordly goods, and who with their families struggled for life upon the prairies during the awful destruction, and bravely endured it all, could not restrain the swelling heart or grateful tears when they read what the noble people of the country had done for Chicago; how the rich and the poor, whites and blacks, all—men, women, and children—had done something to alleviate the distress and mitigate the suffering of fellow-beings in far-off Chicago. How true is it that 'one touch of pity makes the whole world kin.' In some cities the contributions have exceeded an average of a dollar for each member of the population, and in the abundance that has been given unto us the aggregate is largely made up from the prompt offerings of the humble and the poor as well as of the rich. Future statisticians may compute in tabular array the commercial value of the donations to Chicago; but only in the volume of the recording angel will be known the inestimable blessings of that merciful, generous, humane charity which this calamity has kindled in the hearts of the whole American people.

"In due time there will be a formal and complete acknowledgment of donations, public and private; but in the mean time let the nation rejoice that underneath all the conflicts in which men are forever engrossed there is a latent spark of universal brotherhood, which needs but the occasion to develop into the most genial warmth. Property may be lost, wealth may be obliterated; but that people must be great who have hearts in which charity for human suffering cannot be stifled in any event."

It was felt to be a most appropriate recognition of God, and His mercy, and of the goodness of our fellow-men to us, when the following proclamation appeared:—

"In view of the recent appalling public calamity, the undersigned, Mayor of Chicago, hereby earnestly recommends that all the inhabitants of this city do observe Sunday, October 29, as a special day of humiliation and prayer; of humiliation for those past offences against Almighty God, to which these severe afflictions were, doubtless, intended to lead our minds; of prayer for the relief and comfort of the suffering thousands in our midst: for the restoration of our material prosperity, especially for our lasting improvement as a people in reverence and obedience to God. Nor should we ever, amidst our losses and sorrows, forget to render thanks to Him for the arrest of the devouring fires in time to save so many homes, and for the unexampled sympathy and aid which has flowed in upon us from every quarter of our land, and even from beyond the seas.

"Given under my hand this 20th day of October, 1871.

"R. B. MASON, Mayor."

The day was generally observed and the churches were filled. The writer preached on a theme appropriate to the former part of the proclamation in the morning, and in the evening on Good Deeds, to be Held in Everlasting Remembrance. *Mat.* 26:13. "Verily I say unto you, wheresoever this Gospel shall be preached in the whole world, there shall also this that this woman hath done be told for a memorial of her."

This prophecy and command illustrate the divineness of our blessed Lord, because He predicts the world-wide spread of His Gospel, and stakes his reputation upon it; and because He exhibits so delicate and perfect an appreciation of the generous care which this woman offers Him. The event has justified His grand prophecy, for the aroma of that noble woman's name has spread throughout the world. The recognition of her offering by the Saviour, and His award of praise, have given us an example which is equivalent to a rule, that we should treasure in grateful remembrance, and also commemorate the good deeds of our fellow-creatures.

He has also further said, "And whosoever shall give to drink unto one of these little ones a cup of cold water only, in the name of a disciple, verily I say unto you he shall in no wise lose his reward." An act of kindness to God's people, however common and simple the deed of mercy, bestowed in the spirit of Christian love, shall be rewarded by Him.

Possibly there may be some intimation of that Great Day when Christ shall judge men according to their doings, and confer eternal honor on the workers of mercy, saying, "Inasmuch as ye did it unto one of the least of these My brethren, ye did it unto Me."

We read, also, that the works of the blessed dead do follow them, accompanying them into the very presence of God to speak for them and claim the reward. The Bible itself, God's own Word, is a history and memorial of some of the best actions ever performed among men. It is therefore godlike to remember and to celebrate good deeds, especially when we ourselves are the objects of beneficence. Ingratitude is the foulest, basest of sins; gratitude the fruit of a noble nature. It is most becoming in us, who have been recipients of the charity of the world, to manifest our appreciation, to dwell upon the benevolence, to magnify the bounty, to love the donors, and glorify Him who is the Great Author of all good in man and the universe.

1. Let us notice the *spontaneous* overflow of sympathy and beneficence. Scarcely had the tidings gone forth to the surrounding country, and the extent of the evil become known, when we heard that car-loads of cooked provisions were on the way to our city; that women sat up all night preparing food for our homeless thousands; that the dépôts were full of supplies; that distant cities were filling their trains with necessary articles for our comfort; that corporations and communities were raising moneys for our relief; that England was moving to our rescue, and Germany, and all Christendom, indeed, had been touched, and the lines of communication were given up to the Chicago relief-work. Never in history was there a calamity so great and sudden, and never an uprising of mankind so gene-

rous and spontaneous. Unforced as the light, free as the crystal flood from the mountain-spring, gracious as the perfume from the flowers, came all the sympathy and all the help we could possibly receive and use.

2. We may dwell upon the *magnitude* of the world's charity toward our suffering people. Whatever we had need of poured in upon us without measure, and the quality was unexceptionable. The poor never lived so well as during the first few days after the fire; at least we may reasonably suppose that they seldom had bread so white, biscuit so light, ham so sweet, preserves so rich, and everything eatable in such abundance. The munificence of the people at large provided all that heart could wish of food, bedding, clothing, and household furniture. The railways were taxed to their utmost capacity, the churches were filled with material, and all the dépôts of supplies testified to the magnanimity of the American public. Immense contributions of money followed upon the heels of these gifts for immediate use. God opened wide men's hearts and unclasped their purses in our behalf. Across the water our necessities appealed to the generosity of foreigners and strangers, so that quantities of money will flow to our relief from lands beyond the sea. Churches gave, after the general fund was raised in popular assemblies, their collections, and gathered their boxes and bundles, much of which will be privately disbursed to the actually needy in the various Christian congregations. Farmers and merchants came in to open their houses to the homeless, and doors everywhere stood wide to welcome those suddenly left without a roof. Instances might be named and incidents given of the most interesting nature, all of which reveal a humanity and philanthropy which shed glory upon the age, and show the power of Christianity upon the world. "For this is the Lord's doings, and it is marvellous in our eyes." He has made all this overflowing beneficence possible, and to Him be the glory! Our thanks must be given

to the railroad corporations for their nobleness in these times of distress. They have done everything in their power to mitigate and relieve the horrors and evils of our situation. We must not say, henceforth, that corporations have no souls. Our own citizens have shown a magnanimity worthy of all praise, in opening the churches to the homeless, distributing with what care they could exercise in the press of need the public bounty, offering hospitality and sympathy to the sufferers, to their own discomfort, inconvenience, and loss; cheering and helping one another by brave words, kindly offices, and lenient treatment, insomuch that there never was such a calamity accompanied by less actual suffering, or followed by such ample relief. The immensity of the loss was met by prompt and efficient assistance, unexpected and unparalleled in history.

The offers of pecuniary aid to men crippled in business were on the largest scale, as if men rose to the height of the emergency, under the inspiration of the Almighty. The Alabaster box was full of costly ointment, and when it was broken upon us, the fragrance filled the world, and will perfume the age. Its sweetness ought to possess mankind with a sense of brotherhood, and draw them into closer fellowship. It is here most fit to mention the boundless charity of cities heretofore our rivals; instantaneous and magnificent was their response to our deplorable need, and never can we cherish anything but gratitude to their warm-hearted, generous people. All feelings of bitter rivalry must die and perish forever, and only a lofty emulation characterize our mutual endeavors. Let the memory of their good deeds live in our hearts, and be transmitted as a precious inheritance to our children and the generations that follow.

The considerate action of our Governor and Legislature deserves from us a particular recognition, and must knit our people more closely to the mass of our fellow-citizens in other sections of the commonwealth. And doubtless the magnitude and far-reaching extent of the public charity will never be known until the Books are opened at the great Day of Accounts. Nor can our gratitude and thanks be too comprehensive and deep, too constant and fresh, towards our Heavenly Father, and those whom His grace prompted to unexampled works of mercy.

3. Now, again, to heighten our conception of obligation, we must reflect from what possible evils we were saved by the spontaneous and magnanimous action of the American people and the civilized world.

The scenes of Sunday and Monday, during the conflagration, were often of such revolting depravity as to remind us that a portion of our population were fiends incarnate, or beasts in human form. The dregs of a great city contain elements of destruction that rise to the surface when any storm or convulsion shakes it. Nothing is then safe from their raging frenzy. The helpless community become their prey; and they especially attack the better classes, because from them they expect plunder, and their envy of the more fortunate satiates itself in their ruin and distress.

Besides, when disaster is abroad, and riots occur, a demoniac passion for devastation seizes on the ignorant and excitable, and they assist the elements in their fatal sweep. When law and its restraints are thrown off suddenly, it is like unchaining and unloosing a menagerie of wild animals and serpents. This is not too much to affirm; because history confirms the statement, and shows bad men the worst at the very time when they should be most gentle, considerate, and kind. People without roofs, or raiment, or food, would not long brook the sight of comfortable homes and abundant supplies, without forcibly compelling a division. We shudder to think what might have been, without the ample bounty of which we were recipients.

And again also the suffering that would have occurred but for this speedy and gigantic provision for all the homeless multitude! We could scarcely have cooked and dealt out the food needful to prevent starvation; nor would it have been in our power to furnish money and clothing, bedding and furniture; abject poverty would have overtaken and swallowed us all down into a gulf of hopeless misery; famine and death would have held sway over this proud metropolis. If we have thus far happily escaped, and feel measurably secure, let us praise God, for this unstinted liberality, and all the blessings it has insured us,—especially deliverance from dangers of unseen horror and magnitude.

4. Again, let us hold in grateful remembrance what has been done for our relief, that we may act worthily before our benefactors. It would be a shame for us to be avaricious and narrow, from this time forth. "Freely ye have received, freely give." The world expects every man to do his duty in this emergency. Cowardice or meanness now and henceforth must appear doubly degrading and despicable in a citizen of this city.

> "I will live so they shall remember me
> For deeds of such Divine beneficence
> As rivers have, that teach men what is good
> By blessing them."

There are some persons, who sit down and fold their hands in idleness, eating the bread of charity till such time as it shall cease to be given out. They are an excrescence upon society, a burning disgrace to humanity; such men discourage benevolence, and thus curse the deserving. Any one who in any manner imitates them, must share their deep damnation. This, also, is no time for despondency, but rather for heroic action, in view of a helping world, whose aid cheers us to greater exertions than ever, and lays us under solemn obligations to prove our manhood. And it is one of the best things in life, that "a man's life consisteth not in the abundance of the things which he possesseth." The poor are happy, they are often great; their deeds live when pelf is burned or wasted upon folly and sin; and if we were to take a survey of history, it would be found that, what men have nobly done, not what they have gotten for themselves, makes them remembered as a blessing to the world. Never mind whether you succeed in hoarding again, or in regaining your former position. Do not fall down in the dust and cry, or hesitate to do your duty, because all is swept down to ashes, and flung to the winds in smoke.

"Nay, never falter; no great deed is done
By followers who ask for certainty;
No good is certain, but the steadfast mind,
The undivided will to seek the good.
* * * *
The greatest gift the hero leaves his race
Is to have been a hero. Say we fail!
We feed the high tradition of the world."

Our names will brighten the list of those who have suffered patiently, toiled manfully, and sought, through misfortune and trial, a higher and better destiny.

I have sometimes dreamed of the days of old, when our fathers were alike poor and struggling, and had little time for frivolous amusements. They were happier and truer then than people are now. And if there is any life that seems to me loathsome and detestable, it is that of the mass of the population of great towns. The very high are all gayety, fashion, folly, and luxurious vanity; the very low are given over to cheap amusements, vile pleasures, and empty nothing. The large middling class are industrious, earnest, useful persons, who form the balance-wheel of the machinery, the conservative element in society. Reduced as we are to a level, and brought back to first principles, we must humbly confess our indebtedness to our generous helpers, and order our future to please the Great Giver, and to honor those who have saved us from total wreck. Piety, prudence, industry, charity, and fidelity are the cardinal virtues, whose exercise will form the best memorial we can raise to the remembrance of the world's great beneficence.

5. Finally; the offering of that precious ointment was love's gift to Jesus Christ; and the Christianity of the Bible made men's hearts so tender, that when our calamity smote upon them, they broke and gave forth the generous offering, whose odor smells sweet in our nostrils. Christian brethren, be it ours to promote this same holy, humane religion, of which we have been made to partake, and whose fruits in a thousand ways we enjoy, and shall enjoy forever. We seem to labor sometimes in vain. But by patient kindness, bold persistence, and earnest fidelity, we make impressions which affect the deepest elements of society, and mould the public mind. We must be true and energetic; and the ever present recollection of Jesus' love in dying for us, and of his latest exhibition of the influence of His example and spirit upon the race, will especially spur us to new exertions, inspire constancy and zeal, and enable us to give a good account to Him, and to Christendom, of the stewardship with which we are entrusted. As Mary was reproached for her beneficence, as Christ was crucified for his mission of love, we shall not find the path of benevolence one of flowers. We shall meet opposition and many a rebuff; but looking unto Jesus, let us go forward doing with our might whatever our hands find to do, and His recognition and approbation shall be our exceeding great reward; for no well-doing shall fail of His well-done. Amen!

Most happy are we to bear testimony that the sentiments of this discourse accord with those of the people at large. And while there may be difference of views respecting the administration of affairs and the disbursement of funds, there is a unanimity of gratitude. This variance of opinions, and occasional asperity of temper concerning the disposition of moneys and supplies, arises from the extreme generosity and eagerness of some, and the corresponding conscientiousness and practical wisdom of those actually at the helm. Men of power and men of benevolence are guiding the relief work, and the people will yet admire the tact, courage, and self-sacrifice of these men. With the supplemental offices of the good Samaritans in private life, and individual local societies, there will be no great amount of suffering, unless the winter should be unusually long and rigorous.

It is gratifying to know that the people who have aided, are satisfied with the manner in which their bounty has been bestowed. We enjoy the ring of the following paragraph from a city paper, where the gifts have mounted up into the millions:—

"No clear-sighted observer can have read the record of the weeks first following the great Western calamity without feeling that the effect of the great outburst of sympathy for the outcasts of Chicago has been most wholesome and elevating upon the national temper. We had all begun to look at human nature too much through the medium of Tammany thefts, Ku-Klux Klans, and trials for adultery and murder. They had almost put out of our sight the actual framework of social and domestic life, its silent modesties, and pure affections, and the myriad unselfish ties which in real life bind men together. Only such a disaster as that of Chicago could call this hidden ground of humanity to light in its most generous work. The country has had her moments of justifiable pride before now, in the display of her strength, or wealth, or success of arms; but she was never so great as when in the spirit of her Master she went into the highways and byways and compelled the homeless and destitute to come into her royal feast—be warmed and clothed and fed. It will need many years of squabbles and thefts and international jealousies to blot out this glimpse of the substratum of manliness and kindliness in ordinary human nature, or to make us forget how from every nation came the quick response when the great city sat in ashes, and cried aloud, like Job, 'My bone cleaveth to

HAULING SAFES FROM THE RUINS.

my skin and my flesh. Have pity on me, O ye my friends, for the hand of God hath touched me.' "

We close this division of our subject with regret, because so much is left unsaid of necessity, and here we leave a theme of the sweetest and most absorbing interest.

If to give is more blessed than to receive, then indeed has there been a wave of joy rolling over the great human soul; and the experience of this century shall be illumined by a light above the flames of Chicago's burning. As they paled before the sun, so has our gloom fled from the sunburst of a world's beneficence. In the language of Tiny Tim, in Dickens' Christmas Carol, "God bless you every one!"

" Yet, from the grave Chicago's wondrous spirit
Comes forth all brightness, o'er the darkened town,
To say again: " Lo! I am with you, brethren;
With all my thorns, I wear my civic crown."

" To die is sweet embalmed in your compassion;
Your oil and wine make life in every rent.
Oh! let me lean a little while upon you,
And walk to strength in your encouragement

VI. —CONVALESCENCE.

CHAPTER XXXII.

Chicago 's been burnt down in timber to-day,
Chicago 'll be built up in marble to-morrow;
Chicago has capital losses to pay,
Or Chicago has credit her losses to borrow.

No fabulous Phœnix, with flames circled thick,
Give us henceforth, as swift resurrection's *imago:*
In its stead paint up, heralds, an Illinois Chick,
With the legend in gold letters tacked to it—" *Ago.*"

For this Illinois Chick, from her circlet of flame,
Looks calmly and coolly, victorious o'er ruin,
And this word has a right to, in more than in name,
For *Ago* 's " I do," and Chicago is *doing*.

—Punch.

The reputation of this city for boasting was such that people always allowed a margin for exaggeration in statements made by our citizens. It was usual to observe an air of incredulity upon the countenances of those who listened, when Chicagoans told of their exploits and advances. Yet underneath all this apparent doubt, and mingled with this idea of vaunting, there was a growing sense of the amazing energy of the western people. Other cities in the same region reproached one another with want of enterprise and spirit, and pointed hither for an example of what was needed to give them equal or greater prosperity. The world, too, had begun to realize that the Young Giant was a power in the realm of commerce and of all activity. A lady said of her own great city, "If we had been burnt out as you have, our people would have sat down with folded hands and made no effort to recover. Or if they had done anything, they would have waited till spring before they commenced." A St. Louis party tried to induce a friend on his way to Chicago, the day after the fire, to wait a little for them. "No," said he, "those fellows will have it all built up in less than twenty-four hours, and I want to see the ruins." And on he went. This revealed the real reputation of the city among those who knew, in the clash of contest for trade, the stuff of which our merchants were made.

The London *Spectator*, looking at Chicago after the fire, speculates in an interesting way on the elastic energy displayed by business men:

Not a little of the surpassing energy and spirit displayed by individuals after the fire may be traced to the absence of that appreciation of the weight of circumstances which, like his liability to the laws, presses so heavily upon the Englishman. Mr. Joseph Medill, for example, is one of the proprietors of the Chicago *Tribune*. It was thought that the *Tribune* office, a huge block of marble, might resist the fire; the neighboring journalists sent in their presses, and the staff seemed to have waited for the flames as they would for an enemy's attack. Despite the strength of the building, however, the flames "licked in," and Mr. Joseph Medill walked out, to purchase there and then a store at some distance, and a couple of machines, with which, before his old office had grown cold, he was circulating *Tribunes* to the public. It is impossible not to admire such energy, and impossible not to suspect that one source of it was indifference; that Mr. Medill did not really care, as an Englishman would have done; that his heart was not choking, or his brain bursting, with a sense of defeat and pain, as an Englishman's would have been. There is something of "What does it signify?" in it all, as there is in the Mayor's vigorous and benevolent leap through the laws. A merchant, hurrying back to Chicago to see what had become of house and home, is said to have met a friend and asked him of their fate. "House burned, wife safe at our father's, papers all right," was the reply, whereupon the merchant remarked, "Well, when a man has his wife and his papers, what more does he want?" "Heroic stoicism," says the listener, and there is heroism, and stoicism too, in the speech; and so also there is indifference, easiness, fluidity of feeling on points which would have touched an Englishman very deeply. The American cared about his wife and about his papers, but about his house and its associations, and their sudden disappearance out of his life, he did not care at all. Even the burnt-out multitude seemed after the first shock to have turned to work again with an ease which is in itself admirable, but which would, we suspect, be impossible if the chances of life weighed there as they do here. Life, as well as the law, presses more lightly across the Atlantic, and men struck by misfortune turn to work again, not with the dogged resolution of the Englishman, not by a supreme effort of the will, but with a light elasticity and heartiness which resemble frivolity, even while they have with frivolity nothing in common.

It would be a benefit to mankind to ascertain, if only such ascertaining were possible, how far this elasticity is due to American institutions. If it is due to them, that would be the best argument ever advanced in their favor, one object at least of human institutions being human happiness, and there is something to be said for American theories on the subject. The American social system is a result, in part, at all events, of the American political system; and its tendency is to lighten life by increasing sympathy, and diminishing that sense of isolation which so greatly intensifies the impression of any calamity, and which is, we suspect, one of the greatest causes of the depressed tone visible in English life. But we believe that a much stronger cause is one with which institutions have very little to do, the visible presence of innumerable chances in life, the sight, as it were, of endless potential wealth besides that which has been destroyed. A great English peer is not very heavy-hearted if one of his houses is burnt down and no life is lost, and that is very much the American feeling about a similar calamity. The house he lives in is only one of his houses. He has no other just at present, but he will have, and in that certainty he loses the sense of the irreparable character of any loss not involving a human life. Prosperity is sure to come back to Chicago, or if not, then to Milwaukee, and Milwaukee will do just as well as Chicago; and the American, as certain of that as he is of to-morrow's sun, feels misfortune not as a wound, but as a grain of sand in his eyes, annoying, no doubt, but sure to be out in a minute. It is not the present men really fear, but the future; and to the American, taught from childhood to appreciate the vast and certain reversions which belong to him, the future is always pleasant, and life therefore never without light. The burning of his house or of his city matters no more to him than the wearing out of his furniture to the English rich man; he has only to get some more. If his cheque-book is right, all is right; and to Joseph Medill his paper is his cheque-book, and the grand office old furniture soon to be replaced. Americans have not developed a new strength, they only exert the strength they have through a lighter medium.

The London *Times* closed an article with these words:—

"When Mr. Cobden complained that English school-boys were taught all about a trumpery Attic stream called the Ilissus, but nothing of Chicago, it should have been remembered in fairness that at that time Chicago had hardly existed long enough to be known by any but merchants. It will now not soon be forgotten. We may be confident, however, that the natural resources of the place and the native energy of the Americans, will more than repeat the marvels of the original development of the city.

The novelty and rapid growth of American civilization render the people far more indifferent to such calamities than dwellers in older countries who are conscious that their possessions are the accumulation of centuries. At the same time with the news of the fire the telegraph informed us that its mercantile effects were already being discounted in New York, and we have no doubt there are numbers of enterprising speculators who see their way to fortune through the speedy reconstruction of the city. The most cordial sympathy will be felt in this country with individual sufferers, and we can only wish the great mercantile community of the West the prompt recovery which their energy deserves."

The *Daily Telegraph*, in a characteristic article, says:—

It is idle to suppose that such a city is destined to become a Tadmor in the wilderness, or to sink into the chronic decadence of Sebastopol after the bombardment. "Resurgam" might be written upon every brick of the burned-up houses of Chicago. It will rise again, and with a vengeance. Luckily no venerable cathedrals, no historic palaces, no monuments of art, no hoary relics of antiquity, have perished in the colossal fire. Chicago has blazed away with the rapidity of lace curtains, or of ornaments in a drawing-room grate. The articles were handsome and expensive, but they can be replaced. To repair the injury done, all that is wanted is a certain amount of resources, energy, and pluck; and in pluck, energy, and resources the American people will never be bankrupt.

The London *Daily News* has a two-column editorial on the fire, in which it says:

"Nowhere in the world—not in Manchester, not in London, not in New York were busier streets to be found. A river, hardly better than the Irwell, flowing through part of the business quarter of the city, and spanned by innumerable drawbridges, did, indeed, make hideous some of the city scenes, which showed like an uproarious Rotterdam or a great commercial Königsberg

But the streets of shops and banks and theatres and hotels might stand a rivalry with those of any city in the world. Enormous piles of warehouses, with handsome and costly fronts; huge 'stores,' compared with which Schoolbred's or Tarn's seem diminutive, hotels as large as the Langham or the Louvre; bookshops which are unsurpassed in London or Paris; and theatres where Christine Nilsson found a fortune awaiting her such as the Old World could not offer—such were the principal features of that wonderful quarter which has just been reduced to ashes. Nor was Chicago wholly given up to business. Her avenues of private residences were—some, we trust, still are—as beautiful as any city can show. Michigan avenue and Wabash avenue were the streets where her merchant-princes lived; and there is nothing to be seen in Paris, London or New York to surpass either avenue in situation or in beauty. Michigan avenue is a sort of Piccadilly, with a lake instead of a park under its drawing-room windows. The other great avenue was distinguished from almost any street of the kind in Europe or the United States by the variety of its architecture. Mr. Ruskin himself might have acknowledged that in this civilized and modern street, at least, the curse of monotony did not prevail, and the yoke of the Italian style was not accepted. Let it be added that Chicago, having the advantage of newness, and the warning of all the world before her, had but few narrow streets and lanes. The thorougfares were, as a rule, nearly all of the same width. The inexperienced traveller often found himselt sadly perplexed as he wandered through a city of broad white streets, each looking just like another, and any one seeming as well entitled as its neighbor to claim the leadership in business or fashion.

"Chicago will not remain in her ruins as an ancient city might have done. Already in the thick of all the wreck and misery we may be sure that active and undaunted minds are planning the reconstruction of many a gutted and blackened building, the restoration of many shattered fortunes. It is only a few years since the city of Portland, in Mainé, was destroyed by fire; and the traveller to-day sees there a new, busy, and solid town, where the story of the conflagration has already become a tradition. The people of Illinois are still more energetic and fertile of expedient than the people of Maine, and they will not long leave the city, which was their pride, to lie in her smouldering ruins. The claims which Chicago used at one time to urge for the transference of the national Capital to the shore of her lake are, indeed, put out of court for the present; and her rival, St. Louis, will, for some time to come, have the advantage of her in the race for commerce, wealth, and population. But the city whose rate of growth distanced that of any other on the earth, will not be long in recovering the effects even of the present calamity. So much at least of consolation may be found. Before the widows and orphans, whom this catastrophe bereaves, shall have put aside the robes of mourning, Chicago will be rising from her ruins, perhaps more magnificent than ever. Her restoration, we may feel assured, will be in keeping with the marvellous rapidity of her rise, and the awful suddenness of her fall."

While these generous words were heard from across the water, and we knew what men really thought of us, like one who reads his own obituaries, there was no lack of similar expressions from our fellow-citizens. The language of the New York *Tribune* was:—

Chicago may be taken as a fair type of American material energy. We are proud to claim her as a representative city, so far as vigor, boldness, self-poise, industry, and far-reaching enterprise are the characteristics of the American Republic. The destruction of three hundred millions of substantial property is a lamentable disaster; and we shudder at the statement that hundreds of human lives went out with agony in the midst of the fiery furnace; but the indomitable energy of the great community still survives. As Chicago was a representative city in the nation, so it shares in all the recuperative qualities of the Republic. The city which has been laid waste was not alone that of the three hundred thousand people who inhabited it; it was the city of many mighty States whose messages of cheer and trains of relief are this moment speeding to it from every quarter of the Republic. A nation that has survived a great rebellion, and has grown stronger and mightier in the work of replacing the wreck of a four years' war, has an interest in rebuilding Chicago, and in making it stronger, nobler, and more admirable than before.

Though this is a great calamity to the City of Chicago and to the whole country, we shall doubtless be surprised to see how soon both city and country will recover from it. The elasticity of a community which built a city by the Lake within the limits of a brief lifetime, raised its foundations again and again from the morass, drove a tunnel under Lake Michigan, and turned the course of a river against its natural flow, will be equal to even the present emergency. There will be no panic, but the audacious and cheery confidence of the people—not of Chicago alone, but of the United States—will sustain the enormous burden; and mutual forbearance, help, and co-operation will tide over the disaster. Already there are comfortable indications that the Insurance Companies will weather the sudden storm; and that the two hundred millions of dollars which are represented in the risks in Chicago may be forthcoming when the recovering city shall demand this prudent provision. For a time, of course, trade will suffer, and the multitudinous interests inwrought with the prosperity of Chicago will languish. Rival cities will divide among themselves much of the business which Chicago has heretofore absorbed.

Then the St. Louis *Democrat*, a few weeks after the fire, thus recognized the recuperative force of the smitten Giant of the West:

The funeral sermon over the remains of Chicago may be postponed for the present, owing to unmistakable signs of animation on the part of the corpse. If dead, she yet speaketh, and that, too, in the loudest and most understandable Saxon. Through the columns of her leading newspapers she tells the great Northwest—and is careful to make herself heard in bailiwicks which nature and art seem to have set apart for St. Louis—that her merchants are ready with larger stocks of goods than ever before, and that they are prepared to sell cheaper and deal more justly with the general public than any other city, especially St. Louis. All this is done at a cost to the merchants and a gain to the newspapers of many thousand dollars per diem. The merchants of Chicago have a lively faith in the efficacy of printer's ink. They recognize it as an unquestionable truth that the long columns of advertisements, for which they have so liberally paid, have had more to do in giving to Chicago her proud commercial position than any other instrumentality whatever; and, so believing, they make the investment with a cheerfulness which sometimes quite overpowers the facilities of the newspapers. In Chicago the advertising merchant is the rule; in St. Louis he is the exception. Up in that big city on the Lake the merchant has read and believes what Thomas Jefferson once said of the Richmond *Enquirer*, when it was published by his friend Ritchie—that a man who put down a newspaper without reading the advertisements often missed the best part of it. And so they do not coincide with their wiser brethren of St. Louis, who seem to think that men's wives are more interested in police items than in discovering where they can find the cheapest and best silks and shawls, and other indispensables of the female form divine. They never made a greater mistake in their lives. Bless their unsophisticated souls, let them follow the female eye as it traverses to-day's *Democrat.* First, marriages and deaths—with a smile for the first and a tear for the last; then the latest fashion notes; then an elopement, if there be a first-class one; and then a careful scrutiny of the advertising columns, to see who has the largest and best stock for to-morrow's shopping. *The country merchant living near Cincinnati*, Terre Haute, Indianapolis, or other point within trading distance, reads the market reports first, and then turns to the advertising columns to see from whom he can get what he wants. If he can find more information on this subject in the Chicago papers than in the St. Louis papers, he will be very apt to patronize Chicago merchants in preference to those of St. Louis. Chicago understands this and acts upon it. Her business-men keep themselves before the people in flaming capitals on the first page of her newspapers.

In all this, we discover a sidewise blow at the dilatoriness of the citizens of the rival metropolis, the Queen of the Rivers. A gentle rebuke was administered, here and there, to those who looked chiefly on the retributive aspects of the calamity, and recalled the peculiar sinfulness of our way. As *e. g.* the following paragraph:

People who see a Providential judgment in the conflagration of Chicago, have very limited knowledge of Divine economy. God helps those who help themselves, and if two elements of nature—fire and wind—have torn down a mighty city, He who masters these elements and moves the seas and keeps the prairies fertile, has resolved that the city shall be rebuilt. The prophets should seek another occupation.

In the next chapter we shall see how the kind opinions entertained concerning us were fulfilled and verified, and how "Chick—Ago" is "doing" according to *Punch.*

CHAPTER XXXIII.

"The street shall be built again, and the wall, even in troublous times."

DANIEL.

THOSE who have kept the thread of this story have become impressed with a sense of the perfect desolation of the scene, and conceive the horrors of our situation when the enemy finished his victory.

The water was everywhere about us, yet we were destitute of the precious element. Hydrants were dry, reservoirs and cisterns empty, and the atmosphere still parched and the wind raging. People resorted to the Lake and parks with tubs, buckets, pails, pitchers, and cups, a motley array, for enough to prevent thirst and filth. This continued for a week and more, until the water works and machinery were restored. The hungry and homeless were all about us by tens of thousands. Dread winter, a stern foe in our northern climate, stood near with menacing aspect. Men were in an agony, lest banks and all associations should fail, and they become totally bankrupt. Business, too, was imperilled and might be lost forever. It was a season of Egyptian gloom. Houses stood with furniture packed and doors ajar, ready for another alarm of fire. All was confusion and uncertainty. Some said, we must leave the city, as there can be no more to do here for years. Chicago is ruined and lost.

But this was not by any means a general feeling, or one that received encouragement. "The strange people that built Chicago," as some one terms them, were not daunted by adversity; neither did they believe that God had any plans of destruction to execute, by which the site should become a desert. They accepted the situation with better grace than could have been expected. They did not attribute the disaster to anybody's malice. Some, indeed, said the guerillas have done this, and some charged it on the Mormons; but sensible people all scouted any thought of incendiarism, and looked on it as a great, mysterious dispensation of Divine Providence, which was permitted, and occurred in accordance with well-known laws of nature. There they left it for the time being, and turned their attention to the sublime charity of the hour—care for the poor—and then to the work of reconstruction. As in Nehemiah's day, when the fallen wall of Jerusalem was rebuilt, men worked with a weapon in one hand and a tool in the other; so now a part of the day was devoted to benevolence, and a part to recovery from commercial ruin. It was a sad but noble spectacle!

The fires were not extinguished when some men had rented new places for the transaction of business, and advertised themselves as prepared for customers. Others began to clear away the débris for new foundations. "Already," wrote one, "from the smouldering embers, the city is gathering strength for a renewed career of prosperous activity."

But that which was everywhere apparent in spite of the horrors of the scene and its sad hopelessness, was the indomitable pluck which the Chicago men showed, all which no losses could damp and no wretchedness subdue. "Chicago will be hard up for a time," said one; "but we must try and pull through." "It will take a long time to build all this up again," said another. "I thought I was pretty well off yesterday," said a young man, cheerfully smoking a cigar; "now all I have in the world is the suit of clothes I have on." "I think I might have saved my law library," said a rising young lawyer, "but, by Jove! it did not seem the thing to do, when everybody else's property was burning up; so I picked up a few papers in my office, took some volumes of Kent given me by a friend, took off my hat to my old books, and left them to burn."

The spirit of the people is shown by the tone of the press, which gave no uncertain sound, but spoke confidently of the resurrection in these words:—

Whoever has permitted himself to think that the great calamity which has befallen Chicago would paralyze the energies of her people, and check her rapid march to the commercial greatness which is her destiny, has taken but an imperfect measure of the character of this city and of the men who, with nature's aid, have created it. The men who built Chicago still live. The city was not their inheritance; it was the work of their own hands. What they have achieved they know they can achieve again. They have not to wait to consider how to begin; they are beginning. They have already taken off their coats and commenced the work of rebuilding Chicago.

The foundation upon which they have to begin exists in the remaining value of the land. It is impossible, of course, to express anything more than opinion as to depreciation in this value which will be the result of the conflagration. A number of facts and circumstances combine to render it probable that no ruinous depreciation in prices will be witnessed. One of these is the value of insurance. If the value of good insurance should prove equal to one-half the aggregate loss, then the actual loss to the owners would be reduced from $150,000,000 to $75,000,000. But probably the whole amount of insurance is under rather than over fifty per cent. of the total loss, so that if fifty per cent. should be realized upon the total amount of insurance, the loss to owners would still amount to over $100,000,000. The indications certainly are that, upon the average, considerably more than fifty per cent. of the insurance will prove good; it is even hoped that seventy-five or eighty per cent. may be realized. Whatever the amount may be, in reducing the personal loss to owners it becomes an element of strength in the value of the land.

The land value is still more strengthened by the existing city improvements; the sewers, the water and gas mains, the pavements, etc. Twenty years ago none of these necessaries of a great city existed; all had to be built. Now they are all finished and in readiness for use.

But more than by all else the land value is strengthened by the fact that here, in the "burnt district," was the business heart of Chicago, and here, in the very nature of things, it must be again. Here, in the region bounded by the river, the Lake, and the southern limit of the conflagration, is the locality where convenience and accessibility for all parts of Chicago, and for all parts of the country, meet in a common focus.

Here the commercial heart of the city has been fixed by nature; and here it must and will remain in spite of fire and in spite of every adverse influence. And here it is the duty and interest of every citizen to concentrate all his influence and exert all his moral as well as physical force to lift up Chicago from its ruins.

This is the feeling and the common sentiment among all classes of men who "take stock" in Chicago, and among none has it been more promptly or vigorously manifested than the railway companies. All the railway companies having their termini in the South division are making preparations to rebuild immediately upon our old foundations. The companies on the Lake shore desire, indeed, to proceed at once to enlarge their facilities to three or four times their former capacity. The Illinois Central company effected a contract on Friday for bricks to rebuild their freight houses. They have already advertised for 400 brick layers to commence work immediately. The work of clearing the ground is already begun. At present all their trains, as well as those of the Michigan Central and the Chicago Burlington and Quincy, start from the Twenty-second street station; but their plan is to immediately erect a temporary roof upon the walls of the Union depot at the foot of Lake street, and return with their passenger trains to the old premises. So soon as it shall be possible to get new freight houses under roof, their freight trains will do the same.

Again the voice of prophecy was fortified by such living facts as the following paragraphs describe:—

Persons travelling upon the prairie have noticed the mounds thrown up by the ants, and have wondered at the incessant activity of the multitude of laborers. Hardly less activity is to be witnessed among the ruins and upon the streets of the burnt district in the South Division. Never in all the previous history of Chicago was such a scene of thriving activity witnessed. Even those most familiar with the wonderful resources of this city are forced to wonder at the multitude of wagons which are employed in hauling off the débris to make room for the workmen putting up the new structures. Workmen are everywhere at labor, delving amid the ruins to reach the old foundations, that the masons may set to work. Thousands of men and boys are cleaning, wheeling, and piling bricks, while hod-carriers are supplying them to the masons. Teams loaded with lumber and lime throng the streets, carpenters and masons are working bravely; the gatherers of old iron are busily employed collecting their material and carting it away. The removal of safes has ended; every safe has been opened; those which were really safes have been carried off, the others abandoned to the purchasers of old iron. Broken walls have been levelled, and the tottering fragments of once stately buildings have been overthrown. But amid the smoke, the dust, the rain and the fog, there is an incessant throng of busy men, boys and teams, working as energetically as if the whole burnt district was to be restored before Christmas, and they were charged with the duty. The days seem all too short, and work goes on long after dark.

An idea of the number of teams and men employed may be had from the fact that 5,000 loads of débris are emptied into the Lake basin daily, and this work can continue all through the winter, giving continuous labor to the thousands now employed. So great is the demand, that hundreds of boys from fourteen to

THE FIRST BUILDING ERECTED IN THE BURNT DISTRICT.

eighteen years of age are hard at work wheeling, cleaning, and piling bricks. All honor to the brave men who have met misfortune by resolutely beginning the work of reconstruction, and all honor to the men and boys who have gone to work, preferring to earn the bread and the shelter they enjoy, than to compete for the same with the sick and helpless at the churches. The man who thinks Chicago has been destroyed, has only to cross the river into the burnt district to be undeceived. Labor and skill, directed by energy and enterprise, are working like bees in the hive, and, when the spring comes, the desolate places will be desolate no longer, and from the ashes will have arisen new monuments of industry and faith.

Where the proud miles of white marble once extended, there are now ghostly and tottering walls, and a chaos of infinite ruin. One hundred thousand of our people have been rendered homeless, and men who were yesterday princes are to-day beggars. But, in view of this tremendous transformation, there is no faintness, no cowardly disposition to yield the battle. We have here won one of the grandest conflicts known to history, and although our defeat is without parallel, we shall marshal the remnants of our routed but not demoralized armies, and shall march once more to victory. Chicago may be beaten, but it cannot be conquered. In a week, or a month, or three months, may be, we shall be once more in line, shoulder to shoulder, and the world shall see us marching on as cheerily and determinedly as though naught save victory had ever perched on our banners.

Seven days after the fire a gentleman wrote to the New York *Evening Post*, assuring the public that our debts were to be paid, and said:

We are coming on well. There is a lull after the storm; all eyes are now on the future, and our city is a scene of activity unusual even for us. Residences are converting into offices and stores, temporary buildings are erecting, and all is hurry and bustle.

In one form or another, in municipal bonds, in mortgages or in commercial accounts, we owe a large amount in the Eastern States, and especially in your city. A word of assurance to our creditors:

Chicago abhors repudiation. Our citizens detest that word. I attended a meeting of bankers and merchants in Standard Hall three days after the fire, when an insurance seemed worthless and our complications ruinous. The situation was looking desperate, and the matter of a general stay-law for the relief of debtors came up. It met with a burst of opposition that was electric. It was affirmed, amid rounds of applause, that the business men of Chicago would tolerate no such relief, and, whenever it became necessary for the payment of their debts, their remaining property also should go. Men lately of large fortunes declared they could not yet see in what condition their present troubles would leave them, but they were resolved every penny they still had should be turned over to meet their liabilities. Then the applause would be renewed; and this in a room where scarcely a man believed himself to be solvent. Everywhere in the city you meet with but one sentiment among our crippled and ruined men—that they may have to go down, but if they do it shall be honorably and with their colors flying.

To the large holders of our city and county bonds let me say—have no fears about Chicago or Cook County. Every dollar of the principal will be paid, and the interest as fast as it falls due. Our people are already inquiring about this indebtedness, and affirming that whatever else is delayed, the interest on borrowed money must be paid the day when due. We are heavy losers, we are poor, but we have some money left in our city and county, and we will tax ourselves down to the last shirt sooner than have our public obligations dishonored. Even if our commercial honor were not what it always has been, any other course would be suicidal, for we shall want more money and must protect our credit.

This was an expensive city to build. Materials and labor were cheap enough, but our site was a swamp. The business portion of the town was many feet below its present level. The buildings there are gone, but every dollar expended for street elevation remains. There are the heavy curb-walls, the graded roadways, and the long miles of Nicolson pavement; there, too, are the costly sewers, and gas and water-mains. We still have our river-tunnels, the lake-tunnel, and the water-works, the expenditure of many millions, almost unharmed. Had our buildings remained, and what is now left been taken, the loss would have seemed ruinous.

No man is to be reckoned out of the fight until his spirit is broken; and to-day we are more full of energy, hope, and confidence in ourselves than in our most prosperous times. Emerson speaks of a high order of courage which is attracted by opposition, and which is never quite itself until the hazard is extreme. I am not boasting, but you ought to know we have some of that courage here. You might walk about our streets for hours and never read in men's faces a word of our hard story. The lines about the mouth are stern, but the eyes are bright and hopeful. I am proud of our city in its meeting with desolation. On the firelit avenues in that early Monday morning I saw gentlemen stop in the hurrying crowd and salute their lady friends with a word of cheer and all the formalities of a promenade, while they responded with eyes as bright and cheeks as unblanched as they had ever shown there in the sunny afternoons. Through the terrible hours until dawn, and amid the hurrying thousands in the long burning day, I saw but one woman in tears. The men saved theirs until the telegraph told us how the news was receiving elsewhere. Your sympathy was the only thing to unman us. Since our visitation I have mingled with all classes, in public meetings and in private intercourse. I have not heard one word of complaint.

We are face to face with our ruin; we owe you eastern men much money, and I am writing to let you know how we feel. The day after the fire I determined to open a new office at once, so as to do what little I could by example to restore public confidence. After a long search, I was unsuccessful, because everything suitable had been already taken since the fire, the landlords said. No one can understand Chicago who does not remember that we have few old men—least of all among our prominent business men. The capital and influence of the city are in the hands of young men, or men in the prime of life, and these can face beggary more courageously than if their steps were feeble and their best working-days gone. Do not say we are still resolute because we do not realize our misfortunes; we feel what none can feel who have not seen our ruins; but we think that with unbroken courage, untarnished honor, and God's help, we can do again what you saw us do before.

Now, for our commercial liabilities we ask no releases, no stay-laws, no compromises. We are honest, we are energetic, and we have an enormous trade already established. Give us a little time, that is all we ask. The election is with you. If you do not press us, we can pay you, we hope, every dollar. If you do press us, you shall have what is left.

And the editor responded cordially, recognizing the situation, and acknowledging the splendid fortitude and recuperative energy displayed in all our departments of enterprise and service:—

Those Chicagoans are people to be proud of—they are essentially American. The indomitable pluck they show under their calamity, and the manly cheerfulness they display amid the wreck of worldly fortunes, are grand. They have as good a right to sit down and grieve as ever Caius Marius had to mourn over the ruins of Carthage. But there does not seem to be a Caius Marius in all Chicago. Nobody thinks of sitting down; and as for grieving, they haven't time. They are burned out, but they refuse to continue so. They are impoverished, but they won't stay poor. On all sides they are up and doing. The activity with which they are covering the blackened, smoking plain with fresh frame buildings; the vigor with which they proceed to dig bank vaults from the hot ashes, and resume payments out of them before they are cool; the philosophic composure with which laboring men go to put more money into the savings bank, instead of beginning a "run" on it; the prompt decision with which the millioniare of yesterday, beggared to-day, resumes business by writing his name on a shingle and hanging it outside of his shanty; the resolute energy with which the wholesalse merchant, finding his store gone, opens his parlor windows and announces his readiness to retail goods there at the usual prices—all these are illustrations of a spirit which no misfortune can appall.

Mr. Bradish, one of our artists, after describing in eloquent language the burning of the Academy of Design, exclaims: Thus perished the Academy.

But, thank God! not the courage or the hopes of the Chicago artists. For the moment they are disheartened,—they are not dismayed. The great calamity has destroyed their art business. Many have families, and the citizens of Chicago are not able now to buy pictures. But the artists do not ask for charity; they need and will accept orders. There can be no more suitable occasion to promote the cause of art than liberal offers to Chicago artists. This winter will be a severe one for those who must remain there. But already the burnt districts are alive with the pleasant sights and sounds of busy artisans. A great city still exists; another one, as imposing as the first, will soon occupy the desolate places. Within the past month, more than 3,000 buildings have been erected.

A generous people, enterprise, genius, credit, indomitable spirit, the free flow of Eastern capital, the outburst of universal sym pathy,—all these give assurance of the rebirth of Chicago. And speedily will be seen a new edifice, a new Temple of Art, not less beautiful, that shall continue to be, for the coming years, the home of art, and the cherished abode of the stricken artists of Chicago.

There was fear of a rush on the Savings Banks, and the police were guarding faithfully the avenues of approach, and all, with ludicrous gravity, awaited the coming of the deluge of excited depositors to clamor for their money. But, as the day wore on, now and then one straggled in to claim the proffered twenty per cent., but the number who came to deposit was altogether unexpected. There was no run on any bank, and every one of these moneyed institutions commenced doing business within a few days of the fire, and all stand on a permanent basis for the future.

Rents advanced to very high figures on account of the immense demand, and some men re-rented at an advance of five hundred per cent. A shrewd man hired a place, after he saw his building going into ashes and smoke, for twelve hundred, and leased it again for twelve thousand dollars. A correspondent of the New York press said:

As early as Wednesday morning, when the fear of further danger had ceased, the work of reconstruction began. On the smoking ruins of their great edifices these unconquerable people set the signs of revived industry. The needs of so vast a body, homeless as they are, make a great market, and the thriving trade of old times commences at every uncovered corner where a temporary roof can be raised. Inspired by the opportunity, the thriving Shylocks came out Wednesday, resolved to turn the misfortunes of the city to golden account. Bread went up to fabulous rates. All sorts of provisions, though by no means scarce, were put up to extravagant prices; the remaining hotels doubled their former rates, and general dismay fell upon the helpless community; but General Sheridan fell upon the vampires with a general order, and routed them with real live words. To the baker he proclaimed cheap bread. To the hotel men, living rates, or he would run the machines himself. This restored the natural state of things, and the city under the new impulse fell into a more healthy attitude. Presently the newspapers, *The Journal* first, *The Tribune* and *Republican* following, came to life again, and a glimpse of what the country was doing for Chicago reached the suffering people. The splendid record of beneficence aroused a new hope, and the people give evidence in unmistakable ways that they are neither crushed nor disheartened.

After struggling through the mob of newsboys who were besieging newspaper offices, I met the Hon. N. B. Judd, who was returning home after an unavailing search for an insurance company in which he is interested. He spoke lightly, after the Chicago manner, of his losses, but indulged in some enthusiastic expressions about the beauty of the ruins on the South Side. The front façade of the Bigelow House gives an exquisite hint for a triumphal arch, and the south angle of the Palmer House looks a little like the Campanila of the Duomo at Florence. I checked his flow of artistic appreciation long enough to ask him about the prospects of the situation. He answered with hopeful but seasonable words: " The city will be rebuilt; its removal from the sphere of the commercial activity of the age is not possible, in view of its geographical position; it is yet too early to predict with absolute certainty whether the future fortunes of the city are to remain in the hands of those who have so long controlled them, or whether new men are to guide the new destinies. There will be ruin of individuals; whether of classes or not, is as yet unknown; but the commerce of the world demands that there shall be a city here, and, by the hands of one and another, the city will be rebuilt.

"It was only yesterday that I spoke of the desolation of that beautiful line of palaces called Michigan Terrace; to-day the garden and residence is covered with a crowd of mechanics, and the air is filled with the sound of hammers and chisels. The indefatigable owner is everywhere present, ordering and directing everything, and shedding about him a fresh and breezy atmosphere of hope and energy. His losses, of course, are enormous; but he owes nobody, and everybody owes him, so that there will still remain a large balance of this world's goods to one of the men who best know how to use them. He is building three houses for business purposes on his vacated lots, and has contracted to have them ready for their occupants in a week."

While business men were providing for the resumption of trade, or were renewing it, in twenty-four hours, some had to furnish shelter for their families. All things had to be done at the same time. Within a month there were five or six thousand houses, if such the extemporaneous tenements can be called, in course of erection or occupied by families. There were also contracts for several thousand permanent buildings for business purposes, while hundreds of temporary structures rose like mushrooms on every side, for the accommodation of those who were determined to retain their trade by supplying their customers at the earliest possible moment. And the people outside came to the rescue like true brothers in adversity. They proffered help in every form, promised to stand by the merchants and manufacturers, and gave their orders as freely as though nothing had occurred. Our misfortune was felt to be theirs, and they made it as light as possible upon us by receiving a portion of it themselves. It was interesting to see how the marriage statistics showed convalescence. The young people were not to be daunted by so small an obstacle as the Great Fire, and hundreds took the yoke upon them, in order to prove whether two were not better than one to pull a load.

"A Chicago girl wrote to her lover in Springfield, Massachusetts, just after the fire, saying: 'Our wedding was set for next week, and if you will stand up with a woman dressed in a cotton skirt and her father's overcoat, come on.' The brave youth telegraphed in reply, 'Get ready; I'll be with you.'"

Another of our ladies, when offered a velvet cloak by her mother at the East, replied that she would be ashamed to wear one this winter, when economy was the necessity and watchword of the hour. Not display, but work, frugality, charity, are the offices of our noble women, till Chicago is redeemed and our debts are paid. In accordance with this purpose, the papers warned off concert and theatre managers, and summoned the lovers of pleasure to seek cheaper amusements.

The Christians also resolved to restore the lost edifices, by an appeal to the public at large, and exhibited a heroic spirit in undertaking to go forward with their Master's cause in undiminished efficiency and enthusiastic earnestness. The universal watchword was *Resurgam;* and the world's answer is Resurget.

Already, seeing that we mean to rise again—and the country means that we shall rise again—multitudes are flocking hither, to enter upon business with their capital, to invest money in real estate, and to join in rebuilding our city. Thus the wall rises, by the blessing of God, even in troublous times; fear has given place to hope, and convalescence is written on every feature and movement of the Young Giant.

As a matter of history, it is necessary to record that the poor people in their hasty dwellings were made as comfortable as circumstances would permit. An unusually cold winter would entail much suffering, as many of them lack the fertility of invention and enterprise of the genuine American, who is not content to live in squalor and discomfort, when tact and industry can give relief and better his condition.

We close this division by quoting from a Liverpool paper, whose prognostications and comments have been evidently justified to the fullest extent :—

If anything could be more remarkable than the rapidity with which Chicago sprang into existence, it was its sudden destruction. There seems every probability of its resurrection being more remarkable than either. The recuperative power already developed is unequalled by anything in ancient or modern times. No sooner are the flames of the burning city extinguished, than workmen are busily engaged in clearing away the smouldering cinders, and making preparations for the erection of buildings as magnificent and costly as those which have been swept away. One is reminded forcibly of a colony of ants, which, when disturbed by the ruthless passer-by, no sooner recover from the panic of the moment than they set to work to repair the damage which has been done. The Chicago disaster was assuredly enough to have appalled the bravest, and disheartened the most sanguine. Such a calamity, breaking with such abruptness on a community, might well have paralyzed their efforts, and led to their practical annihilation. But there is about these mushroom cities of the West an energy of which we in the Old World know nothing. While Englishmen would be stopping to discuss the rival plans for rebuilding, and schemes for raising the money, and wasting time in long-winded speeches, America would have the whole thing done. This extraordinary energy and elasticity which enables its people to rise like giants refreshed from every disaster, is one of the most wonderful characteristics of the New World. It was developed to a remarkable extent after the War of Independence; it was developed to a yet more remarkable extent after the lamentable civil war of a few years since. Losses, both in men and money, which would have broken the credit of many countries, were to the Americans only stimulants to call forth their extraordinary qualities. Chicago is a splendid example of this splendid energy.

VII.—THE FUTURE.

CHAPTER XXXIV.

MEN said at vespers: "All is well!"
In one wild night the city fell;
Fell shrines of prayer and marts of gain
Before the fiery hurricane.

On threescore spires had sunset shone,
Where ghastly sunrise looked on none.
Men clasped each other's hands, and said:
"The City of the West is dead!"

Brave hearts who fought, in slow retreat,
The fiends of fire from street to street,
Turned, powerless, to the blinding glare
The dumb defiance of despair.

A sudden impulse thrilled each wire
That signalled round that sea of fire;
Swift words of cheer, warm heart-throbs came;
In tears of pity died the flame!

From East, from West, from South and North,
The messages of hope shot forth,
And, underneath the severing wave,
The world, full-handed, reached to save.

Fair seemed the old; but fairer still
The new the dreary void shall fill
With dearer homes than those o'erthrown,
For love shall lay each corner-stone.

Rise, stricken city!—From thee throw
The ashen sackcloth of thy woe,
And build, as to Amphion's strain,
To songs of cheer thy walls again!

How shrivelled in thy hot distress
The primal sin of selfishness!
How instant rose, to take thy part,
The angel in the human heart!

Ah! not in vain the flames that tossed
Above thy dreadful holocaust;
The Christ again has preached through thee
The Gospel of Humanity!

Then lift once more thy towers on high,
And fret with spires the western sky,
To tell that God is yet with us,
And love is still miraculous!

JOHN G. WHITTIER.

There were predictions of ill omen concerning the probability of resurrection within a brief period. To many the very removal of the wreck seemed an insuperable obstacle. The view of such gigantic ruin overwhelmed them; and it is true that there is much that years alone can reproduce. The beautiful trees, that had slowly rooted and grown to towering majesty, cannot be soon replaced. For years the newness and rawness of a primitive city must again be suffered. Yet so much remains uninjured as to give us ground to expect that the resurrection may be far speedier than the first upbuilding. The representative energies of the great North-West still hover amid the crumbling ruins of what but yesterday was Chicago; and as busy hands are already effacing the scars which now disfigure the site, so surely will they make for this noblest exponent of the free, elastic growth of the North-West a Future more brilliant than her past.

There is not the remotest probability of our sinking back into insignificance, or dwindling into extinction. The voice of the people is the voice of God, and they have said, by their capital, their charities, their grand utterances, that here must stand a great city, whose future no mind can fitly conceive. After visiting the Golden City of the Pacific, and riding through the region traversed by the new railroad that bound East and West into closer fraternity, the Hon. Benjamin F. Wade, the Statesman of Ohio, said in 1866:—

"Again I say to you, that the importance of this location transcends what most now think of it. It will never have but two rivals. San Francisco, on the Pacific, may contest the palm of greatness with it, and New York has got to run fast to get out of its way. You may deem that an extravagant expression, but recollect that New York had to struggle for one hundred and fifty years before she had the population and wealth Chicago has to-day. No people of this country have more of intelligence, more of enterprise, more of the American Yankee go-aheadativeness than the people of Chicago. I say again, that there are but two cities on this continent that can compete with it for the palm of greatness. Thirty-two years ago it had a few rude buildings, and I have been amazed to-day, as I passed through and viewed the wonderful progress that has been made; I am sure I have had no conception of the importance of this point, and, what is still more important, of the vastness and richness of the great country that lies West, and which is bound to contribute in the future so much to build up the second, if not the first, city on this continent."

If this man could have looked upon our city five years later, he would have seen more to admire, and to fortify him in his lofty expectations. And now we are to forecast the future in the light, not of blazing destruction, but of the glorious past and the actual present.

CHAPTER XXXV.

CHICAGO must be great, yea, far transcend all former greatness, because of several reasons; among which is the marvellous faith which inspires those who have had the largest experience, and occupy posts of influence and power. It was, of course, a playful remark which an old gentleman made after visiting New York City, upon being asked what he thought of the metropolis. "Why," said he, "it is a fine place, but it lacks one thing. That is the only fault I find with it. It is too far from Chicago." Here was the spirit that gave us our prominence, gone to seed. But the real creators of this amazing prosperity were animated by an intense conviction that here was the central focal point of America, and they must not rest until manifest destiny was consummated.

There seemed something almost irreverent in the confidence which men cherished and expressed. Says an eminent clergyman: I recall the conversation of a leading citizen of Chicago with me at my last visit there, and his words still ring in my ears: "There is no possibility of checking the growth of this city; its future is as fixed as God's throne."

This language was not intended to be considered boastful, nor did it deserve to be termed blowing: to the minister it seemed extravagant, inasmuch as things had been slower in New England under his eye, nor did he see, as the enthusiast saw, the immense resources upon which the city would build its future.

I recollect a similar remark made to me by a gentleman connected with the railroad interest, as I was returning home after six months' absence: "Nothing can stop Chicago now."

Such was the belief of influential men, and they naturally imparted their zeal and hopefulness to others; so that the entire population were combined in a mighty effort, not to inflate public expectation, but to give the city a position worthy its advantages.

"All things are possible to him that believeth," is the solemn statement of Holy Writ. And our Saviour said, "According to thy faith be it unto thee." Of faith there was abundance and of works no lack. For in no city were men of ability and earnestness worked harder, and nowhere did talent and industry reap quicker and larger rewards.

But is there the same firmness of faith in the future, since the sudden arrest of its onward career? Do the wise and far-seeing men anticipate a growth like that of the past? Doubtless there was anxiety in the minds of many lest the crown should be plucked from the brow that wore it so proudly. But that soon gave place to the same marvellous confidence which made every man a hero, and banished slavish, enervating fear. The language of the press was like this which follows:—

"Away with despondency! With a world to comfort us, why should we not hope? Fire has destroyed one-half of our substance; but twenty-five years ago one-hundredth of that substance did not exist, and every cause which contributed to the making of our wealth then, exists in an improved form to-day! We then had a marsh, with malaria and fever; we now have high-graded streets and pure, bracing air. We then had ox-carts and canoes; we now have railroads and a mighty merchant fleet. We then had a foul river, whose stench was in our nostrils, and a short and shallow canal that half defeated its aims; we now send a tide from the Lake to the Gulf, and our clear rolling river runs to the sea, while a lake tunnel fills our reservoirs with sweet water. We then struggled with Nature to gain a little by Art; now Art and Nature have become one in the physical advantages which no conflagration can destroy, and Chicago, with her great business division in mournful ruin, is greater, in the resources of regaining what she has lost, than any city ever built by human hands on a site possessing the greatest possible advantages of nature Moreover, there is in the history of all great fires a lesson whose unbroken force bears mightily upon our bewildering present. It is this: From the débris of all great conflagrations has sprung a sequel greater in everything that constitutes human good than was that which preceded the ruin. The great fires of London, without a single exception, increased the city's population and swelled her commerce. Every burnt portion of Constantinople, the city of fires, has been rebuilt so much better, that fire in the East is looked upon as an agent of civilization; the new in every case is greater and stronger than the old.

"New York, by her great fire, has gained ten times more than she lost. Portland, which lost $9,000,000 worth on July 4, 1866, now considers that conflagration a blessing, for the number of her people, their commercial prosperity, and her home and foreign trade have been enhanced in five years as they could not have been save by the occurrence of so tremendous a catastrophe.

"Already the ring of the carpenter's hammer and the click of the mason's trowel tell of renovation. The ruins are being brushed away, and marts where busy trade will reign before a month has passed are springing up as if by magic. The marvel of Chicago's growth has been equalled only by the magnitude of her downfall; but both will be surpassed by the miracle of her resurrection. We have lost, it is said, more than $200,000,000. All that we had a week ago was made by these agencies:

"1. Individual energy, pluck, and enterprise.

"2. The Lakes.

"3. The Railroads.

"Is any one of these three agencies destroyed? NOT ONE!"

JOHN M. VAN OSDEL.

CHAPTER XXXVI.

THE men of nerve and brain, of energy and courage, remain to guide the destinies and uphold the character of the city. In this the fire was merciful. Had a plague or other epidemic cut down our men, and decimated our population of leaders, this would have been a worse calamity than loss of property.

The press is still in the field, unconquered, and binds its magnificent powers to the re-creation of trade and confidence.

The architects, builders, merchants, manufacturers, mechanics and artisans are working together manfully, and the future seems big with promises of superior excellence and grandeur.

Several of these founders have already been named and outlined on former pages of this book; those who are to follow are but specimens of hundreds equally deserving as models in those qualities and deeds which have raised our city to its princely eminence, and shall lift it out of ruin into increased glory and greatness. These memoirs we have compiled from "Biographical Sketches," published in 1868.

Mr. J. Y. Scammon, whose name has occurred in connection with the burning of his mansion in Terrace Row, was a Maine boy, and after finishing his studies, he left his native State for a tour of observation.

"In the course of this journey he reached Chicago, in September, 1835. He made the voyage on a steamer from Buffalo *viâ* Green Bay, and the passengers were landed at Chicago by means of small boats, the steamer being unable to enter harbor. He put up at the old Saugauash Hotel, which was reached from the landing by a devious path through prairie-grass and deep mud. The hotel was crowded, the weather horrible, and large numbers of the people were sick with bilious fever. Chicago presented no very inviting prospect to the stranger. At that time the late Colonel Richard I. Hamilton was Clerk of the Court of Cook County, and

Mr. Henry Moore, an attorney, was his deputy. When the weather had improved sufficiently to justify his travelling, Mr. Scammon made ready to depart; but on the very eve of his leaving, Mr. Moore called upon him, stating that the Circuit Court had commenced its session; that he could no longer serve as deputy; that the person employed in his place had been stricken down with fever, and therefore he desired Mr. Scammon to assist Colonel Hamilton during the term. The request was complied with. In the rooms of this building Mr. Scammon performed the duties of Clerk of the Court, received his clients, and lodged at night. In 1836, he entered into partnership with B. S. Morris, Esq., in the law business, which continued for eighteen months. A year later, he formed a law partnership with Norman B. Judd, which continued until 1847. At that time Mr. Scammon had become largely interested in the Galena Railroad enterprise, and devoted his time principally to that business.

"The men of the present day can hardly be expected to comprehend fully the courage and enterprise necessary at that time to keep alive the project of a railroad extending westward from Chicago. The construction at the present day of two or more railroads across the continent, with branches and cross-roads, is not one-half so imposing and startling an enterprise as that which in those days was projected by Messrs. Ogden and Scammon. When these gentlemen came to Chicago, Illinois was in the full glow of excitement upon the grand question of internal improvements. This system, which, so far as railroads were concerned, excluded Chicago, culminated in 1837, and sunk rapidly. A most disastrous torpidity of enterprise followed. Capitalists avoided Illinois, and the hope of any railroads was abandoned by even the most sanguine. Messrs. Scammon and Ogden stood almost alone amid the ruins, unappalled by the overwhelming disaster. The Michigan Central Railway eventually extended its line to Lake Michigan, at New Buffalo, and there it had stopped.

Messrs. Ogden and Scammon, after a long effort, succeeded in reviving an abandoned Indiana charter, giving the exclusive right to construct a railroad from Michigan City to Chicago, and to this law is Chicago indebted for its first continuous railroad communication eastward.

"Previous to this, these gentlemen had travelled repeatedly from Chicago to Galena, holding meetings in every village and at every cross-road, urging the people to a united effort to secure a railroad communication from the Mississippi to Chicago, and thence east. They both had invested largely in the enterprise, and they, by personal pledges, eventually succeeded in obtaining subscriptions to stock to an amount sufficient to authorize the commencement of the railroad, being the pioneer railroad in the vast combination of roads which now bring the treasures of the West to the lap of Chicago.

"One of the early settlers of Chicago, he has been one of the early founders of many of its institutions. He was the first of the New Church or Swedenborgian body of Christians in Chicago. He and his wife and one other person were the founders of that body of Northern Illinois, and he has lived to see himself surrounded by a numerous circle of religious associates, and worshipping in one of the finest church buildings in the city. He organized the Church of the New Jerusalem in Chicago. He was also the first man of any prominence in Chicago who favored the practice of the medical school of Hahnemann. He was, as we have seen, a pioneer in the railroad system; he established the first bank under the general banking law of the State; he was one of the original founders of the Chicago Academy of Sciences, and of the Chicago Astronomical Society, and is the President of the Board of Trustees of each of those societies. The Dearborn Tower, the western tower of the grand edifice of the Chicago University, in which is placed the Alvan Clark Telescope, the largest refracting telescope in the world, was built at his expense, and named in honor of his deceased wife, whose maiden name was Dearborn. He was elected one of the Trustees of the Chicago University on his return from Europe, and one of its professorships was endowed by his munificence. The family of Mr. Scammon consists of one son and two daughters."

This man still lives to consecrate his genius and experience with undivided earnestness to the rearing again of our fallen metropolis. From this sketch we can see through what difficulties Chicago rose to power, and easily believe that the present obstacles are far less imposing than those which he materially helped to overcome. He may have less to give than before the fire, but having experienced the blessedness of large liberality, he will not be backward in renewing his labors in behalf of the institutions imperatively demanded by a vast population.

The future appearance and character of the buildings of the new city must depend on its architects, among whom, representative names are William W. Boyington and John M. Van Osdel. When we have looked on the ruins of the superb edifices designed and erected under their supervision, we have felt sympathy for these men, whose monuments seem to have crumbled into dust. In the old world we see the names of architects imperishably connected with the massive and elegant structures which are historic; when these fall, the work of the builders passes away, and appeals no longer to men's admiration and reverence. It is the fortune of our architects that they live to renew these memorials of their skill and power, and to write their names indelibly upon the future of Chicago; and, already, their heads and hands are full of plans and contracts for buildings which shall rival those that have melted and perished. The men who have means eagerly place them again in brick, iron, stone, and mortar, demonstrating their unshaken faith in the inevitable greatness of the new city and in the ability of the architects whose work the fire destroyed.

"Prominent among the architects of the city of Chicago stands the subject of this sketch—William W. Boyington—a true representative of his class, and an acknowledged leader in that great architectural reform which, during the fourteen years of his residence here, has been in progress in Chicago, appropriating her waste places to occupancy by the busy multitude, and changing her shanty dwellings to palaces, wherein operate and dwell the real kings of the Great West—her business men. He has been a power in shaping the destiny of Chicago in its external aspect. From him has gone forth the fiat which has set at work and kept busy thousands of intelligent workmen, whose every movement was in harmony with the one great idea of the author, and ever tending to its completion. Dozens of draughtsmen and clerks have detailed his conceptions on paper, and thousands have given them more enduring form in wood, brick, cement, or marble. A vast number of our largest, most stately, and most useful edifices are the realizations of his thoughts on architecture.

"In the spring of 1853 Mr. Boyington came out to Chicago, to see the chances offered in this city, which was then just beginning to be talked about in the East. He returned home, and after some months' delay, wound up his business in Massachusetts, and in November removed hither. His first work here was to make out a plan for Charles Walker, Esq., of the ground on which the great Central Union Depot now stands, showing the character of the buildings which could be placed upon it, the Railroad Company being then about negotiating for the site for the depot grounds. He has been ever since that period most prominently identified with the history of our civic growth, as the city was just ready for architectural style, finding ample scope for the exercise of his talents, and generally meeting with the recognition which his ability deserved, especially after the first few months, by which time he was generally conceded to be a man of extraordinary talent in his profession. His success during the subsequent thirteen years is scarcely equalled in the history of any architect in the whole of the United States.

"Up to the year 1853, when Mr. Boyington came to Chicago, the city could boast of but very few buildings worthy of note in an architectural point of view. Here and there a structure was visible possessing some claims to notice, but, with a limited range of exceptions, the buildings in the city were little better in appearance or comfort than the old log-house, and not one-half so substantial. How wonderfully the scene has changed! The revulsions of commercial panics, the universal suspension of banks, the almost entire stagnations of trade, the terrible excitements of war, none of these have stayed the successive piling of bricks, the aggregation of slabs of marble, and the rearing of massive timbers, to form our city into one great system of architectural beauty."

Mr. Boyington is not past his prime, and with all his immense experience he can outdo his former achievements, and leave himself many monuments by which his fame will be transmitted to posterity. Fortunate for the city is it that he lives, and men like him, to reconstruct the public and private buildings, which shall be our honor, our profit, and our joy.

His colaborer and competitor is also in the vigor of an energetic maturity, and has plunged anew into his profession.

"In the autumn of 1836 Mr. Van Osdel formed the acquaintance of Hon. William B. Ogden, of this city, which resulted in his removal to Chicago. Mr. Ogden at first engaged his services simply as a master-builder; but soon found that he was every way competent for the responsibilities of an architect, and engaged him to design as well as construct a residence for him in this city. The house which he built on Ontario street, the following season, was for several years the best in the city, and is still occupied by Mr. Ogden." We have read the sad story of Mr. Ogden's attempt to find it after the fire.

"Mr. Van Osdel also turned his attention to ship-joinery, and to him belongs the honor of having done the finishing of the first vessels that were built in Chicago, being the two steamboats 'James Allen,' and 'George W. Dole.' Our lake commerce was a mere trifle at that time; but it had begun to give promise of its gigantic future. In 1838 he constructed several large pumps on the Archimedean-screw principle, for the purpose of lifting water out of the excavations then in process for the Illinois and Michigan Depot. During the following winter Mr. Van Osdel invented a horizontal wind-wheel, which was extensively used in working these canal-pumps. The first important work in which he engaged on his return to Chicago, which was in the spring of 1841, was the erection of grain-elevators. Here, too, he was the pioneer.

"In 1843 he entered into partnership with Elihu Granger, in the iron foundry and machine business. This partnership continued until February, 1845. His wife dying at the time, and his own health being impaired by over-work, he was advised by the leading builders to devote his time to architecture, they pledging him their support. He therefore opened an office on Clark street, over Mrs. Bostwick's millinery store, precisely where is now the main entrance to the Sherman House. His receipts during the first year were only five hundred dollars, although he did all the business of the kind which there was to be done in the city. As the city grew, and his skill as an architect became more widely known, his business increased, until his net profits for the three years ending in 1859 were thirty-two thousand dollars.

"To enumerate all the public buildings, private residences, and extensive mercantile blocks which were designed by Mr. Van Osdel, and built under his superintendence, would be to give a long list, including many of the best edifices, not only of Chicago, but of Illinois. We will only mention as specimens, the Cook County Court-House, the Chicago City Hall, the Tremont House, all the five-story iron-front buildings in the city, being over eleven hundred lineal feet of such frontage; the residence of Peter Schuttler, corner of Adams and Aberdeen streets, Chicago; the residences of ex-Governors Matteson, of Springfield, and Wood, of Quincy—the three finest residences in the State. Mr. Van Osdel has accumulated an ample fortune; he has not suffered himself, however, to be placed upon the retired list, but is to-day one of the most active men in the city. He is at present architect for the completion of the State Penitentiary. His report on the progress of the work, with estimates of work done and to be done, received the unanimous approval of the last General Assembly of Illinois, which pointed him out as the architect best deserving a place among the Trustees of the Illinois Industrial College, located at Champaign. He was elected by the Board as a member of the Finance and Executive Committees, also of the Committee on Buildings and Grounds, three of the most important committees of the Board. Mr. Van Osdel was mainly instrumental in having a Polytechnic School established in Chicago as a branch of the Industrial University, of which he is Treasurer." Since the above notes were made, he has built the most magnificent stores in Chicago, and is Potter Palmer's architect for the new hotel, which is designed to eclipse anything in the world. These men are prominent supporters of the church, and eminent for their liberality.

From them we turn to the merchant princes, and select Mr. John V. Farwell and Mr. C. T. Bowen as men of whom the city is justly proud. The name of "Farwell Hall" was a fitting recognition of his generosity towards the Young Men's Christian Association, which had its headquarters in that building. And Mr. Bowen had begun a foundation for a Youth's Library in that Hall, which would have been as useful to the generations to come as it was honorable to his head and heart.

"In the spring of 1845 Mr. F. left off his books and came to Chicago with exactly three dollars and twenty-five cents in his pocket, working his passage on a load of wheat. The road was a canal of mud. Driver and passenger frequently had to put their shoulders to the wheel, or their hands to the lever. They made their ninety-five miles in four days, without losing their temper or calling upon Hercules, who, if he were a witness of the spectacle, must have wondered afterwards, as he saw in the affluent merchant the youth who pried the load of hay out of the prairie-mud. Reaching Chicago, he drifted into the City Clerk's office, and got employment at twelve dollars per month.

"His aptness for business was soon apparent. He had skill in trading, in managing and in planning, and energy adequate to the carrying out of his plans. Besides this, he was one of the few who realized the possibilities of the Northwest, and fully foresaw the destiny of Chicago. While others conjectured, he was convinced; while others stood by wondering whether to invest, he went forward and proved his faith by his works, and a great, high faith he had in this city and this section when he became a partner in the firm he had served as salesman. His hand was felt upon the helm immediately, and his word had weight in the councils of the concern. That was in 1851, when the house did a business of about $100,000 per annum. Its business now foots up $10,000,000. The entire dry-goods commerce of the city had a new impetus under the leadership of Mr. Farwell. For lead he did, with such boldness as to confound the wisdom of the wise in trade, and to make the most enterprising among them shake their heads in an admonitory fashion.

"In 1856, through Mr. Farwell's irresistible persistency, the wholesale mart on Wabash avenue was built, now occupied by the firm of John V. Farwell & Co., which, after several changes, came to be the name of the firm in 1865. The enterprise was stoutly opposed by the oldest member of what was then the firm, and was set down by the longest heads in the city as a project that must bring its owners to ruin. But time has demonstrated the wisdom of the undertaking. It was to the wholesale dry-goods cause of the Northwest what the memorable raid of Sherman was to the cause of the National Government. If it was daring to look forward to, it was grand to look back upon.

"The men who built a commerce are to be honored with those who found a commonwealth. Commerce is the corner-stone of the commonwealth. First ships, then schools; first trade in corn, then in books. What are dwelling-houses without warehouses?

"But for commerce there had been no Chicago. Once a commercial capital, and Chicago became a seat of learning and of literature, a market for knowledge as well as for breadstuffs and dry-goods. This is the metropolis which the man of this sketch helped mightily to build by his enterprise, and then to adorn with his philanthropy. And such men have a fame which Chicago will never let die. Their renown is indissolubly linked with hers. And as we ramble through this buzzing and busy dry-goods hive on the Avenue, with its hundred of men and its piles of fabrics from every part of the commercial world, we cannot but feel a thrill of pride in the man who founded and built it all. But we have a livelier and a nobler satisfaction when we contemplate this man as "the servant who was found faithful' to his stewardship, as well as the merchant who was found equal to every exigency. Prosperity did not quench the ardor of his convictions, deaden his sensibilities, nor blunt his moral sense. When poverty departed it did not carry conscience away with it; when riches came they did not bring penuriousness along, but openhandedness instead. The merchant had an end beyond his merchandise, the tradesman was not content with trade. Affluence was made no excuse for self-indulgence. The miserable cupidity which brings a man to his knees before the golden calf was held in scornful detestation. The grovelling avarice which makes a business man a slave to his business was equally despised. The love of Christ constrained the love of money. The love of God induced the love of man, and the love of man was shown by deeds and devices for his amelioration and elevation. Mr. Farwell increased in philanthropy as he increased in means for exercising it. The world which lieth in wickedness, and the church which is as a net to save it, are the objects of his alert solicitude and unremitting liberality." His brother, C. B. Farwell, is the member of Congress from Cook county, and is a man of brains and energy. They have gone forward with accustomed sagacity and pluck in the restoration of the commerce of Chicago.

Mr. Bowen was born in New York State, and at the age of seventeen was a clerk in Little Falls. "From there our future merchant came to Chicago, and entered the service of Mr. Wood, who was the first to introduce into the village the system of trading on strictly cash principles. Never was a clerk better suited for his position, and the duties which devolved upon him were admirably adapted to fit him for the part he was afterwards to sustain in the commercial development of this city and the Northwest. Before he had been in Mr. Wood's employ three months, he was placed at the head of the establishment. The proprietor was absent the greater part of the time, and the whole responsibility rested upon the shoulders of young Bowen. He gave his personal attention to every department of the business. He was at once cashier, bookkeeper, and head salesman; the first man at the store in the morning, and the last to leave at night. But his labors were not confined to the counter and the desk. Not content with seeing that customers were well served and books accurately kept, he added largely to the custom of the establishment by pursuing a system of advertising and 'drumming' peculiarly adapted to these pioneer days. At that time it was the custom of the farmers from the country to come to Chicago with their produce, and camp out for the night in what was then the southern suburbs of the town, in the vicinity of Eighteenth street, and it was Mr. Bowen's practice, mornings, before it was time for trade, to go the rounds of the camp and distribute advertising circulars among the campers, setting forth the superior inducements of 'The People's Cheap Store.' Not content with merely scattering these, he would, by a few words fitly spoken, win upon their personal favor. In that way he became widely and always favorably known to a large circle of customers, whose trade added materially to the profits of his employers. The personal popularity of young Bowen was very great. The farmers liked to trade with him better than with a kid-glove counter-jumper, who fancies the condition of mercantile success is good clothes and fastidious drawing-room manners. And we may add that the same good sense which characterized Mr. Bowen then, has ever since. Not only so, but he has been careful to surround himself with associates and assistants similar in character. At this day there is no one connected with his establishment, from the senior member of the firm to the porters, who does not by his works show his faith in the dignity of labor, of whatever kind.

"Mr. Bowen's theory in regard to advertising was then, and always has been, that no promises in regard to quality of goods or their price should be made that he could not fulfil. Enterprise may reap an ephemeral reward, even when dishonest; but great, lasting success is conditioned on probity.

"Mr. Wood was not slow to testify his appreciation of the services. The salary for the first year had been fixed at two hundred dollars, but at the end of the year Mr. Bowen found six hundred dollars credited to his account, without anything having been said by either party upon the subject. At the same time his salary was, without solicitation, raised to one thousand dollars. This was nobly generous of Mr. Wood. Yet he could richly afford to do it, for the young man's services, even then, were remarkably cheap, considering the amount and kind of service rendered.

"In 1853, Mr. Wood retired from business. He was succeeded by Mills, Bowen & Dillingbeck. The members of the firm were

D. H. Mills, George S. Bowen, Chauncy T. Bowen, and Stephen Dillingbeck. The business continued to be conducted on the same plan as before, only on a much larger scale, and even more profitable. This firm was in 1856 succeeded by the famous house of Bowen Brothers, of which George S. and Chauncy T. were the co-partners. In July, 1857, their oldest brother, James H. Bowen, came on from Albany, New York, and joined them.

The business of this house during the last ten years has been immense. There is not a merchant in the West who has not heard of Bowen Brothers, and the majority of those who have been in trade any length of time have, doubtless, had more or less dealings with them. The enviable reputation of Chicago as a centre for wholesale supplies is largely due to the enterprise and scrupulous honesty of this house. Its sales for the last three years amounted to more than fifteen million dollars. Neither St. Louis nor Cincinnati, cities which once looked down in disdain upon Chicago, has a house that can make any such showing into several millions. About a year ago the firm of Bowen Bros. retired from business, and erected one of the finest mercantile blocks in the city." Men who began at the bottom and climbed to the top, know well how to repeat their efforts when reverses come. Along with these pioneers and founders have come up a host of young men of sterling stuff, whose opportunity has now arrived, and who will have before them a grand career, worthy of being written out for the admiration and encouragement of succeeding generations

---

CHAPTER XXXVII.

"The West cannot exist without Chicago; her natural situation and advantages give her control of the Western States, and in future times, seated in all the majesty of empire, on Lake Michigan, she shall look back, as to a fearful dream, upon the recent conflagration, and in all probability remember it as a blessing rather than a curse."—ANON.

FAITH, existing in the hearts and throbbing in the pulses of great or common men, could not produce grass on the top of a bare rock, or create a city without people and without the co-operation of nature. We have no king to dictate where towns shall be established, and compel them to be inhabited. In this free country, men gather into the best locations, as naturally as water runs down hill. And probably there is no spot more adapted to the wants of the commerce of the Northwest, in all its vast domains, than that on which Chicago stands.

The energetic and wise master-builders have planned well and executed firmly, and they need profound wisdom and limitless energy to fulfil all the hopes of mankind; but it would seem as though Providence had provided a place for the confluence of nations, and given every possible suggestion to guide and encourage our citizens. The fire has altered no natural laws, changed none of the causes of growth that inhere in the geographical position, left materially uninjured much that has been done to promote the transaction of business and the comfort and ease of living; while it has advertised us to the world, and opened to us their sympathies, as nothing else could have done. True, there is much latent selfishness in the world, and men will go where they can buy for the least money what they want—in short, where they think they can do the best for themselves. If we keep the largest stocks, and sell for the smallest profits, and convince the world about us that this is true, our city will continue to command the attention of buyers and sellers, and grow into increased greatness. These results are possible, because of our magnificent harbor for the Lake shipping. Already we have thirty-four miles of dockage, with opportunities for indefinite expansion, by means of slips, and are at the head of the grandest inland navigation on the globe.

"During eight months of the year there is an average daily arrival and departure of some fifty sailing vessels and steamers.

These bring coal, iron, wood, lumber, and heavy goods. Of these Lake craft, three hundred and ninety-eight are owned in Chicago. These are of an average capacity of 214⅓ tons; the exact aggregate is 85,313 tons. Vessels bringing coal and iron from Buffalo and Cleveland are much larger. The entire fleet entering and clearing from the port of Chicago average 239⅔ tons; and the total number during the eight months of 1870, from April to November, both inclusive, was 12,546; while the arrivals and departures, during the same eight months at the ports of New York, Philadelphia, Baltimore, New Orleans, San Francisco, Mobile, and Savannah, were 12,259—287 less than at the single port of Chicago. It is true, the sailing vessels and steamers entering New York are much larger—averaging 599⅔ tons; but even their aggregate tonnage is far less than the port of Chicago! The fleets of deeply-laden vessels that daily arrive and depart from our youthful city would greatly surprise even a resident of New York or Liverpool."

The deepening of the canal to the Illinois River, which is thus connected with Lake Michigan, must greatly increase the already enormous trade which floats on its bosom. This all remains intact; and five of the grain-elevators stand erect and ready for business. So that the great expectations of Chicago cannot fail, unless nature fails, and the world burns up. Our system of parks will be needed in due season; and the continuous drive through parks and boulevards of twenty miles, all round the city, will yet be thronged with the prosperous and happy people of our reconstructed metropolis.

The railroads of the Northwest are all vitally concerned in the new Chicago's prosperity, and each added mile is directly or indirectly tributary to the markets of this city. Besides the great rival lines to the Pacific now in operation, the Northern Pacific commands an immense territory, a vast empire of fertile lands rich in mineral resources; and this must become a source of wealth, also, to us, bringing not only that splendid region to our doors, but aiding us to control the Oriental trade which has enriched so many cities of the world.

Our manufactures have begun to acquire root and strength, and are destined to develop indefinitely, because the labor, coal, and materials are all here waiting the call of capital to convert them into immense profit. The following table shows what nationalities are here, and suggest the various industries they pursue, and the ties which bind us to the Old World. The numbers represent families:—

United States	28,839
Ireland	19,145
Canada	3,167
Norway	2,910
Austria	1,426
Denmark	655
Poland	370
Italy	275
Germany	28,870
England	4,947
Sweden	2,940
Scotland	1,750
France	722
Holland	527
Switzerland	337
Wales	247

Besides these, there are families from twenty-seven other countries. Whatever business any man may wish to follow or carry on, which appertains to civilized communities, he can find the artisans here who know how to aid him. These people are mainly inhabitants of Chicago still, and their presence will attract and employ the capital of the world.

The actual indebtedness of Chicago is under fifteen millions

JOHN V. FARWELL.

and the largest part of the improvements remain for which this money was expended. According to the best estimates, our total loss will amount to two hundred and ten millions. A portion of this will be recovered as a basis of operations, so that we are better off to-day than we were fifteen, or possibly ten, years ago.

The instantaneous leap forward, after the first touch of the hand of a world's charity, and the successive strides of the Young Giant, towards former supremacy, assure us that the future shall be as the past, and increasingly glorious.

AGE AND POPULATION OF WESTERN CITIES.

	Settled.	Population, 1870.
Detroit	1700	79,580
St. Louis	1764	310,864
Pittsburgh	1784	86,235
Louisville	1785	100,754
Cincinnati	1789	216,239
Chicago	1830	70
Chicago, present population	1871	334,270

Facilities for the cattle trade have been furnished by the Union Stock Yards, which the fire did not harm, and are of such importance as to justify the insertion of the following description:

Probably no enterprise in the history of Chicago has combined so many corporations and capitalists together into one great company as the Great Union Stock Yards. Railroad companies that have heretofore been rivals for the live stock trade of the West, and often at war with each other upon this subject, are now a unit, working together as architects of this great undertaking. Their tracks have been extended to a common centre, and nine of the former competing roads now connect directly with the Great Union Stock Yards. The broad prairie that stretches southward from the city is now traversed and retraversed by their different branches, all tending toward the great bovine city of the world.

Packers and commission dealers, whose extensive establishments have heretofore demanded their entire attention, are now found at this nucleus, prospecting upon the results of the enterprise, laying plans for the future, and prognosticating the prosperity that is to follow the opening of this great cattle mart. Their estimates for the future might be considered chimerical by the Rip Van Winkles of other and less go-ahead cities; but Western men know the extent of the broad prairies of Illinois and neighboring States, which stretch away like the pampas of South America, yielding pasturage for innumerable herds of cattle, found nowhere else in the country.

Among the first business transactions of the hamlet, now grown into this great city, was buying and selling cattle and swine, large herds of which were easily driven to market here, slaughtered, and shipped to other points. The packing business was only another branch of the trade, and beef packed in Chicago was to be found in the marts of Liverpool long before the growing Western town from whence it came had a "local habitation and a name" among the cities of the continent.

At the World's Fair, held in London several years ago, the attention of Queen Victoria and Prince Albert was called to several tierces of beef from the packing establishment of the Houghs, in Chicago; and they were awarded a premium. Thus the produce of the new city began to grow in the estimation of foreign dealers, and an impetus was given to the trade. Steadily advancing, the exports from our harbor began to look like those of much older cities; and St. Louis and Cincinnati lost their laurels—the latter ceasing to be the recognized "porkopolis" of the land. Reaching out like a young giant, the new commercial port seized upon the produce of the prairies of Illinois and the West, and put an embargo upon the growth of older towns less centrally located. Dealers in live stock soon left their old landmarks in Cincinnati, St. Louis, Louisville, and established themselves in the Garden

City; the places that had known them knowing them no more, unless it was to hear of their prosperity and increasing wealth. Railroads sprang into existence, and cut the prairies in every direction, while the lakes were whitened by the unfurled sails of thousands of vessels; and the great rush of business which now blesses Chicago as a metropolis was established permanently, upon a basis having for its foundations millions of acres of productive lands, great natural resources, and untold commercial advantages.

On the first of June, of the present year, ground was broken for the new yards. The first thing to be done was to drain the land—a work of no small importance. An immense box sewer was constructed along Halstead street, to serve as a main discharge for the drains and sewers. This structure is half a mile in length, running north and south, and four feet in the clear. Constructed on the most improved plans, these drains and sewers, underlying the yards in every direction, perform their work in the most admirable manner. The soil is now in good condition, and no inconvenience will be experienced from wet land or standing water. In this particular, the great bovine city will be far ahead of the populous and crowded human city which it adjoins, and of which it is destined to become an important part.

The total length of the drains and sewers is about thirty miles. They have caused a wonderful transformation in the level, wet land on the prairie, which it has heretofore been considered impossible to drain. The argument deduced from this is, that all the low land surrounding Chicago is valuable for building purposes, and that it can be thoroughly drained, so as to afford a solid foundation for structures of any size.

The tract of land selected as the site of the yards was now thoroughly drained, and what a short time before was a marshy prairie, covered with rank grass, appeared dry and firm, admitting of the passage of loaded wagons, and the laying of railroad tracks over it. Lines of rails were soon constructed, leading from different railroads, which were to transport the immense amount of lumber required for the construction of the yards to the spot. Large sills of timber were placed upon the ground, across which were laid three-inch joists. Upon this foundation the planking was commenced. That portion of the yards to be used for cattle pens was planked with three-inch pine planks, placed firmly upon the joists and nailed thereto. Two-inch plank was similarly placed upon those portions where the hogs are to be kept. The planking being raised from the ground, affords the water and refuse from the yards an opportunity of draining off to the ground, where it immediately finds its way into the drains and sewers which underlie the whole, thence into the main sewer on Halsted street, and into the river. The entire planking, like the draining, was done in the most substantial manner, no expense or pains being spared to make it firm and solid, so that no accidents might result in the future from its sinking or breaking through, beneath the tread of the herds destined to pass over it. A portion of the planking was done by contract, and the remainder by the company. As many as 1,000 men were employed upon it at one time.

The entire 345 acres comprised in the yards are laid out into streets and alleys, in the same manner as a large city. Through the centre, from north to south, runs a broad avenue, which has been named E street. This great central thoroughfare is one mile in length, and seventy-five feet broad. It is divided into three sections, like a bridge, to facilitate the driving of cattle through it. Droves passing to the south will take one section; those passing to the north, another, meeting on the way without the slightest inconvenience or stoppage. The drover's whip will not be called into requisition in passing through this avenue, as all will be "fair sailing." This street runs through the entire grounds, and is paved with Nicolson pavement; the blocks used being the refuse ends of plank, etc., which economy greatly reduced the expense. There is not a finer or smoother drive in Chicago than this well-paved and finely rounded street. Running parallel to Avenue E are other streets, leading to the railroads that surround the yards on all sides but the south.

These streets are crossed at right angles by others, running east and west. The principal one of these passes by the Hotel, and has been named "Broadway" by the workmen. It is, indeed, a broad avenue, and will probably retain the name, as it leads from the Hough House to the Bank and Exchange building, where the life and excitement of the yards will centre. It is sixty-six feet wide, planked with heavy timber, and traversed on the south by a raised sidewalk.

There are five hundred of these enclosures, all lying on the different streets, like the buildings of a city, and all probably numbered. In size these enclosures vary from 20 × 35 to 85 × 112, while others are precisely the size of a car, calculated to hold just one car-load of stock. The cattle-pens are open, but those designed for hogs are covered with sheds, and so arranged as to prevent the hogs "piling," which they are inclined to do in cold weather.

The yards are provided with six hay-barns and six corn-cribs, situated on different parts of the enclosure, convenient to different sections of pens.

Perhaps the greatest feature of these yards is that of the different railway accommodations. Nine of the principal railroads of the West find a common centre here. There have been constructed fifteen miles of track, as branches, which connect these roads with the yards, besides many switch-tracks and side-runs. Upon the north are tracks of four railroads—the Great Eastern, the Michigan Central, the Michigan Southern, and the Pittsburg and Fort Wayne. These roads all run in from the east, and their tracks are so arranged by the side of "shoots" that whole trains can be unloaded at once. On the north, and parallel to the "shoots" belonging to these roads, are others running nearly parallel. They are for the accommodation of two roads, the Chicago, Burlington, and Quincy and the Illinois Central, which also approach the grounds from the east. The east and west sides of the yards describe an inward curve, along which are platforms and "shoots." The Chicago and Rock Island Railroad owns those upon the east, and the Chicago and Northwestern and the Chicago, Alton and St. Louis those upon the west, where their tracks are constructed. By the act of incorporation all the roads have the privilege of running over each other's tracks, but so ample are the arrangements that this will seldom, if ever, be necessary. The yards are provided with water-tanks for the engines, wood-yards, turn-tables, and everything that is required at a great depot, which in fact these grounds are—the greatest in the world.

The facilities for loading and unloading cargoes of cattle at these yards are unsurpassed. Each road has 1,000 feet of platform, which is provided with "shoots" leading directly into the yards and pens of the division appropriated to the use of such road. When a train of cars loaded with live-stock arrives, it draws up in front of the "shoots." Gates are so arranged that they open across the platform extending to the cars, and thus form an enclosure through which the stock passes directly into the yards. These gates enable a whole train to unload as quick as one car. Several of the "shoots" are made double, so that the upper and lower floors of a carload of hogs can be passed out at the same time. This arrangement is so perfect that there is little chance for an accident to happen to the stock as they pass down the avenue formed by the gates, and are thence driven into the pens. As many as 500 cars can be loaded or unloaded in this manner at the same time, the whole operation occupying only a few moments. This fine arrangement is considered one of the greatest features of the yards. Water is furnished in ample abundance by Artesian wells on the place.

Since their opening, these yards have verified the opinions expressed above, and constitute a system as nearly perfect as human skill can construct. The Hough House has become the Transit House. A neat chapel, erected by the Second Baptist Church, and patronized and sustained in part by that body, opens its doors and affords religious privileges to the residents and visitors there, who take a deep interest in the exercises of the Sunday-school and the preaching of the gospel. Who shall deny that Chicago must go forward more swiftly in the future than the past, when these vast facilities for business which we have named, and these surroundings of country and population are fully considered?

The testimony of our rival city, St. Louis, is a generous recognition of our geographical supremacy. Said the *Missouri Republican*:—"Chicago, though stricken in purse and person as no other city recorded in history ever has been, is not crushed out and destroyed, and her complete restoration to the place and power from which she is temporarily removed is only a question of time. It would be sad, indeed, if a conflagration, though swallowing up the last house and the last dollar of a great commercial metropolis, could fix the seal of perpetual annihilation upon it, and declare that the wealth and prosperity which once were should exist no more forever. Such might be the case, perhaps, were there none other save human forces at work; but into the composition of such a city as that which the demon of fire has conquered, enter the forces and the necessities of nature. Chicago did not become what she was, simply because shrewd capitalists and energetic business men so ordained it. That mighty Agent, who fashions suns and stars, and swings them aloft in the boundless ocean of space, marks out by immutable decree the channels along which population and trade must flow. When the first settlers landed at Jamestown and Plymouth, and began to hew a path for civilization through the primeval forest, it was as certain as the law of gravitation, that if this continent were destined to be a new empire, fit to receive the surplus millions of the eastern hemisphere, and contribute to the progress and enlightenment of mankind everywhere, there must and would be a few prominent centres, so to speak, around which the vast machine could revolve. Those centres were determined by the geography and topography of the country; and when the advancing tide of immigration touched them they began to develop as naturally and irresistibly as the flower does beneath the genial influences of sunshine and showers. For practical purposes neither Jamestown nor Plymouth were of any special consequence; therefore the one has ceased to exist altogether, and the other remains an insignificant town. But the inner shore of of Boston harbor, the Island of Manhattan, the site of Philadelphia, Baltimore, Cincinnati, New Orleans, St. Louis, and San Francisco, furnished the required facilities, and we see the result to-day. Nature declares where great cities shall be built, and man simply obeys the orders of nature.

"The spot where Chicago river empties into Lake Michigan belongs to the same category as those we have mentioned. It was designed and intended for the location of a grand mart to supply the wants of the extreme north-west—that portion of the central plateau lying on the line and to the north of the Union Pacific Railway, and the Western part of the British possessions. The trade from these sections seeks an outlet there, and finds it better and more available than anywhere else. This fact was settled before the first brick was laid in Chicago; was settled when Chicago rose to the rank of the fifth city in the republic, and is settled just as firmly now, when, to all human appearances, her destruction is wellnigh accomplished.

"Natural advantages, then, must compel the reconstruction of Chicago, even though every foot of its soil passes out of the hands of the present proprietors. And if we examine what the fire has spared, it will be found that the nucleus of a new and rapid growth is not wanting.

"If we add to these resources the railway lines converging to that point, which represent an aggregate capital of $300,000,000; and remember that every railway is directly interested in the process of reconstruction, and will aid it in all possible ways, it may not be difficult for even the most incredulous to see why and how Chicago must grow again. That she is absolutely ruined or permanently disabled is sheer impossibility, which no sensible person will for a moment credit."

CHAPTER XXXVIII.

If it were possible to obtain a commission of true, honest, public-spirited citizens, to whom all general affairs were entrusted, with power to make laws for the city, and determine the character of its future, they would doubtless greatly change many things, and introduce reforms and establish customs of incalculable benefit to all coming generations. In the nature of things this is impossible, and all Utopias exist but in the brain of enthusiasts, never probably to issue into living realities, while men are prone to error and sin. Education should be the right and duty of every child of the city; in other words, all persons should enjoy, either freely or compulsorily, the advantages of learning. The principal temptation of city life should be put away by the prohibition of all sales of intoxicating liquors, and by such careful legislation as to prevent any drunkard from existing among us, or any dram-seller plying his trade among our citizens openly or by stealth. Gambling should be made a crime and absolutely crushed out, and forever prevented. Harlotry should be trampled under foot, and kept down by every resource of law. Honesty should be encouraged, and justice magnified by the officers and judges, whose example should be above reproach. The Sabbath should be regarded as a sacred institution of universal obligation, and defended from the encroachments of power or the perversion of selfishness and ignorance. Religion should be the voluntary choice of all men; and its ordinances and machinery, simply protected from the rude hand of violence, should be given free scope in the improvement and satisfaction of the people. Literature, science, and art should enjoy every encouragement, and be made to minister, not, as in Paris, to the worse portions of our nature, but to the ennobling, gratification, refinement, and culture of the whole community.

In the material improvements there should be care exercised to guard against the recurrence of fires. In the French capital, the man in whose house the fire begins that consumes property, recovers no insurance. The buildings also are constructed slowly, and with such regard to the destroyer's ravages that, in the last ten years before the Commune, the damage by fire did not amount, according to the testimony of an American merchant there, to two hundred thousand dollars. It should, therefore, be an offence to build, in the heart of the city, of anything combustible. Let a city grow to stand half or twice a thousand years. This would be economy, and our liberty should not be construed into license to prepare, under our neighbor's eaves, a tinder-box to burn him down or do him damage. Water arrangements should be made as perfect and safe as ingenuity could devise and money procure. Immense engines and accessible reservoirs should be provided, by which whole blocks could be flooded and placed beyond peril, as gigantic barriers against the progress of conflagrations, however furious. The Fire Department should be organized and drilled to an efficiency like that attained among soldiers of the regular army.

How magnificent might be the future of our city under a system like this! Our influence would extend on every railway and highway, borne by the billow and the breeze to remote districts, and wherever it was felt, the tone of public sentiment would be exalted, and men would turn to us as the mariner to his compass or chart, for laws, sentiments, principles, and fashions, and the whole conduct of life. Our example would be such that the Republic, energized and purified, would pulsate with new life, and her glorious career would prolong itself to the end of time.

We close with this pleasant picture drawn by another's hand:

"Then shall our fair city rise out of her ashes, and sit beside this lake for ten thousand years to come, beautiful for situation, the joy of the whole earth. The fair Garden City, the centre and glory of the garden of the world, one of the fairest jewels in the diadèm of America, the strong right hand of our noble nation, where our children will live in great peace and prosperity when we are dead and gone, where starvation and squalor shall be known no more, where the poorest home will be filled with plenty, and the poorest child have an equal chance with the richest to come to the knowledge of the truth, as it touches our whole life here and hereafter. Where all homes will stand close to all temples, and all temples near all homes—a city like that John saw in his great vision, that standeth four square, and the height, and the length, and the breadth of it are equal."

THE FIRES IN WISCONSIN.

CHAPTER XXXIX.

WHILE we were struggling in our agony, neighboring States and communities were also visited by the raging monster, and suffered a scourge as keenly felt and more destructive of life. The drought, whose pernicious influence had desiccated the air in our own vicinity, and parched everything to a state of preparation for fire, was very general in the western country. Water, for ordinary purposes of family use and for cattle, had become a luxury in many places, and even an expensive one. The streams and springs were dry in large sections, and the people unprotected from such a foe as charged down upon them. Occasional conflagrations were occurring in the woods of Wisconsin and Michigan, caused by the hunter's carelessness, or as a natural consequence of his sport. In this way, the wadding lodging in the dry grass, prairie fires have originated which desolated the fairest regions of our country, year by year. But upon the blackened soil there appeared again in the vernal season a fresh growth that made all look fair when summer came. So we may hope our desolated regions will bloom again when the forces of nature and the energies of man combine in harmony to develop the seeds and roots of beauty and wealth that now lie dormant. The smoke from these stray burnings increased until the bosom of the Lake was veiled, and the country inundated by its volume. These things were of common occurrence, and did not seem to be precursors, as they were, of that devastation which has befallen northern Wisconsin and western Michigan. In actual loss of life we suffered less than the people of those districts; while the protracted nature of their visitation and their remoteness from lines of travel made the individual suffering more keenly felt. Here we had every comfort that a sympathizing world could provide speedily brought to our doors; but there aid came more slowly, as the tidings of their calamity lingered longer on the way to the ears of the world.

WIND, FIRE, AND SMOKE.

It is difficult to apprehend vividly enough the rush of the wind over our prairies, and especially at such a time when the fire drew the air towards itself with accelerated velocity—each developing the force of the other. Accounts inform us with uniformity concerning the density of the smoke-cloud, and the intensity of the fire torrent.

A friend, who was in a sailboat on Little Sturgeon Bay, describes the fire blowing off the shore as terrific, so much so that trees on an island about half a mile from shore were set on fire, and the island burned over. He says that after the fire he could have picked up a yawlboatful of birds in the bay, that had got burned in their flight, and dropped into the water. A passenger on one of the Lake boats running across to Green Bay gives some facts of interest which serve to confirm the dreadful nature of the time we are describing.

The boat was greatly detained on her upward trip on account of high wind and smoke, and the latter was so dense that the boat had to be steered entirely with the compass. The fire on the east side of the Bay extended in an almost unbroken line from the eastern shore of Lake Winnebago to the northern extremity of the Eastern Peninsula, fully 150 miles, burning up in its course fences, barns, houses, and an endless quantity of cedar telegraph poles and tan bark, the latter of which was piled in immense heaps on the docks. So deep and dismal was the darkness caused by the immense volume of smoke, that the sun was totally obscured for a distance of 200 miles. This midnight darkness continued for a week. The boat, of course, was delayed, but she left Escanaba for Green Bay on the fatal Sunday night at twelve o'clock, but only made her way twelve miles out when forced to return on account of the stormy sea beneath and the sea of fire overhead. The air was red with burning fragments, carried all the way from Peshtigo and other places along the shore, a distance of nearly fifty miles. The boat laid in Escanaba harbor until six o'clock A.M. Monday, when she was again started, the storm having but slightly subsided; but the course was pursued, and Menomonee was reached with great difficulty. As they approached Menomonee they passed vessels loaded with furniture, etc., all being ready to leave if the place took fire. It was here the passengers learned of the destruction of Peshtigo.

A correspondent writes, twelve days after the event:—

This letter, to give it a local habitation and a name, is dated where Peshtigo was. In the glory of this Indian summer afternoon I look out on the ghastliest clearing that ever lay before mortal eyes. The sandy streets glisten with a frightful smoothness, and calcined fragments are all that remain of imposing edifices and hundreds of peaceful homes. This ominous clearing is in the centre of a blackened, withered forest of oak, pine, and tamarack, with a swift river—the Peshtigo—gliding silently through the centre, from northeast to southwest. Situated seven miles from the Green Bay, on the Peshtigo River, the town commanded all the lumber trade of the northern Peninsula, and grew rapidly into importance as a frontier mart of Chicago. Built by an enterprising but lately singularly unfortunate Chicago sufferer, William B. Ogden, the town has had but one purpose, to make money for its founder and keep up the lumber interests. But one industry breeds many, and in time a railroad running seven miles to the bay, connected the little city with the great chain of lakes. Great foundries and machine-shops rose on the banks of the river, and a busy mill stood in ceaseless operation in the centre of the town. The banks of the Peshtigo teem with a rich and various growth of timber, and a trade of years stood always in prospective to her busy people. The great Northern Pacific Railroad was to be tapped by a road even now building to the place where Peshtigo was, and every hamlet and town in Northern Wisconsin envied and admired the wonderful little city.

The keen eye of trade and speculation was not deceived; population flocked in amain, and fully 2,000 people had established permanent homes. The site was well chosen for beauty as well as business; the river at this point runs through a slight bluff, which breaks into a low flat before the stream escapes from the borders of the town. The excellent water-power as well as the lumber interest had determined the spot, and a mill was one of the first establishments in operation when the walls of the village began to rise. Below the mill the ground on either bank sloped gently into low, pebbly flats, which joined the water's edge a few rods from the centre of the town. The business and residence streets were wide and well laid out, the houses prettily built and carefully painted, and little ornamental gardens were frequent.

The river cut the town pretty fairly in twain, the works and shops of the Peshtigo Company covering most of the northeastern shore, while trade and business for the main part held themselves on the southwestern bank. The site was, and is to this day, unmistakably a clearing. A solid wall of pine, oak, and tamarack hedge in the desolate waste, even now. As it stood, the pretty bustling village combined the orderly enterprise of New England and the irrepressible vigor of the typical Western "city." Roads cut through the forest communicated with a long line of prospering lumbering hamlets and thriving farms, to the west and south. The surrounding woods were interspersed with innumerable open glades of crisp brown herbage and dried furze, which had for weeks glowed with the autumn fires that infest these regions. Little heed was paid them, for the first rain would inevitably quench the flames. But the rain never came, and finally valiant battle was waged far and near against the slowly increasing fires. In this, as in other towns, the danger was thought well warded off by the general precautions. The fire had raged up to the very outskirts of the town weeks before that fatal Sunday, and the fires were set outward to fight the enemy. Everything inflammable had apparently been taken out of harm's way on that memorable Sunday. One careful citizen traversed the western outskirt, and assured his people that no danger could come from that quarter.

The sharp air of early October had sent the people in from the evening church services more promptly than usual, although numbers delayed to speculate on a great noise and ado which set in ominously from the west. The housewives looked tremblingly at the fires and lights within, and the men took a last look at the possibilities without; for many it was truly a last glimpse. The noise grew in volume, and came nearer and nearer with terrific crackling and detonations. The forest rocked and tossed tumultuously; a dire alarm fell upon the imprisoned village, for the swirling blasts came now from every side. In one awful instant, before expectation could give shape to the horror, a great flame shot up in the western heavens, and in countless fiery tongues struck downward into the village, piercing every object that stood in the town like a red-hot bolt. A deafening roar, mingled with blasts of electric flame, filled the air, and paralyzed every soul in the place. There was no beginning to the work of ruin; the flaming whirlwind swirled in an instant through the town. There is no diversity in general experience; all heard the first inexplicable roar; some aver that the earth shook, while a

THE BURNING O

:HTIGO.

credulous few avow that the heavens opened, and the fire rained down from above.

Moved by a common instinct, for all knew that the woods that encircled the town were impenetrable, every habitation was deserted to the flames, and the grasping multitude flocked to the river. On the west the mad horde saw the bridge in flames in a score of places, and turning sharply to the left, with one accord, plunged into the water. Three hundred people wedged themselves in between the rolling booms, swayed to and fro by the current, where they roasted in the hot breath of flames that hovered above them, and singed the hair on each head momentarily exposed above the water. Here despairing men and women held their children till the cold water came as an ally to the flames, and deprived them of strength.

Meantime the eastern bank was densely crowded by the dying and the dead. Rushing to the river from this direction the swirl ing blasts met the victims full in the face and mowed a swath through the fleeing throng. Inhalation was annihilation. Scores fell before the first blast. A few were able to crawl to the pebbly flats, but so dreadfully disfigured that death must have been pref erable. All could not reach the river; even the groups that fell prone on the grateful damp flats suffered excruciating agony. The fierce blaze, playing in tremendous counter currents above them on the higher ground, was sufficiently strong to set the clothing aflame, and the flying sand, heated as by a furnace, blistered the flesh wherever it fell. All that could break through the stifling simoon had come to the river. In the red glare they could see the sloping bank covered with the bodies of those that fell by the way. Few living on the back streets succeeded in reaching the river, the hot breath of the fire cutting them down as they ran. But here a new danger befell. The cows, terrified by smoke and flame, rushed in a great lowing drove to the river brink. Women and children were trampled by the frightened brutes and many, losing their hold on the friendly logs, were swept under the waters.

This was the situation above the bridge; below, a no less harrowing thing happened. The burning timbers of the mill, built at the edge of the bridge, blew and floated down upon the multitude assembled near the flats, and inflicted the most lamentable sufferings. The men fought this new death bitterly; those who were fortunate enough to have coats flung them over the heads of wives and children, and dipped water with their hats on the improvised shelter. Scores had every shred of hair burned off in the battle, and many lost their lives in protecting others. The firemen had made an effort to save some of the buildings, and the hose was run from the river to some important edifice. The heat instantly stopped the attempt, but not before the hose, swollen with water, had been burned through in a hundred places. Although the onslaught of fire and wind had been instantaneous, and the destruction almost simultaneous, the fierce, stifling currents of heat careered through the air for hours. These currents were more fatal than the flames of the burning village. Ignorant of the extent of the fire, and the frightful combination of wind and flames, many of the company's workmen, some with wives and children, shut themselves up in the great brick building and perished in the raging heats of the next half hour. Others on the remote streets broke for the clearing beyond the woods, but few ever passed the burning barrier. Within the boundaries of the town and accessible to the multitude the river accommodation was rather limited, and when the animals had crowded in the situation was full of despair. The flats were covered with prone figures with packs ablaze and faces pressed rigidly into the cooling moist earth. The flames played about and above all with an incessant, deafening roar.

The tornado was but momentary, but was succeeded by maelstroms of fire, smoke, cinders, and red-hot sand. Wherever a building seemed to resist the fire, the roof would be sent whirling in the air, breaking into clouds of flame as it fell. The shower of sparks, cinders, and hot sand fell in continuous and prodigious force, and did quite as much in killing the people as the first terrific sirocco that succeeded the fire. The wretched throng, neck deep in the water, and the still more helpless beings stretched on the heated sands, were pierced and blistered by those burning particles. They seemed like lancets of red-hot steel, penetrating the thickest covering. The evidence now remains to attest the incredible force of the slenderest pencils of darting flame. Hard ironwood plow-handles still remain, perforated as though by minnié balls, and for the main part unburnt. When the hapless dwellers in the remote streets saw themselves cut off from the river, groups broke in all directions in a wild panic of fright and terror. A few took refuge in a cleared field bordering on the town. Here flat upon the ground, with faces pressed in the sand, the helpless sufferers lay and roasted. But few survived the dreadful agony. The next day revealed a picture exceeding in horror any battlefield,—mothers with children hugged closely lay in rigid groups, the clothes burned off and the poor flesh seared to a crisp. One mother, solicitous only for her babe, embalms her unutterable love in the terrible picture left on these woeful sands. With her bare fingers she had scraped out a pass as the soldiers did before Petersburg, and pressing the little one into this, she put her own body above it as a shield, and when the daylight came, both were dead,—the little baby, face unscarred, but the mother burnt almost to cinders.

The hardy lumbermen are not wont to exaggerate, and the perfect accord of every story and incident confirms every episode of this tragedy. Faithful to the helpless, a stout woodman carried out on his shoulders one deadly sick of fever. He burrowed for the helpless body a sepulchre, and then began the struggle for his own life. He had lingered too long, and his scarred body was found near the refuge of the man his heroism had preserved.

The tornado played through the desolated streets, and swept the river and the low land adjoining. The timber of the mill floating down among the people, made additional labor and danger, and daylight broke terribly on the saturated survivors before they dared drag their cramped limbs from the icy waters. The mingled crowd of men, women, and children, cows, and swine had held this watery refuge since 10 o'clock of the night before. Of the hundreds of human beings that entered the waters not all escaped; the frightened cows trampled many under the waters; the blistering heat blinded many who groped hopelessly about in the current, and finally sank. To this day none can tell how great was the slaughter in the waters. After the burning heat of the night, a numbing chill followed, and the water-soaked group crawled over dead bodies and hot sands to the only blazing building in all the waste about them. Groups of dead bodies were found within a stone's throw of the water; families rushing down for a breathing place had been blown upon by the rushing blast and struck lifeless. The ghastly throng huddled shrieking and bewailing about the flaring embers, and the terrible roll of the missing was soon called from end to end of the ashen waste. No vestige of human habitation remained, and the steaming, freezing, wretched group, crazed by their unutterable terror and despair, plead with each other to restore the lost ones. The hot blasts of the night had blinded them, and they could but vaguely recognize one another in the murky light of the new day.

Long after the flames had died out, when there was no more to feed on, the hot sands rendered moving about an exquisite torture, and long into the dismal mid-day the survivors were confined to the narrow circuit near the river. As the day wore on, help came in slowly from the northward. Several railroad gangs had escaped annihilation, and one gang, led by an ex-prize-fighter named Mulligan, came with promptness and efficiency to the rescue, through miles of burning prairie and blockaded roads. On Sunday night something over two thousand people were assembled within the confines of this industrious, prosperous city ; the dreadful morning light came upon a haggard, maniacal multitude of less than seven hundred. When the work of rescue began it was found that a great number had escaped by the bed of the river and the northern road to the port, and as the day advanced, half-naked stragglers, unkempt and blackened, began to stream into the sparse settlement. As the molten sands cooled off, the woful work of recognition began. Peering into blackened faces, mothers, fathers, brothers tremblingly sought out missing ones.

Some, in the immeasurable anguish of the moment, had dashed themselves against the sands and let out the life with their own hands that the licking flames coveted. Men too distant from the river to hope for rescue or safety, had cut the throats of their choking children, and were found in groups sometimes unscarred by the flames.

In the streets, full twenty corpses were found with no apparent injury or abrasion. Fatuous tradesmen, in the sudden rush of flame, had thrown their valuables into wells for security. Every well in the city was turned into a flaming pit, and the very waters half evaporated by the heat. Survivors attest that women and children, cut off from the rivers, were put into wells and covered with bedding. I have looked into every well in the ash-covered clearing, and there is no possibility that a living thing could have endured the flames that boiled and seethed in them.

For hours the unreasoning search was continued by the famish ed-dying remnants, but to little avail ; the dead, when recognizable, lay where they had fallen in the streets; where the houses stood, the ground was whipped clean as a carpet, and all hope of identifying human ashes was idle. The next night the long-prayed-foi rain came, gratefully to the living, and ki dly to the fleeting ashes of the dead. The great dread that hovered over the bay cities and towns was allayed, and the threatened danger nearly gone. Be·

fore dark, help came to the perishing sufferers from the neighboring villages. The wounded were taken by boat to Green Bay, whence some were forwarded to Milwaukee.

From 9 o'clock Sunday night until dusk of Monday may be taken as the time of the main action of this terrible drama. By Tuesday the sweeping miles of fire had been quenched by Monday night's rain. A slight drizzle still further aided the work of rescue. The ravages of the one night's tornado left unmistakable traces on every hand. Through the solid growth of timber a clean swath of blackened stumps and roots marked the course of the fiery tempests. The roads were cumbered with roasted cattle, and frequently with the carcases of bears and deer, while the ditches and cleared fields were strewn with smaller game and wild birds. Nearing the vicinity sadder relics were found, for those who penetrated eastward through the wall of flame met equally fierce flames in the clearest places. Remote dwellers on the highroads, warned of the great danger, with their families safely packed on their great farm-wagons, made northward through the highways for security; but the flames engulfed them in the heart of the woods, and the fragments of stout vehicles, burned to the irons, now strew the road hither from Marinette, the last town on the Northern Wisconsin border. The highroad enters Peshtigo from the north, through a break in the encircling belt of woods, where the pretty Episcopal Church stood—the last to burn in the fatal place. Even before this was reached, a putrid hecatomb of dead cattle cumbered the wooded street. Among the pines, scores lay, not burnt, but smothered to death. Through this underbrush thirty bodies of men and children were picked up, more or less injured by fire. In a great many instances the human remains were distinguished from animals by the teeth alone. One horror-stricken relative recognized the relics of his nephew by a pen-knife imbedded in an oblong mound of ashes. What does it avail to narrate circumstantially the inexpressible horrors of these succeeding days. What good to tell of the dead faces staring upward through the calm waters; or the piteous circumstances of a hundred heart-wrenching tragedies during and following that treacherous Sunday blast? No moral underlies the terrible story; all that frightened human nature was capable of came into play that direful night; the slaughter resulted from no sin of omission or commission on the part of man. No unseemly panic aided natural causes in achieving, comparatively, the completest devastation in human annals. On the contrary, superhuman daring and energy were put into active operation to mitigate preternatural horrors. The immensity of visible destruction at Chicago surpasses the completeness of this devastation, but Chicago, with all its woes, has not two-thirds of its citizens to deplore as dead.

With one of the men who passed through that night of destruction, I wandered over the pretty rising plain where Peshtigo spread its thriving stores and handsome houses. Save where the houses were built with cellars, which was very rare, there is no trace of a former habitation. Here and there are metallic remnants of sewing-machines and cracked stoves. The hardware and drug stores leave almost the only reminders of things that were,—a blackened mortar stands idly in a wild confusion of melted glass and lead, with the pestle ready for a new decoction.

Two or three men with troubled faces were moving about putting up a shed for the Relief Committee. They answered civilly and sadly that they had been in the fire, but saved themselves and nearest kin. They should have starved to death if the outside world had not stepped in, and now hoped to be shortly on their feet again. They despaired of the bright cheery little town ever being again as it was, but complacently "reckoned," if the scared ones didn't drive newcomers away by their silly stories, a new people would make a new Peshtigo. If you ever walked over the ground where a camp had been burned, and there are few that served during the war that have not, you found there as much semblance of a substantial city as now marks the spot where Peshtigo's 2,000 people carried on the business of life a few days ago. On the bank of the river fish killed by the lusting flame are still to be seen, which the day after the fire were soft and white and unwounded. Crossing the frail remnants of the bridge on timbers charred and fragile, my neighbor said, "It was as like the Judgment Day as I can imagine. Friend Hansen, with his wife and four children, believed firmly that it was, and while the fire rained down he began to walk composedly up and down his parlor with his family about him, and I have never seen him since."

The material loss is estimated at $3,000,000, the greater part of which falls on William B. Ogden, who suffered simultaneously greater losses in Chicago. But undaunted by his accumulating misfortunes, that energetic man instantly sent an agent on to rebuild the mills and shops, and gather a new people in the place if possible. There are 400 dead authentically accounted for there, besides half as many missing who cannot be accounted for, and probably never will be. Many of the mill hands and company's employés were utter strangers in the place, and the majority of them, something like 100, trusting to the stout walls of the company's building, perished *en masse.*

It is a significant fact, as showing the character of one of Chicago's noblest and most valued citizens, that, in the midst of his immense misfortunes, and his extended and complicated business responsibilities, Hon. William B. Ogden could lay aside all these for the sake of writing the following letter, in the hope of assisting in the recovery of the lost child of one of his Peshtigo people:—

To the Editor of The Chicago Tribune:

DEAR SIR:—Frank Jacobs, a Hungarian of Kossuth's party, and now about eighteen years in Peshtigo, escaped, with his wife, the death that overtook so many on the night of the dreadful fire here; but his wife's sister, Miss Charlotte Seymour, who left his house in advance of him, taking his only child, was drowned in the Peshtigo River, and the child, a little boy about two and a half years old, has been missing since.

The boy's name was Frank, and they called him "Frankie."

It was reported that some one who saw Miss Seymour go down between the logs she was clinging to, failed to save her, but did seize and save the child; and that it with a great many others was taken to Menomonee on the morning after the fire, and there took the steamer for Green Bay.

Mrs. John De Marsh, of Peshtigo, tells Mr. Jacobs that she saw the child on the steamer, on its way to Green Bay, in the care of some kind gentleman passenger on his way to Chicago, who asked when she told him that she knew the child, and whose it was, if its parents were still living. She did not then know whether they were or not, and very likely left him with the impression, at that time of great distress and confusion, that they were probably not living, as she had not seen them since the fire.

The gentleman caring for the child said to her that he should take it to Green Bay, and to Chicago, if he did not find its parents, and should take care of it until he did find them. Mrs. De Marsh does not know who this gentleman was, but if she is not mistaken it would seem that the child was saved and is still alive.

The object of this letter is to ask, if admissible, that *The Tribune* publish these facts, in the hope that the child may be found thereby, and his very distressed parents relieved from their painful state of uncertainty and suspense about their boy and only child.

The child was perhaps two and a half years old; has light-brown hair and blue eyes. The only word he spoke plainly, his parents say, was "Ike," the name he gave to a favorite uncle; and if asked "Where Ike is?" he will show, his parents think, an interest, and that he understands the meaning.

Frankie wore on the night of the fire a black and white checked flannel shirt, a red flannel dress, and red, brown, and white checked apron, with a band of purple and white check.

In the hope that this child may have escaped death on that fearful night, and that this statement of the circumstances of the case, as related to me, may lead to his discovery and return to his unhappy parents, I remain, dear sir, with great respect, very truly yours,

W. B. OGDEN.

A citizen of Green Bay who passed through the fire at Peshtigo, who saved himself and a woman and children he happened to meet, by getting on a low spot of ground or in a ditch, and covering over with wet blankets, tells this story:—They had got well covered up in this burrow, when a half-frantic woman rushed along with a great bundle in her arms. She had been well dressed, but her clothes were half burned off. She stopped and deposited her bundle, which consisted of a child and a lot of clothing, and then shrieked, "Great God, where is my baby!"

At this the narrator sprang up, and saw, a few rods off, a baby in its night-clothes, lying on the road and kicking up its heels in great glee, while a billow of flame rolled over it, striking the ground beyond and leaving the baby in the centre of a great arch of fire.

The baby had slid out of the bundle, unperceived by the mother in her haste. He immediately sprang for the child, and with difficulty rescued it. It is no wonder that the mother fainted when she secured the child.

Wal. Heath was one of the proprietors of the Peshtigo House. When the fire occurred, his family, with the girls employed in the house, escaped from the hotel by a team, and were saved on the low land below Ellis's house. Heath got into the river on the west side of the bridge and clung to the centre pier of the bridge. The wind blew the fire from the hotel to where he was. The hotel was near the south end of the bridge and on the west side of the street. At the north end of the bridge and east end of the street was the Peshtigo Company's water mill, and the flames from that also blew directly to his position. Thus it seems that the wind on the two sides of the river blew in exactly opposite directions. Heath was saved from the fact that being on the west side of the pier, the flames from the water mill divided at the pier and passed him on both sides. The bridge being on fire, he dared not swim through with the current, but when the fire on the bridge had got uncomfortably close, he took off his coat, pulled off his boots, and swam up stream to a place of safety. He had a very narrow escape from death, and has not yet recovered from breathing the hot air and smoke.

He tells us that the most vivid imagination cannot picture the scene of the calamity as bad as it actually was. In his opinion as many as 1,000 people lost their lives on the Peshtigo; that 752 bodies have been buried, and that many were entirely burned up. The names of half the dead will never be known. They are buried all over Peshtigo, and the boards that mark their graves are marked "2 unknown," "3 unknown," etc.

Much has been said of the intense heat of the fires which destroyed Peshtigo, Menekaunee, Williamsonville, etc., but all that has been said cannot give the stranger even a faint conception of the reality. The heat has been compared to that engendered by a flame concentrated on an object by a blow-pipe, but even that would not account for some of the phenomena. For instance, we have in our possession a copper cent, taken from the pocket of a dead man in the Peshtigo Sugar Bush, which will illustrate our point. This cent has been partially fused, but still retains its round form, and the inscription upon it is legible. Others in the same pocket were partially melted off, and yet *the clothing and the body of the man was not even singed.* We do not know how to account for this, unless, as is asserted by some, the tornado and fire were accompanied by electrical phenomena.

The house, barn, and fences of Mr. Hill, of the upper Sugar Bush, were burned, and Mr. Hill and his family all lost. By the side of the family was a narrow alley, just wide enough to drive through. In this alley stood a wagon, and while the barn and fence were entirely destroyed, the wagon box was not even singed.

Alf. Phillip's house, in the upper Sugar Bush, was destroyed, but the family escaped. They state that two opposite currents of air apparently struck the house, which was 16 by 24 feet, and carried it bodily into the air, they think about 100 feet. In the air it burst into flames, and in a few minutes was entirely destroyed. The house was not on fire when it left the ground.

We do not believe that any other explanation of the great calamity can be made than that it was caused by fire, wind, and electricity.

Another correspondent says:—

The story of the Wisconsin fires has never yet been told, and from the suddenness of the calamity and the intensity of the clouds of fire, its fullest extent can never be known. From Mr. J. Harris, chief messenger of the House of Representatives, who was an eye-witness of the fury of the flames on the fatal Sunday night of the 8th of October, and who has just returned from Green Bay, Wis., we are enabled to give to the public a statement of the number of the families who were totally swept away in that terrible holocaust of fire.

In the towns of Brussels, Union, Nasewaupee, and Gardner, in Door county, on the east shore of Green Bay, six families, numbering forty persons, were burnt to death; also, thirteen other families, of from two to four in each family. In these four towns in Door county, the total is 117 persons burned to death, besides a large number severely and slightly burned, and there are 167 families rendered homeless, with loss of all they possessed.

In the town of Peshtigo, in Oconto county, on the west shore of Green Bay, twenty-five families, numbering 157 persons, were burnt to death; also forty-six other families, of from two to four in each family. At Peshtigo, including the upper, middle and lower Sugar Bush farming settlements, the number of bodies identified and buried was, at the last account, 530, while the number of bodies reduced to ashes by the intense heat can never be known, and the destitute persons number over 1,200 souls.

The foregoing statements apply only to Door and Oconto counties, and do not include the destruction of life and property in Kewaunee and Brown counties, from which only partial lists of the losses have as yet been gathered up; but the loss of life there is said to exceed one hundred, and there must be many hundreds of destitute persons to provide for.

Many persons express surprise that so many human beings perished in the flames, and ask, "Why could they not escape?" Mr Harris, who resides in that part of Wisconsin swept by the fire, explains this by showing on the map that the Sugar Bush settlements, where the greatest loss of life occurred, were located in the dense forests, entirely surrounded by timber of large growth, and are situate from three to five miles away from the Peshtigo river, which was the only chance for escape. The fiery tornado came upon these people with such fury, and so suddenly, that there was no escape, and this fact accounts for so many families being swept away. The same holds good for the frightful loss of life at Williamson's Mill, in the town of Gardner, in Door county, near Sturgeon bay, where the sawmill and other buildings, with about seventy men, women and children, were located in the woods, surrounded by the dense hemlock forest, through which there was but one road, and that choked up with *débris* from the shingle-mill of the most combustible kind, leaving no chance for escape when the tornado struck them, as is shown by the horrible fact that, of the seventy persons hemmed in there by walls of fire, sixty-one, including the Williamson family, were burned to death. The village of Peshtigo being built on both sides of the Peshtigo river, the latter afforded means of escape to several hundred of the inhabitants, of whom, but for that opportunity, numbers more would have perished. So, too, at the village of Menekaune, the whole place, including the splendid new saw-mill of Jesse Spalding & Co., just built at a cost of $80,000, was swept clean in less than an hour; but being located on the shore of Green Bay, the inhabitants all escaped by taking to steam-tugs, sailboats and other craft, and no lives were lost. Our informant, Mr. Harris, was at Menekaune on the night of the fire. He visited Peshtigo after the fire, and subsequently the other burned towns on both sides of Green Bay, and says that no pen can describe what he saw on that fatal night, nor can tongue tell of the sufferings of the hundreds who perished in the flames, or of the thousands who escaped. In going through the burnt towns back in the woods, and witnessing the wide-spread desolation, the wonder is that so many escaped alive, and many of those who did escape were saved by throwing dirt on each other to keep from burning.

Another reason for the great destruction of life and property in Wisconsin, is found in the fact that no rain had fallen for three months before the 8th of October; the clearings, and the woods, and even the swamps, were as dry as tinder, and the people had been fighting the fires in the woods for several weeks before; all that part of Wisconsin was full of fire and smoke, and when the tornado came upon them it swept all before it, travelling faster than the people, who were struck down while running for a place of safety. The intensity of the fire may be gathered from the statement made by Mr. Harris, that when he was at Peshtigo, after the fire, his attention was called to a mass of ashes at the spot where was the kitchen of the large boarding-house. It appears that before the flames reached that house numbers of people fled there for safety. It is said that the kitchen was full of people, and that an immense sheet of flame struck the house so suddenly that none are known to have escaped. All that remains is a pile of human ashes, from which can be picked out pieces of human bones, the largest not two inches long, and these split and broken. The village of Peshtigo, with its saw-mills, factories, hotels, churches, school-houses, stores, and fine residences, were all in flames in half an hour, and the whole is swept clean as a prairie.

As soon as daylight came on the Monday morning after the fire everything was done that could be done for the sufferers by the few surgeons living there, but to add to the anguish of the hour the telegraph poles were burnt down, and they were sixty miles from Green Bay City, the nearest telegraph station in operation. The Lake Superior steamers not running a boat on Monday, it was Tuesday forenoon before despatches could be sent away for help. Mr. Harris sent despatches by that boat to Governor Fairchild and to the Mayors of the cities of Green Bay, Oshkosh, Fond du Lac, and Milwaukee, calling for medical aid and supplies for the survivors, but the despatches did not reach those gentlemen until Tuesday night, and it was Wednesday noon before the surgeons and nurses and supplies arrived, and it was near the end of the week before the outside world knew the extent of the calamity and the sufferings of the people of Wisconsin.

The great calamity at Chicago was known wherever the wires penetrated by the Monday night, while the fires were yet burning; and towns, and cities, and States vied with each other in forwarding instant relief of all kinds to the sufferers, while to Wisconsin's fearful and greater sacrifice of human life was superadded the loss of the telegraph: the loss of which prevented a large amount of relief being forwarded to Wisconsin, so much needed by her suffering people. Large amounts of money were instantly raised for the relief of Chicago by benevolent and wealthy individuals, by corporations, and by cities and legislatures, a portion of which would doubtless have been given to Wisconsin had her calamity in all its fearfulness been known earlier throughout the country. Chicago will need all that is being poured into her lap; but while this generosity is flowing in upon her, do not let the sufferings of Wisconsin be forgotten.

Governor Fairchild has visited the burnt towns on Green Bay, and gathered up the facts of the condition of the sufferers, and says that all are now supplied with food and clothing, but that from 3,000 to 4,000 people will have to be housed and fed for the next six to nine months, until they can get back on their farms and raise the next year's crop. While the supply of food and clothing forwarded to Wisconsin has been generous and ample, the amount of cash funds in the hands of the Relief Committees will be wholly inadequate to meet the demands made upon them. Further contributions of money will be needed to feed these people, to rebuild their houses and furnish them with farming tools, furniture, seed, and feed for the cattle saved from the flames; and looking to what has already been done for these suffering people by the outburst of American generosity, we have full faith that it will be as generously continued until the afflicted are able to help themselves.

What a period of terror and destruction for the North-west, unparalleled in our history, and quite unexampled in the annals of time, if we consider the brevity of its duration and the immense losses of property and life!

While our blinded eyes witnessed the destruction of our homes and business in the Garden City, the same heart-breaking scenes were transpiring in other places on either side of Lake Michigan, in Indiana, and Ontario. There was a carnival of death. A Chicago man, who lost heavily, had a small farm in Michigan, and there were his wife and son. The forest igniting, fire drove

A WISCONSIN HOUSE ENVELOPED IN FLAMES.

through his beautiful timber land, roared around his dwelling, almost compelling the desertion of all to the flames. It was saved only by heroic exertions, and the farm was a waste. It seemed as if sorrows were never to cease, and yet he held up his head like a Christian hero, trusting in God the good provider. More fortunate he than thousands whose all was stripped from them as the autumn winds disrobe the trees. Like these, thanks to God, the miserable victims will put forth life and vigor, and yet stretch out their thriving beauty to Heaven, and bask in the summer of His mercy who heals and restores whom He has smitten. The accounts of whole regions smoking like a volcano are not exaggerated, as no pen can fitly describe the occurrences of that memorable week from October 7th to the 14th.

SUMMARY OF WISCONSIN LOSSES.

CHAPTER XL.

"After making a deduction for exaggeration, I supposed that 500 would cover the number of the dead on the west side of the Bay. I now learn from reliable sources that the actual number of interments up to Monday night counted up to 504. Add another hundred for remains of ashes and charred bones at Peshtego, and I think we have not far from the true number on the west side. Add one hundred and fifty for the east side—making 750 in all—and the death roll is nearly complete.

"It is impossible to figure the aggregate losses of pine timber and farm property with any degree of closeness. It is to the interest of mill-men to underrate the amount of fallen pine that must be secured this winter to save it. A medium estimate of damage to pine lands in the Green Bay region is $400,000. The damage on the Wolf is figured at $300,000. The loss of the fifteen saw-mills burned is put at $225,000. The loss of cord-wood, ties, hemlock bark, etc., is set at $200,000. The loss of fences, buildings, wagons, cattle, crops, among the six hundred farmers cannot be less than $600,000—making a total aggregate of more than $3,000,000, aside from those at Peshtego.

"The country through from Brown county north to Big Sturgeon Bay, for four hundred square miles, is utterly devastated. At least four hundred farms in this tornado section alone are left desolate—stripped of every improvement. Fences, barns, dwellings, implements, furniture, wagons, harness, and crops, all went up in a whirlwind of fire. It will take thirty years in that cold, hard soil, for their timber to grow again. In the aggregate, their losses must foot up to one thousand dollars a family. Farmers here have saved half of their teams, that were let loose in the woods, and a third of their stock. But they have no hay, straw, grain, or feed of any sort—not even the poor chance of browse in the woods. Nearly all, with large families, have lost their last cow and pig. In a ride of six miles, on nearly a straight line, I saw but three hens and a fanning-mill—the only farm implement left in the town. In the Belgian settlement, on Red River, sixty-two families were burned out in a row! Not a house, not a shed, not a coop—not one fence rail left upon another. The families had fled, almost naked and breathless, to the few cabins on the outskirts that were saved.

"There are three hundred, or more, wounded sufferers remaining in hotels, boarding-houses, and hospitals about the Bay. Fifty of the Peshtego sufferers were at the Dunlap House, Marinette. Half of them were able to be about. Burned ears, faces, hands, and feet were common to nearly all. Many in rooms could hardly stir in bed. There were women with great burns on the sides and limbs, with faces like kettles, and hands like claws, burned to the bones.

"Men could fight better, and dare more than women. Most of them perished by suffocation. Little children are sadly maimed in their feet and faces. I saw one with a heel gone, and another with an eye. Nearly all will recover without loss of sight or limb. I could fill a book with stories of the hospital. Most of them suffer more from hurts of mind than body. I have a sad memory of a poor widow who lost her crippled boy who went on crutches, and a sprightly little girl who fell between the burning logs. They were all of her family. 'The screams of both,' she said, 'seemed forever sounding in her ears.' There is a future, and, no doubt, compensations for all these suffering ones.

"Most of these cabins that are left are crowded with two and three families each. I saw one with four men, five women, and sixteen children—two of them suckers. They had just received an outfit of clothing—warm stockings, knit hoods, thin shawls, thin gaiters, and light-colored dresses for the women and girls; odd-fashioned hats, bursted boots, thin jackets, and summer coats and pants for the men and boys. There were some occasions of laughter, but none of ridicule; all were glad and surprised at getting what they did. I saw no immediate want of provisions. Flour, pork, and hard bread are distributed to all, packages of tea and coffee to most. There are nearly potatoes enough in the country, if distributed. Their stock that is left has been driven off to meadows and fields not burned over. One large-hearted old farmer was keeping eighty odd cattle belonging to his unfortunate neighbors. Without stopping to consider the ways of Providence, or the uses of philosophy, these simple-minded people seem to have understood the art of helping one another."

Captain W. R. Bourne, who has recently returned from a visit to the Wolf River pineries, tells us that, in his opinion, the pine damaged by the recent fires on the Wolf River and its tributaries (the Shioc, Embarrass, and Red Rivers) will amount to 50,000,000 feet. If this pine is all cut the coming winter and got into the streams, the damage to it will be about $1.00 per thousand feet; but if not cut the loss will be almost total, as it would be bored by worms another season and destroyed for every purpose but fencing. There is very little pine injured above Keshena. Thus the lightest estimate of damage to the Wolf River pine is $50,000.

The pineries along the bay shore have suffered to a still greater extent, but the damage is easier repaired, provided the lumbermen can raise the means to put into the river two or three years' stock of logs, for a much greater amount of logs can be put into the Menomonee, Oconto, and Peshtego Rivers, which are not navigable, than into the Wolf, where they would interfere with navigation. One lumbering firm down the bay estimate their loss at $50,000.

Lucile Mechand tells the story of the family's adventures in thrilling words :—

On the morning of the 11th of October, just as we were sitting down to take breakfast, Mr. Richardson, a neighbor of ours, came running into the house and told Mr. Mechand that he must come out immediately and see what could be done. During the night the wind had risen, but not so greatly as to amount to anything like a gale, but rather did it resemble the ordinary fall wind. Mr. Mechand did not seem at all uneasy, and leisurely swallowed his breakfast before following Mr. Richardson, who had disappeared as soon as he had stuck his head into the room and called my husband. Mr. Mechand went into the woods and stayed till about noon, when he came running back and said that he climbed up to the top of Brown's Hill, where the wind was blowing a gale, and from there had seen the fire, which was coming toward us at a tremendous pace. Indeed, I had feared as much, and had been exceedingly uneasy all the morning, for the smoke which for days had been in the valley where we lived had become more and more dense, and occasionally hot puffs of wind had blown down over the hills, driving the smoke in a dense cloud before it. I asked my husband if he thought there was any danger to be feared ; he shook his head and answered, "No ;" yet I knew by his face that he was far from being devoid of fear. He ate his dinner hastily, and then ran out again, and was met at the door by a neighbor, who said that the fire was advancing with frightful speed. Indeed, the air had now become sultry as it never had been before, except on some hot days in summer immediately before the coming of a thunder-storm. The air was stifling, and the smoke got into one's lungs and nostrils in such a way as to render it exceedingly unpleasant. Mother sat in a corner holding little Louis in her lap, and I noticed that she seemed restless, and that her eyes shone with a light such as I have sometimes seen in the eyes of a wild beast, and had only seen in hers in the old days when she was about to have an outburst of fury.

I was frightened and fidgety, and didn't do anything in the right way. I went and took the boy away from mother, who relinquished him readily; and then, as I had afterwards terrible reason to remember, although I hardly noticed it at the time, she went to the cupboard and secreted something in the bosom of her dress. Mr. Mechand stood at the door speaking hurriedly with the man whom he had met, when a burning branch of pine fell at his feet. Instantly the air darkened, a violent puff of wind rushed upon us, and smoke poured in volumes about the house. Then, following the gust, a bright sheet, or rather wall, of fire seemed to be pushed down almost upon us, and instantly everything was in flames. Mr. Mechand cried out to me to bring Louis with me, and seized mother by the hand, and we all four ran in terror out into the woods ahead of us. I ran on blinded and choked by the smoke, and carrying Louis in my arms. He was pale with terror, and did not utter a single cry, but clung to my neck as I hurried on, stumbling and tripping almost at every step. So suddenly had been the rush of the fire that we had no chance of saving anything but our lives, even if we had cared to do so. I kept calling to my husband to keep in sight, but, poor fellow, there was no need of doing so, for I could see that mother was a great worry to him, and that he had almost to drag her along. She kept looking from side to side, and trying to break away from him; even then I thought how terrible it would be if she should become furious again. What on earth could we do with her.

We must have gone on this way for at least three miles, and I was almost exhausted, for Louis was a boy six years old and large for his age, and I had been carrying him all the way. The trees were compact, and in some places the undergrowth was close and stiff as wire. Mother kept getting worse, and Mr. Mechand, who was a short distance ahead of Louis and me, had the greatest difficulty to make her obey him. Presently he stopped, and evidently was waiting for me to come up. So I put Louis down and told him to keep alongside of me, at the same time taking him firmly by the hand. The fire had come much slower than we, and I believe we must have been at least two miles ahead of it, although there was no telling, for I could see nothing behind or far before me but smoke curling like a mist in and out of the trees. Behind us, indeed, it was heavier and looked a sullen, dirty white.

We could not have been six feet from my husband when my mother broke away from him, and with a loud cry darted off into the woods, and then I knew that what I had dreaded had indeed come to pass, and that excitement and danger had brought back an old sickness upon her. She was a maniac. Mr. Mechand darted after her, and in the terror of the moment I forgot all else, and I followed him, leaving poor little Louis behind. I must have been crazy to do so, but on I rushed, and soon saw that mother was cunning enough to attempt to escape by doubling on her tracks, for I saw her dress dart past the bushes at my side as she ran diagonally away from me. I sprang after her, and after running for about five minutes found to my horror that I had not only lost her, but Louis and his father. Madly I tried to retrace my steps, but there was nothing to guide me—no path, no blazes on the trees. The wind shook the trees and almost bent them double; the sultry air filled with smoke, and all the horrors of my terrible condition made me frantic. I rushed about helplessly, crying and screaming, "Louis!" "Louis!" "Father!" But that last word made me calm for an instant, and I felt that I was not alone—not utterly lost in the burning woods, for the spirit of my dead father was near, and there were guardian angels. I knelt on the ground, took my crucifix from my neck, and prayed. In kneeling down I found to my great joy that my dress was wet. I had knelt near a spring. I bathed my face and hands, and soaked my hair and the upper part of my dress. But then my boy—my little Louis! I sprang to my feet, and calling on the Virgin to direct me, dashed on in the direction ot the fire. I had not gone more than a quarter of a mile when I found my darling standing with head erect, and flashing eyes filled with angry tears, trying to beat away some wolves, which, hungry though they were, seemed bent only on flight. I cried, "Louis, Louis, c'est moi, ta mère!" and clasped him to my heart. It was my boy, and he was saved. He had not seen his father, though once he had heard a man's voice calling, but the voice seemed to have come from an immense distance. "Oh, Louis," said I, "we are lost unless we find him. We must run for our lives." The boy began to cry, and then I was ashamed of what I had said, and tried to cheer him up. The fire must have been very near us then, for I could not only feel its heated breath, but above my head, among the tree-tops, sparks and firebrands were whirling in the air. I took Louis in my arms, determined that never again should he be separated from me, and pressed onward with some vague idea that I should soon reach Wolf River.

Night was coming on, and since noon we had had nothing to eat. I did not feel hungry, but was tormented with thoughts of what might happen if we should not soon reach a place of safety; for I feared that Louis would give out, and that was one of the reasons which made me carry him. My arms ached, and my limbs were scratched, bruised, and bleeding. Still I made good headway, and soon came to a natural clearing, on the thither side of which we sat down tc rest. By this time night had come out, and what a night! No moon, no stars; but the cloudy heavens lighted up afar with the horrible fires of the burning woods. The clearing in which we sat was the dried-up bed of a stream which, for some unaccountable reason, had no thickly-wooded shores, and we were at least two hundred feet from the edge of the forest in flames. All this time Louis, manly little fellow that he was, had not even asked for food; nor had he cried since I myself foolishly frightened him.

We sat there a long time while I was trying to think where we were, but I could come to no conclusion. I had heard my husband speak of a stream which had run dry, but that was in a northeasterly direction from our house; and notwithstanding the fact that I was lost, yet I had a general notion I was approaching Wolf River. The stars could give me no information, for I could not see them. What to do I scarcely knew; but when the heat of the fire became such that I could not doubt that it was near I determined to press on away from it, and taking Louis's hand I set out. On ordinary nights it should now have been dark; but there was a nameless glare, yet not a glare, a horrible *reflet* which came down from the sky and mingled with the smoke. Hardly had I risen from the ground when in the direction of the woods on the other side of the clearing, I heard a clashing noise, a mingled gnashing and hoarse barking, which I instantly recognized as that of wolves, and I scarcely had time to snatch up Louis and run behind a magnificent pine tree, whose trunk was at least six feet in diameter, before I heard them scrambling up the side of the hill and felt them rush by me. I looked out and could see their eyes coming toward me like the wind. They did not stop for an instant; and when they passed there came in their track a herd of deer, uttering cries that seemed almost human in their intense agony. They ran blindly, for something more terrible than wolves was behind them; they struck the tree and were hurled back by the shock, some of them falling back upon those below. The stampede seemed to last for full ten minutes; and when it was over, and I, trembling with fear, dared once more to emerge from my refuge and look across the clearing, I saw the woods at its edge already burning—saw it lurid through the smoke, and felt its terrible heat upon my face. I turned and fled, in the wake of the deer and the wolves. My shoes were stripped from my feet, and my ankles were torn and bloody. Fallen trees lay in my way, but I clambered over and crawled under them in my desperate fight. I was agonized with terror and despair, and finally sank to the ground with my boy in my arms.

I must have fainted, for I knew nothing of what passed till I was rudely shaken by the shoulder, and heard a wild, gibbering laugh. I opened my eyes, and above me stood my mother with a drawn knife in her hand. The woods seemed all ablaze, although the air was not so intolerably hot as it had been. The forest beyond the clearing must have been burning at its edge, and the strong wind carrying the smoke upward and above our heads. My mother looked down upon me with eyes blazing with that hated light of insanity.

"Ho, ho!" said she, "fine time of night for a mother and child to be running through the woods! Fine night this! Night—it is day! Look at the red light—'tis the light of dawn! Le jour! le jour du jugement est arrivé! And the rocks are burning! Call upon them to fall upon you! The clouds of thunder and the day of doom! The Lord is coming, and the wheels of his chariot burn with his mighty driving. Let us go up to meet him in mid-air. Let us ride on the smoke and thunder, and sweep the stars from the heavens. Come, you shall go with me!" And she seized Louis, who had thrown himself upon me, and was clinging in terror to my breast.

I sprang to my feet and cried, "Mother, mother, what would you do—would you kill me and Louis?"

"Kill you! yes! why wait? The Lord calls and the devil drives. He has let loose his imps against the world. The trees fall crashing in the forest; for all hell's demons pull them down with hooks of fire. I have seen them as I followed you. I have seen you all the way. I rode over on a wolf; 'twas a loup-garou, an old friend of mine, brought me over safely, and kept me from the deer. I will kill you; would you burn to death? You shall go up—up higher than the moon, and beyond the fire. Come, let us go!" and again she seized Louis while the knife gleamed in the air.

I sprang at her, and with all the strength of ten mothers in my arms, I struggled with her. Torn, worn, and bleeding, as I was, the thought of my child and my husband gave me the strength of a giant. I overpowered the mad woman, and forgetting that she was my mother—that she was anything but the would-be murderess of my boy—I seized her by the throat when she was down rolling on the ground, and I would have strangled her. Her insanity had almost made me mad. I felt then what a murderous maniac feels.

But then I thought my mother was lying almost dead and powerless, and the fire would soon advance and perhaps overwhelm us all. My hand was stayed, and when my mother rose to her feet, all of her wildness was gone, and in its place had returned that calmness—almost imbecility—which had characterized her for the last few years. She was ready and willing to do everything that I told her, but I kept that knife fast in my hand.

The wind had fallen, and a slight rain was dropping among the leaves overhead, as we went on for an hour or two longer, and then, overpowered with exhaustion, and no longer greatly dreading the fire, we lay down in a hollow and fell asleep. When we awoke, it was morning. I was sick and completely exhausted, and hardly knew that there were men around us. Yet there were, and good, kind men, too, who gave us food and drove us to a place of shelter, whence, as soon as we were able, we went to Green Bay, where I soon recovered from the sickness and terror of that dreadful night. My mother continues in that same state of imbecility which the doctor says will soon become complete dementia. Louis was not long in recovering, but as yet I have heard nothing from my husband.

LUCILE MECHAND.

The Boston Relief Committee speak as follows of the work of charity and of ruin:—

In this State, as in Michigan, the work of relief is done by two distinct committees, appointed by Governor Fairchild, one at Milwaukee and the other at Green Bay.

We visited the committee at Green Bay, who were in full communication with the committee at Milwaukee. The Green Bay committee consists of six gentlemen, with W. R. Bourne as president.

In general, it may be said that the region burned embraces the lower half of the peninsula between Green Bay and Lake Michigan on the one side and the region on the west side of Green Bay, which extends northward from the Oconto River across the Menomonee River into the upper peninsula of Michigan. The Lake Michigan coast of the peninsula, as far south as Manitowoc County, and the upper peninsula of Michigan, are under the charge of the relief committee at Milwaukee.

Abundant supplies have poured into the Milwaukee committee, so that nothing further is needed by them, as we were informed by the Green Bay committee, who have received a portion of their surplus.

The remainder of the burnt territory, which is situated on both sides of Green Bay, is under the care of the relief committee at the city of Green Bay. The peninsula was inhabited principally by Belgians engaged in farming, which was their only resource, although near the coast were some small towns engaged in the lumber business.

The fires had been burning here as in Michigan for weeks, when on that fatal Sunday night the gale, increasing to tornadoes, swept in currents for one or two hundred miles across farms and villages with fearful destruction of life and property. The whole population of the peninsula was twelve thousand, out of which at least four hundred perished. In the little village of Williamsonville seventy-four out of seventy-eight persons were destroyed. The committee has on its books the names of about four thousand persons who are utterly destitute, who must receive aid until the next harvest.

The committee informed us that a sufficient supply of general clothing has been received, but more is greatly needed of some articles to be hereafter specified. Money is, however, the great want, as much that is both of immediate and future need can only or much more profitably be obtained there.

We visited Peshtigo, on the west shore of Green Bay, where the greatest loss of life occurred. The town of Peshtigo, which has been largely built through the intelligence and capital of the Hon. William B. Ogden, was declared by all to have been the best built, most prosperous and happy town in all that region. It had a population of at least 1,700 in the town itself, and about 1,000 more in the "sugar-bush," or farming region outside.

The origin and progress of the fire is similar to that described in other places, but the devastation was far more complete. In less than ten minutes after the first alarm, which was at nine o'clock, Sunday evening, October 8, the town was enveloped in flames. The people came rushing in from the neighboring farms wild with fright, followed by cattle and horses in a confused rout. The aroused inhabitants ran from their houses and their beds, and attempted to fly before the tornado of smoke and fire, which not only kindled into a flame the houses, but filled the air with flying bricks and timber. The inhabitants believed the last day had come. The tornado swept in currents and eddies of fire, in which many were caught and smothered on the spot, while others with great difficulty worked their way, some to the river and others to an open field on one side of the town. The destruction was complete. Not a building remained except one half-finished frame house, which had been seized, charred, and left. Hundreds remained through that fearful night along the banks of the river or immersed in its waters awaiting the daylight. The scene the next morning exceeds the power of description. Families which had been separated in the fiery darkness sought their scattered members; but hardly a family was left unbroken. The ground was strewed with the charred bodies of men, women and children, and animals.

Outside of the town the devastation was nearly as complete—buildings, fences and trees were thrown to the ground and burned; but three houses remained in the township. The exact number of lives lost can never be known. It is variously estimated at from 600 to 1,000.

The inhabitants all asserted that there were currents of air on fire. The atmosphere seemed saturated with inflammable gases from the pitch-pine forests which had been burning for weeks. The heat was far greater than that of any ordinary conflagration, melting iron and bell-metal at a distance of many rods from any burning buildings. The heaviest loss of property falls upon the Peshtigo Lumber Company, who had here some of the best mills in the world; but the entire property of the inhabitants is swept away, adding absolute destitution to their other afflictions. They fled for refuge principally to Marinette and Peshtigo Bay, where their immediate necessities were supplied. Most of these people are determined to rebuild their homes in Peshtigo.

The men will find employment in the woods during the winter; but your assistance is needed to render their families tolerably comfortable in the shelters which, for the present, they must call their homes. In regard to the distribution of the supplies which we found most needed, we made arrangements with responsible persons on the spot, who are intimately acquainted with all the people; so that you may be assured that they will be bestowed with care and discrimination.

The destitution has been so generously provided for that, on the 1st of November, the authorities were able to publish the accompanying reports:—

Mr. George Godfrey, sent out as a special agent of our Relief Committee, returned from the north on Saturday, and made the following report of his trip and the progress of the work of supplying the destitute with the necessaries of life:—

MILWAUKEE, October 30, 1871.

Chairman Milwaukee Relief Committee:

DEAR SIR:—In accordance with your request I went north on the propeller St. Joseph, on Wednesday, with some articles for the relief of the sufferers by the late fire. Finding Mr. Wing, County Clerk of Kewaunee County, at the boat, in charge of the goods for Kewaunee, it was arranged that I should go on to Ahnepee, which I did, arriving there on Thursday evening, when I transferred my charge to the Relief Committee of Ahnepee, Mr. E. Schwartz, Chairman.

On Friday morning, in company with Mr. Schwartz, I set out to visit the burned district, and we spent the day in exploring the region. The weather was severely cold and flakes of snow filled the air. The timber land, much of it very valuable, had been ravaged by previous fires, and looked very desolate; but the pathway of the great fire, which swept up from the south on that memorable Sunday night, presents a heart-sickening appearance. From two to four or five miles wide in places, and extending north and south indefinitely, so far as I could see, forests, fences, barns and houses were swept away. Farming tools and household furniture, carried out into the fields, fared no better than those left in the house. Pumps in the wells were burned off to the ground. In many instances the wells are artesian, from which it is impossible to get water without pumps. Large quantities of live stock were burned to death. Of the loss of life the public is generally informed. In one spot in the town of Brussels some thirty-six persons were found and buried. Other more isolated instances were discovered; in one place three or four children were found on their hands and knees, with their heads against a large stump, dead in this position. In most instances the victims had apparently died without a struggle, probably killed outright by the first hot breath which they inhaled. In all probably one hundred persons have perished in the towns of Lincoln, Brussels, and Forestville. I have the names of some thirty-four families in the town of Lincoln, aggregating some one hundred and seventy persons—men, women, and children—who survive, but in a homeless and destitute condition. In the town of Forestville, and contiguous to Ahnepee, and looking to that place for relief, are some twelve families, aggregating about sixty persons. There is also a portion of the town of Brussels which is easily reached from Ahnepee, but I did not succeed in getting a list of the sufferers in that locality.

On the county line between Door and Kewaunee Counties, and between the towns of Lincoln and Brussels, goods were arriving and were being stored in the barn of Eugene Naze, in the town of Lincoln. These goods were from Green Bay, and were consigned to Charles Mape. Some nine or ten wagon loads arrived there on Friday, consisting chiefly of flour and provisions, with some clothing. These, in addition to what was already in store at that place, would go a good way to relieving the immediate pressing necessities of the people.

In the way of food they are tolerably well supplied for the present. Of old cast-off clothing there is a good store, but the cry is for blankets, quilts, and bedding. Hay and feed for the surviv-

REFUGEES FROM WHITE ROCK, HURON

ICH., SEEKING SAFETY IN THE WATER.

ing cattle are absolutely necessary. In order that the people may help themselves they must preserve their cattle. There is no grass nor " browse " in the woods. Boards to cover the log cabins which they are now putting up, are indispensable; stoves to warm them and cook their food, are lacking; these, with all their utensils, must be supplied. Many of the inhabitants, astounded and bewildered by the calamity, were about to flee from the country; but, upon learning that relief was coming, they have plucked up courage and are going to work to repair, as far as they can, the great damage, and get upon their feet again. The rational and best disposed of the community are doing all they can to encourage them in this resolution, and, with the aid from outside and their own endeavors, I have no doubt but in a few years smiling plenty and peace will again dwell in that now stricken land.

HEADQUARTERS MILWAUKEE RELIEF COMMITTEE,
MILWAUKEE, November 1, 1871.

To a Benevolent Public:

Through the spontaneous liberality of a sympathizing people, especially of our women, from Maine to San Francisco, we have now on hand and in transit an ample supply of clothing of every description for men, women, and children, to meet the wants of the sufferers by the fires in Northern Wisconsin and the peninsula of Michigan. Money is still greatly needed for purchasing provisions, building material, tools, and farming implements, horses, oxen, cows, hay, feed, etc. The money may be sent to Alexander Mitchell, Treasurer.

H. LUDINGTON, Mayor.
M. P. JEWETT, Chairman.
J. R. DUTCHER,
C. J. KERSHAW,
Executive Committee.

34

THE FIRES IN MICHIGAN.

CHAPTER XLI.

On the opposite side of the lake the forest fires have apparently been quite as bad. We are told that almost every county in Michigan has suffered from them. The lumbering town of Manistee has been nearly consumed, two hundred and six buildings having been burned on Sunday night, with a total loss of over a million dollars. Holland, about twenty miles south of Grand Haven, has been literally reduced to ashes, and the flames seem to have eaten across the whole breadth of the State to the foot of Lake Huron. The disaster, according to our present advices, was most complete on the peninsula between Lake Huron and Saginaw Bay. All that part of the State lying east of the bay and north of a point forty miles above Port Huron has been practically swept bare. This district, covering a region about forty miles square, was the seat of an extensive lumber trade, and all along the shores of the lake and bay were prosperous settlements, large and small, at which lumber was sawed, planed, stored, and shipped, and depots maintained for the supply of the woodmen and other persons employed in the business. The flames, approaching from the west and south, must have hemmed in these villages and cut off all escape except by water. How many of the luckless inhabitants were able to avail themselves of this avenue of safety we do not yet know. Two or three steamers from Detroit have been cruising off shore to pick up the fugitives, and about sixty have thus been rescued, some severely burnt, and all destitute. At one place five children are known to have perished, and there is reason to fear that the worst has not been told. Simultaneously the forests and prairies of Western and North-western Indiana have been on fire, and though there has been no loss of life, so far as reports are yet at hand, the destruction of the harvest has been enormous.

There is something more awful in the thought of a burning forest or a prairie in flames than even in a catastrophe such as that which has fallen upon Chicago. The most stupendous efforts of man seem hopeless of arresting a conflagration which rages unceasingly through two or three entire months, and sweeps in its fierce wrath over thousands of square miles of territory. Nothing checks such a visitation but the exhaustion of the combustible material, or the blessed rain, which at last stayed the flames in the woods of Michigan, just as it quenched the glowing cinders of Chicago. The destruction of the great commercial city of Illinois was a disaster of almost incalculable pecuniary magnitude, but it will be repaired in a few years. The burning of the grand primeval forests means far less to the banker and the tradesman; but it is a misfortune which can never be repaired.

October 11th, a correspondent telegraphed :—

The news from St. Clair and Huron counties of this date is of the most distressing character. All that portion of the State east of Saginaw Bay and north of a point forty miles above Port Huron has been completely swept by fire. A number of persons perished, and it is feared we have not heard the worst. The flourishing villages of Forestville, White Rock, Elm Creek, Sand Beach, and Huron City are entirely destroyed. Rock Falls and Port Hope are partially destroyed. Nothing has yet been heard from Port Austin or Port Crescent; but it is hardly possible they escaped. At all these towns there were large stores, many of which were filled with winter stocks; extensive saw-mills, shingle-mills, and docks covered with lumber, all of which have been swept away. It is stated there is but one dock left on the shore about Forestville. A steamer which left Port Huron last night for the relief of the sufferers returned this evening with about forty men, women, and children, five of whom are severely burned. The revenue cutter Fessenden, which started for Port Austria, picked up a sail-boat on the lake, containing Isaac Green, principal owner of Forestville, together with his family, and eighteen or twenty others, who had escaped the flames at Forestville. The telegraph operator at Forestville escaped through the fire back into the country. All the telegraph offices along the shore have been destroyed, but communication will be restored as soon as the damage done to the lines can be repaired. Five children are known to have perished near Rock Falls. R. B. Hubbard, at Huron City, shot all his fine horses and cattle to prevent their perishing by fire. He loses very heavily, having had a large store, mills, docks, etc. The extensive property of Stafford & Hayward, at Port Hope, is about the only establishment that escaped. $5,000 were subscribed to-night for the relief of the sufferers of this State. The light rain of yesterday seems to have greatly abated the fires throughout the State, and it is believed the worst is passed. There is scarcely a county in the State but has suffered more or less from fire, and the loss will amount to nearly a million of dollars. The damage to the pine land is incalculable.

The town of Bridgeport was saved from destruction by a shower of rain yesterday morning. Charles Chandler's barns, on his farm near Lansing, were burned yesterday, together with several fine horses.

The cutter Fessenden reached Port Huron this morning with seventeen refugees from the lake shore, two of whom are fatally burned. Port Austin escaped the flames.

A Mr. Brady, of Detroit, who was in the village of White Rock, Huron County, when it was burned, says, that after vainly striving to keep the fire out of the town, the inhabitants, hastily gathering the few valuables that came nearest to hand, fled to the most open places away from the houses, and, driven from these, rushed into the water itself, and even here were not safe from the scorching effects of the heated air without occasional plunges beneath the surface, or frequent washings in the surf. Mr. Brady was in the water eight hours, lying part of the time on a log, over which the light surf dashed. About him were men up to to their waste in water and holding children in their arms, women but poorly protected by their clothing from the chill of the water, which was their only security against the burning heat of the air. The inhabitants, of course, saved almost nothing. Not only were their houses, fences, barns, and stocks destroyed, but their furniture and clothing, and even the deeds by which they held their lands, and their insurance papers. From their painful position in the water they were released by the subsidence of the fires; but there was neither food nor shelter within miles, and for many of them naught but beggary apparently remained when shelter should be found. The fire at White Rock occurred Sunday night, and it was not until Monday afternoon that the sufferers were taken off the shore by the steamer Huron, which took them on board, and, coming down the shore, released from similar straits others who had lived in Forestville and Cato, which towns were also burned on Sunday night. The steamer, after taking on board as many as she could carry, left many for a second trip. The sufferers were cared for by the citizens of Port Huron, and the steamer went back for another load. Besides the towns named it is supposed that Center Harbor and many other smaller villages were destroyed, and it is feared that the loss of life has been considerable. The pecuniary loss at White Rock is more than $250,000; that at Forestville is still greater.

A letter from Saginaw City reports that a large fire occurred there on the night of the 8th inst., which destroyed the large steam saw-mill, salt-blocks, and a number of houses adjoining, together with a large quantity of lumber. The loss will not fall short of $400,000. At latest reports the woods were all on fire in the vicinity of Saginaw, and the entire city was in danger. Business had been suspended, and the entire population were doing everything in their power to save their property.

Between Saginaw and Birch Run the loss had been heavy; many cars had been burned and trains delayed.

A letter from Port Huron, dated the 10th, says:—

The fires are still raging on all sides of the city, and a fierce south wind has been blowing for three days. Yesterday afternoon teams were employed in carrying water to the south part of the city, where the most danger was apprehended, and at a late hour last evening the flames were very much checked. Between here and Lexington fires are raging fearfully, and many of the telegraph poles between this city and that place are destroyed. Along the Grand Trunk Railroad, and the Port Huron and Lake Michigan Railroad large piles of wood are on fire, and the fences for miles are consumed. Travelling is very much interfered with, and at some places it is impossible to pass the roads.

Fires were also reported to be raging in every direction around Lansing, and, on Sunday, the students of the Agricultural College were called upon to help fight the flames. They were divided into squads, which relieved each other, and on Monday night the danger was supposed to have passed.

A Detroit report says:—

A score or more of men, women, and children arrived in this city yesterday by boat and rail from the up-country counties, and the statements made by them in regard to the woods' fires are appalling. All of them have suffered the loss of every dollar of property, and some of them show scars and blisters to prove how closely they were pursued by the flames.

John Kent and wife were living about ten miles above Forestville and about five from the lake. He states, as do all the others, that fires have been running in the woods for months, but have travelled more rapidly and have created greater destruction within the last ten days than in all the time before. For weeks the smoke in Sanilac and Huron counties has been so dense that women and children have been made sick, and every human being has been half blind. Fowls were smotherd as long as three weeks ago, and the effect on cattle and horses was to render them unfit for work.

Although Kent had reason to apprehend danger to himself, wife, and two children, he did as nearly every one else did, stood by his little property in the hope to preserve it. He had a considerable clearing around his house, and imagined that the flames would not reach him. He had plenty of water near his house, and filled barrels, tubs, crocks, and everything which would hold water, and placed them where they would be of service in extinguishing sparks and cinders.

Friday last he could hear the roaring of the flames and the falling of the trees from his house. At night the heavens were rendered so light that he needed no lamp in the house. His dog left him early Friday morning, and the house cat disappeared two days before, the animals seeming to have a better knowledge of the danger than tho man. Towards noon the flames appeared on the outskirts of Kent's farm. His children, two little girls, the youngest not a year old, were left in the house, and husband and wife repaired to the fire to try to beat it back. With anything which would strike or smother they fought the advancing flames, and for a distance of twenty rods kept them in check. But while busy here, the flames crept over the dry gound from other directions unheeded and neglected. Fighting with all their strength, father and mother gave no heed to anything but the fire before them, until they were at last startled by a scream from the house. Instinctively they felt that the flames had seized it, and they started to the rescue of the children. But the smoke had settled down so thick that they ran in all directions without finding the house, and knew not its locality until the fall and crash of the roof told them that the little ones had met an awful fate.

"I tell you, mister," said Kent to our reporter yesterday, "it made us crazy. The fire was all around us except to the west, house gone, barn burning; hay and everything destroyed. There was only one thing to do—I got hold of Mary and plunged through the fire and smoke until we got out into the Lake road, and then we had hard work to keep ahead of the fire before reaching the water. It was awful, sir, to hear that screaming from those burning children, and it was dreadful to go away and leave them roasting there."

Many of the others had almost as bad experiences. While some of the farmers left the woods ten or twelve days ago, seeing that nothing could prevent the progress of the flames, others like Kent hoped for a rain, and trusted that the fire would not advance over the cleared lands. There was a family named Cross, a man living about a mile back of Kent's, and as he did not see them come out, and knew them to be at home at the time his house burned, he is certain that every one, five in all, were roasted in the flames.

A Detroiter named J. P. Parker returned to the city yesterday after an absence of ten days in the counties on fire. He states that no one can form an idea of the desolate scene unless he was a spectator. For several days Parker was right on the borders of the fire, being driven back six or seven miles some days, and meeting with scores of people who were driven out of their burning houses. He states that one day, while he was making a great effort to get through the woods to see about the fate of a saw-mill, he encountered an Indian and his squaw, the latter having a papoose strapped to her back. Parker tried to halt them for a moment to make inquiries, but they passed him on the keen run, the Indian yelling out, "No more wigwam!"

The only avenue of escape was the Lake. Many of the settlers along the shore packed their goods on rafts and anchored them out in the Lake, some being thus afloat for many hours before being taken off by steamers. Others who were fleeing for their lives, and had no time to build rafts or hunt for boats, had no other resource but to plunge into the water and wait for the fire to burn itself out.

All the telegrams and letters received from the Lake shore region confirm the statements of these people. There is reason to apprehend that very little property, lying anywhere near the great fire belt, will escape, and that the fire will constantly increase and travel in new directions. There is already a piteous appeal for help, and the cry will continue for weeks to come. Every effort of our citizens and of the people throughout the State must now be directed towards raising money, food, and clothing for our unfortunate sufferers. Detroit has already raised a considerable sum, and more is being raised every hour, while large collections of clothing are being made in various quarters of the city. If any citizen has a shilling in money, a pair of boots, cast-off clothing, a blanket, or anything else that will cover and comfort the poor victim, it is hoped that he will call at the City Clerk's office and leave it.

The steamer Marine City arrived in Detroit about seven last evening, and from Captain Robertson we obtain some further particulars about the fires on the shore. In the counties of Alpena, Alcona, and Iosco the people had, up to the time the Marine City passed, succeeded in keeping the fire out of the villages, though in some cases the struggle was for a long time a doubtful one.

In Alcona on Monday the furniture and other movables were taken out of some houses, in the fear that the efforts to keep out the fire would not be successful, but the houses were finally saved

In Harrisonville the fire approached uncomfortably near to the dwellings, but the people had thus far succeeded in keeping it off.

Between Au Sable and Tawas there was a great deal of fire in the woods, but the villages had not yet suffered.

At Alpena there was a report that there was fire above Devil River, and that it was working its way toward the village.

In all of the villages mentioned the inhabitants felt reasonably secure in case there should be no high wind; but in the event of a gale, without rain, they considered the danger very great that they would suffer the same fate as other villages further down. They had a slight rain on Tuesday, but not enough to afford much security against the spread of fire. All along this coast there was such a dense smoke that it was impossible to see more than a few feet, and the eyes suffered severely from its effects. At Alpena it was necessary to light the lamps in the cabin, as the smoke caused a darkness as great as that of early evening.

At Forrester, where the Marine City stopped, one of the sufferers from the interior came on board in order to go as far as Lexington and procure provisions for his friends in the country. He had walked down to Forrester from a place six miles back of Richmondville, where there were twenty-five families in one house, with only two days' provisions on hand, and he was the only man in the whole number who felt able to get as far as Forrester. The rest of the men were suffering from fire blindness, burns, or exhaustion. The house in and about which they had gathered was the only one saved in that vicinity, and the two days' rations which they then had were all that they had been able to save from the fire. An effort will be made to bring them away as soon as possible, but the means of transporting them down to the shore are very scanty. According to the latest news that the Marine City could obtain, there have been nine villages burned on the coast, viz.:—New River, Huron City, Port Hope, Sand Beach, Center Harbor, Rock Falls, White Rock, Elm Creek, and Forrestville, while nothing is known of many of the villages back from the coast. The only dock left between Richmondville and Point Aux Barques is that at White Rock.

In Huron City there is one public house standing, that of R. Winterbottom, while the remainder of the village is burned.

Port Hope is all burned, except the fine residence of Mr. Stafford, his store, and the Masonic Lodge. Mr. Gunning, living back of Cato, is burned out, and it is feared that the same fate has befallen the Luddington settlement back of Sand Beach.

On Monday Port Austin was all right and supposed to be secure. We get by the way of Saginaw a rumor that it has since been burned, but the rumor comes in such shape as to lead to the belief that it is incorrect.

The tug Frank Moffat arrived at Port Huron Wednesday night, with forty passengers from Sand Beach, five of whom were badly injured. She reports that Verona was entirely, and New River partly destroyed. It was reported that many lives were lost, and also that there was a large destruction of cattle, hogs, and horses. R. B. Hubbard & Co., of Huron City, owners of large numbers of choice imported stock, shot the animals down to prevent them from burning. The steamer Huron, which left Port Huron Tuesday night, returned to that place last evening with another load of rescued villagers, the second that she had brought down. She starts up the shore again this morning to continue the work of mercy which she has carried forward so efficiently, and will be devoted by her owners to this duty as long as there is any apparent need of her services.

From the Saginaw Valley it is reported that everything between Pine Run and Bridgeport has been destroyed. On all sides of Saginaw—indeed, throughout the whole valley—the woods were burning fiercely, and the flames were continually sweeping rapidly onward, carrying destruction to property of all kinds. In East Saginaw five buildings were burned at midnight, Sunday.

The occupants of the building where the fire originated were with great difficulty rescued. Some of them had their feet and hands almost roasted. While the fire was raging a fire broke out in Saginaw City, which destroyed the entire property known as the "Island" and situated between the river and the bayou, on the side of the river. The fire originated in the shingle-mill of Burnham & Still, just above the upper bridge. This mill, drill-house and boarding-house were entirely destroyed; also a house owned by a Mr. Burnham; the saw-mill, drill-house, kettle and steam salt blocks of Chapin & Barber; a lot of lumber and two thousand cords of wood, owned by Chapin & Barber; the shingle-mill of Lathrop & Inscoe, and several dwellings. Total loss, $75,000. The woods between St. Charles and Chesaning are all afire. Nearly a thousand cords of wood were destroyed at St. Charles. Above Midland the telegraph poles are all burned down, and much valuable timber is destroyed. At County Line four buildings have been destroyed and more are in danger. Four or five hundred cords of wood and twelve miles of fencing are burned near Birch Run.

In all the valley cities the most intense excitement prevailed since the terrible fires have given us a foretaste of what the future must inevitably be if rain is not soon graciously vouchsafed to us. So soon as the proclamation of the Mayor was made public the mills were all shut down, and the men sent to such points as they were supposed to be most needed in. It was but a short time after the call of the Mayor was promulgated ere hundreds of men with spades, hoes, axes and rakes were on their way to the various points of danger. Necessarily disorganized as were the men in the first excitement of gathering, it was not long ere they were systematically at work in the thus far successful endeavor to stay the progress of the fire. Clerks in the stores and laborers in the workshops and mills united side by side in preventing the advance of the flames. Yesterday was indeed a day of terror in our val-

'ey. In every direction wagons drawing barrels of water, men with pails and axes and other implements with which to fight the rapacious monster, could be seen eagerly pressing to and fro in their mission. Occasionally a countryman could be seen imploring men to go with him miles into the country to save his property, but almost everybody had too much to do near his own dwelling-place. In the doorways and yards anxious women eagerly inquired for the latest news, and more than once our reporter was asked with apparent earnestness if, in his opinion, "this was the great and final day." Anxiety was depicted on every countenance, and we fear it will but deepen if the much-desired rain does not speedily fall.

From the Saginaw *Enterprise* we take another account of a city saved by courageous and well-sustained endeavor:—

The report that Midland was in flames was not correct. The city suffered no damage; but had it not been for the extraordinary exertions of the people, we would have had a different story to tell. The whole place on Monday night was entirely surrounded with a vast sheet of flame, and the crackling of the fire and the crash of falling trees made the scene a fearful one. Valiantly did the people fight the flames, but so steadily and surely would they spread that all the exertions of the people appeared quite futile. The fierce wind added to the fury of the flames and carried sparks and burning fragments through the air. At one time the whole heavens appeared one mass of fire, and the destruction of the place appeared inevitable, but the unceasing labors of the people kept the fire within bounds, and the city of Midland escaped.

The *Enterprise* also has the following additional disasters:—

Late last night we received the news that four shingle-mills were destroyed between three and four miles from Midland. They belonged to Messrs. George Rockwell, Collier & Garber, Dowlers, and Reardon & Andrews. Besides these mills, all the shanties, boarding-houses, and barns attached, and a large quantity of shingles were also consumed. The flames made a clean sweep, and men and cattle were driven in all directions to seek a place of safety. The total loss is estimated at $50,000. So fierce were the flames and so rapid did they spread, that before the men working at the mills were aware of it, they were completely surrounded by the burning woods, without the remotest hope of escape. They rushed in all directions, but they could not find a way out of their dangerous location. The burning circle around them was gradually but surely growing smaller, and there was no time to be wasted. Some lowered themselves down into wells and others dug holes in the ground, in which they sought protection. In these uncomfortable positions they remained all night, not daring to move for fear of the falling trees. Yesterday a gang of men with wagons was sent out from Midland to their relief, and all escaped without serious injury.

Kawkawlin is entirely surrounded by fire. John Gordon came to Bay City for help. His eyebrows were singed, his hands were blistered, and a piece of his buggy was burned. A hotel half-way between Bay City and Rifle River was in imminent danger from the rapidly approaching fire. It became so smoky that the family were in danger of being suffocated. The proprietor, Barney Shoots, went to the railroad and hired a man for $20 to bring himself, his wife and Mr. Jay's little daughter, who was visiting them, to Bay City. He left a man to take charge of the horses and cattle. The man arrived yesterday morning and reported the horses and cattle safe, and the house still standing.

The railroad bridge across the Pinconning River has been burned. Yesterday morning a train carried out the material and men to build a new one.

On the eastern coast of Lake Michigan half of the flourishing town of Manistee, with 4,000 inhabitants, has also been burned. The loss is computed at $1,300,000.

In Minnesota the loss of life has been least, though the area of destruction has been probably as great as in Wisconsin.

The conflagrations have extended to within sight of St. Paul, and have swept irresistibly over the greater portion of Wright, Meecker, McLeod and Carver counties; and from thence out as far as Breckinridge there is an almost continuous belt of burned country. The towns, however, have generally managed to escape, and it is believed that the worst of the danger is over. The lives lost will not, probably, foot up to fifty, all told. Still, the devastation among the woods has been fearful, and thousands of square miles have been reduced to a charred waste.

SAVED BY HEROISM.

A large portion of Gratiot County, Mich., was overrun by the late fires, destroying dwellings, fences, timber, saw-mills, and other property. The only mill saved in that section was that belonging to D. L. Case and James M. Turner, of this city.

An old Englishman in their employ, named Jacob Laird, saw that the mill was in the range of the fire, and that it was speedily coming upon him, and he made preparations to meet and fight it like an old soldier, for he had served the Union cause gallantly during the late rebellion. Taking all the movable property from the mill, blacksmith shop, and boarding-house, he buried it where it would be safe from the devouring element. Then he dug a series of wells, inclocing the mill and hundreds of thousands of feet of lumber, placing by the side of each well a barrel filled with water, a pail in each well and another in the barrel. These wells were dug thick as needed for the speedy protection of the property. One well he dug deeper than the others, that in case his efforts to save the property should be in vain and his own life in danger, he could jump in as a last resort.

The fire came down upon him like a tornado. With his force of hands he met it, and where it crossed the line here and there, setting fire to piles of lumber, the water ready at hand quenched the flames. At last he came off victorious, for he saved the mill, lumber, and all the property, with the loss of his own hair, eyebrows, whiskers, and even the woollen shirt from off his back. When rallied by his employers as to whether he "didn't find it hot work," his reply was, "It was not much of a soldier who couldn't face the fire, after facing as many cannon as he had."

As a reward, Laird's wages have been nearly doubled, and he was furnished with a fine suit of clothes, and told that he could remain there as long as he chose.

During Monday the city of Grand Haven was full of terrible rumors as to the fire in Holland City, but nothing definite or reliable could be learned until the arrival at two o'clock of a train from the north side of Black Lake, containing several passengers, among whom were Miss Jennie and Miss Clarie Pennoyer, two young ladies who have been engaged in teaching in the doomed city. The statement of these young ladies is nearly as follows:—

The fire broke in upon the city from the woods about three P.M., Sunday, but no buildings of any consequence burned until dark in the evening. No one thought the city was in any special danger until ten or eleven o'clock, but at that time a strong wind setting in from the woods, the fire swept over the city with wonderful rapidity. The main part of the city was soon in flames. The house where the Misses Pennoyer were staying caught fire about three o'clock Monday morning. The ladies had packed their trunks, and, hastily dressing themselves in wrappers, just managed to escape. The Lake View House went next, and then the fine City Mills of Wakeman, Gerlings & Co. The ladies, after leaving the house, ran to a small mound near by, and soon found themselves surrounded by fire. Mr. George Howard, whose efforts were indefatigable, managed to assist them out of their precarious position.

The portion of the city where Professor Charles Scott resided

REBUILDI

IICAGO.

was completely destroyed, and the Professor not being found, it was generally feared that he had fallen a victim to the flames. Mr. Joslin, of the firm of Breynan & Joslin, another of the best citizens, was engaged in rescuing persons from the flames. He insisted on going once more to the rescue; friends advised him not to venture, but he would not be dissuaded, thinking there were still lives to be saved. He did not return, and is believed to have been suffocated and burned to death. The livery stables were emptied of the horses which were taken to the public square, as the only place of safety. Thousands of people were collected there. Women and children were then running about the streets, wailing and crying, unable to find their husbands and fathers, brothers and sisters. Many females barely escaped with their night-clothes. A child ten years of age was picked up on the street burned to death. It is impossible to tell how many lives are lost. Some nine or ten citizens are missing, but some may yet be found.

When the Misses Pennoyer left, men were trying to keep the fire from the College buildings, but the succeeding train reported that these buildings, although of brick, were burning; also the Union School building and all the churches, except the "Seceders" or the "True Reformed Church." One woman in leaving her house tied her baby in a bundle, but in her hurry she took the wrong bundle, and to her dismay discovered her mistake when too late. Of seven children she could find only two. Fortunately, however, the bundle containing the live baby was picked up in the street, and it was believed that the other children were also found.

Mr. M. D. Howard, when he saw the fire rapidly approaching his residence, visited every room in the house for the last time. The house was elegantly furnished, and the fine piano, together with every article of furniture, was destroyed. Nothing was saved but a little personal clothing. The family took refuge on a tug and were taken toward the mouth of Black Lake to a place of safety. George Howard was very diligent in picking up children and women who were running frantically about in places of danger and removing them to a place of safety. In this way he saved numerous lives. He sometimes had to capture them by main force to save them from destruction.

The City Hotel at first was considered by Mr. Myers out of danger, but his most valuable articles he fortunately buried in the ground, and these were all that was saved of the best hotel in the city. The house was in flames when the family and boarders escaped. The other hotels all shared the same fate.

Mr. George Howard, at the very commencement of the fire, took fourteen spades and handed them each to Hollanders, who were standing around, and requested them to use them in throwing sand on the fire, so as to prevent it from spreading to the destruction of the city. They actually refused to work, giving as a reason that it was Sunday, and it would be wrong to do any work on that day. Had they gone to work like men, this terrible conflagration and suffering might possibly have been prevented.

The woods along the line of the Michigan Lake Shore Railroad, between Holland and Pigeon River, were in flames. The miles of marsh were one sheet of flame, and it was with great difficulty the train came through. The heat inside the cars was intense.

A message from Mr. A. D. Howard was received by the train, stating that the people were in danger of starving, as all the stores were destroyed, and asking that a supply be immediately shipped. Mr. D. Cutler immediately called on the stores and ordered a supply of crackers, and all the cooked provisions that could be collected sent to the depot. The train did not get off till Tuesday morning, when rain came and subdued the flames along the line of the road.

We felt profound sympathy for children in these seasons of terror and destruction, for they could not reason, but saw the peril in the skies by day and night, and passed through the bitter struggles for life which so often terminated fatally, and always brought suffering and distress.

A Port Huron correspondent of the Detroit *Post* says:--

You have already been told the story of the little boat-load of children carried from Rock Falls to Canada, and saved in spite of storm and hunger and exposure. I saw Mrs. Mann, the mother of these children, who arrived here yesterday morning on board the Huron. She had given up all of them for lost. But, mother-like, though four were saved, she mourned deeply for the lost one, who, half-clad and shivering in the cold water in the bottom of the boat, sailed away upon an unknown and measureless sea almost in sight of land and deliverance. There were five children in this boat belonging to Mrs. Mann, and four to the owner of the boat who took them away, making nine infant voyagers, who for three days, without food and drenched to the skin, floated across Lake Huron in a boat which was kept from going to the bottom by means of an old boot and a shoe, which were the only vessels for baling that these unfortunate travellers had on board. The mother's heart seemed deeply touched and troubled because no last offices and loving ministries could, in the nature of the case, be paid to the little one whose voyage of life was at once so brief and eventful. When these four children were put on a tug at Kincardine, Ontario, to be returned to their parents, it struck a rock just as it was getting under way and went down. The children were rescued and sent homeward by the cars. They have at last reached Port Huron, after adventures by field and flood almost equal to Othello's, and it is hoped that they will arrive home without further accident.

The babe of Mrs. Shubert, of Paris, one of the Polish settlers, was carried from its burning home by its grandmother, while its mother stayed behind to fight the fire. The grandmother was compelled to lie down in a roadside ditch with twenty others, where they passed the night, it being the only refuge from the flames. The infant was only three months old, and required nourishment. Luckily the fire had driven a cow to seek company and shelter with these human beings. A big tin pan was found in a wagon, and the animal was milked. The baby's aunt took the mushy compound which the flying sand, cinders, and ashes made of the milk into her mouth, and fed the child in that original manner.

During the recent terrible fires in Western Michigan, there were three brothers, owners of valuable mills and buildings, which they and their neighbors (some of whom were Christian men) were defending from the fire until all were exhausted and in despair. One of the owners, a frank, rough, wicked man of huge frame and generous impulses, said many hard words about God's permitting the destruction of so much property for no good to any one, etc., etc. Finally, he gave up and said to his neighbors: "Go home, go home; nothing more can be done for us; God can do as he pleases." Just then a few drops of rain fell; looking up, they saw the cloud, and all redoubled their efforts. A slight rain fell, and the fire was checked, and the mills saved. The rough man dropped upon his knees, great tears rolled down his face, his hands were clasped, head bowed, and he agonizing to express his thanks. Suddenly he sprang to his feet, vigorously swinging his hat, and with the most intense earnestness shouted, "*Hurrah for God!* 'HURRAH FOR GOD!"

ONE COUNTY IN MICHIGAN.

Beginning below Port Austin, Huron County, Grindstone City, a place of three hundred inhabitants, is half destroyed; then follow New River, three buildings burned; Huron City, five hundred inhabitants, totally destroyed; Port Hope, six hundred inhabitants, half gone; Forest Bay, two hundred inhabitants, every house gone; Sandbeach, four hundred inhabitants, all destroyed; Centre Harbor, one hundred and fifty inhabitants, everything gone; Rock Falls, three hundred inhabitants, half of the town burnt; Elm Creek, one hundred and fifty inhabitants, totally destroyed; White Rock, six hundred inhabitants, every house consumed; Verona Mills, three hundred inhabitants, every house in the place gone, except the minister's. Thus were the little villages of this region scourged. The destruction in the country was proportionately great; the farming townships of Sheridan, Bingham, Paris, Verona, Sherman and Sandbeach were traversed by the flames, which lapped up everything in the shape of houses, barns, fences, stock, farming implements, etc.

A more particular account thus locates and estimates the misfortunes of this single county:

The total number of persons and families burned out, and of losses in Huron County are as follows, in the townships named:—

"Verona—Twenty-seven families, one hundred and fourteen persons, loss $42,150.

"Bingham Township—Thirty-five families, comprising one hundred and sixty-eight persons, burnt out.

"Sigel Township—Twenty-one families, one hundred and five persons, loss $6,300.

"Grant Township—Four families, twenty-one persons, loss $1,700.

"Colfax Township—Four families, twenty-two persons, loss $2,000.

"Sheridan Township—Four families, nineteen persons, loss $1,100.

"Sherman Township—Twenty-five families, one hundred and sixteen persons, loss $16,150.

"Huron Township—Twenty-three families, ninety-seven persons, loss $8,550.

"Dwight—One family, seven persons, loss $2,000.

"Mead—One family, six persons, loss $400.

"Hume—Three families, nineteen persons, loss $1,800.

"Mr. W. R. Stafford lost $53,000 by the fire.

"In Port Hope, Rubicon and Gore townships, there are three hundred and thirty-eight persons burned out, with a loss of $176,825.

"The total number in the families that are losers in this county alone is not less than 3,000.

"There have been eleven school-houses burned in this county, as follows: all in Paris township, four in number; one at Vernon Mills, Gibson school-house in Sherman, one at White Rock, one at Centre Harbor, one at Sandbeach, one at Forest Bay, and the Hellem's school-house in Dwight.

"The loss by the burning of bridges across streams and crossways, through swamps, coupled with the almost total destruction of fences, will amount to thousands and tens of thousands of dollars. There is hardly a farm in the county but has had a portion of its fences burned, and this is true even in those sections where but few or no buildings have been destroyed.

"The loss to this county by the burning of pine and other valuable timber is very great. It is too soon to make anything like an accurate calculation of the total loss from this source, but from converse during the week with the Supervisors and others from different parts of the county, we know we are safe in saying that it will exceed $1,000,000."

A gentleman, writing of Tuscola County, gives a forlorn picture:—

The fire through this county has destroyed nearly all the pine, and thousands of acres of hemlock timber. There are a large number of individual cases of suffering by the fire, some losing all they possessed in the world but their land, others losing nearly all their fences, others barns, with their contents of hay, wheat, oats, etc. The destruction of hard timber is also enormous, that more especially on low grounds. Here the fire has burned so fiercely that the roots of the trees have been burned off, and hardly a tree is left standing.

There is a strip of country east of Cass City, embracing a portion of Tuscola County and nearly all of Huron and Sanilac counties, which is probably the worst burnt district in this State. Hardly a building is left, and the pine and hemlock lands are all destroyed. Whole townships of timber are burned up by the roots and have fallen in every conceivable shape, rendering it next to impossible to lumber it without more expense than the actual worth of the timber. I saw and conversed with a gentleman from this district, who was there during the whole fire, and who had several very narrow escapes from being burned alive, being twice carried by others who were with him from the fire in an insensible condition; and finally, after there was no further hope of saving his home, it was then too late to make his escape with his family, and they took refuge in an out-of-door cellar, which is a hole dug into the ground and then covered over with slabs, and dirt thrown upon this.

They were compelled to remain there for twenty-four hours without food, and almost suffocated by the dense smoke. This is but one instance of many equally as narrow escapes. Every particle of anything like hay, corn-stalks, or straw was burned, and many of those who had oxen and cows are now selling them for ten dollars a head, because they have nothing to feed them, and there is not a green thing in the woods or fields for miles. The people of Watrousville have been active in raising supplies and forwarding them to the sufferers. Yesterday Rev. Mr. Goodman, of East Saginaw, was on his way into that county as a committee of one to prepare the way for large supplies to be sent from that place. Business of all kinds has been entirely disarranged by the fires, and there is a general complaint of dull times. But the probability is that the rain, which commenced falling this afternoon at about one o'clock, and at the present writing, ten P.M., continues, will entirely extinguish the still smouldering fires, and give a permanent relief from the anxiety and feeling of insecurity which has pervaded the whole people of this section for the past three weeks, and in a short time business will be resumed with more vigor than ever.

The breadth of wheat sown in this county is very small, probably not one-fourth of the usual amount. But a small portion of the last wheat crop has been disposed of, farmers generally holding it for a higher price. Hay, and all kinds of coarse fodder and coarse grains are very high, and it will require a good deal of economy to get their stock through the winter.

In Chicago a man took refuge in a water-pipe and was roasted; and now we have the story of a man in Michigan, who found his death in a hollow log.

SKELETON FOUND IN A LOG.

Some three months since an Englishman, named Halvry, after a short stay at Quebec, came along to Detroit to visit his brother-in-law here, Mr. John Gloveson, a produce buyer, living on Twelfth street. Halvry left England with the intention of purchasing a farm either in Canada or the States, and when he came to Detroit he left his family at Quebec. As Gloveson was considerably acquainted in the Lake Shore counties, he induced his relative to think of going into some of them and buying him a farm, and agreed to go up with him on a prospective trip. They were both ready to start—in fact, had left the house—when Gloveson was handed a telegram, which called him to go to Jackson, or run a risk of losing several hundred dollars. He, therefore, reluctantly abandoned the trip, and gave Halvry such instructions as induced the man to make the voyage alone.

This was just a week previous to the news of the fires in the woods which created such loss of life and damage to property in

Sanilac, Huron and other counties. Halvry wrote from Forrestville, two days after reaching there, that he liked the locality very much, and had had three or four offers of partnership in business, which he was considering. He also stated further that he was going back into the country to look at some farming lands, and should not probably come down the Lake for several days.

When the fire came, interrupting communication, Gloveson was in Illinois, and he did not return home until several days after the news of the destruction of Forrestville and other towns. Waiting from day to day for news or for the reappearance of his relative, and hearing nothing, he at length decided to go up there, having found by telegraphing to Quebec that Halvry had not joined his family. He accordingly proceeded on the trip, and, after a hunt of thirteen days, returned three or four days ago, bearing only evil tidings. Gloveson found plenty of people at Forrestville who remembered the Englishman, but for three days could not find any one to tell him where the man went when leaving the town. He at length found a farmer, whose property had been swept away, who had shown Halvry around his farm, situated about four miles from the town. This was only two days before the advent of the flames, and the smoke was so thick as to cause many complaints from the new arrival, and he declined purchasing in a locality subject to such a nuisance. Another man remembered meeting and talking with the Englishman in the woods where men were getting out some timber, and the last heard of Halvry was that he was looking at some wild land on the afternoon before the fire, seven or eight miles back of Forrestville. Day after day, until he had travelled hundreds of miles, Gloveson rode and walked over the blackened and desolate country, finding no further news of his relative. One day, when travelling across a bit of forest where the fires still smouldered and flickered in the ground, and where the flames had done great damage, he sat down to rest. In a moment he became aware of a horrible stench, and looking about him, he made a terrible discovery. Fifteen or twenty feet away was a large log, or the remains of one, for the fire had burned up all but the end which had become heavy with water from resting in the neck of a small marsh, dry then, but fed by a creek at other times. Sticking out from the hollow of this log were the feet and legs of a skeleton, nothing but the bare bones left, and beyond the skeleton feet was the roasted body of a man, the flesh cooked and shrivelled down, but emitting a smell which Gloveson could stand only for a moment at a time without retreating. At length he seized hold of the bones and drew the body out, when the sight and the stench were still more horrible. At the shoulders the fire seemed to have stopped, leaving the flesh half cooked, and it was now ready to fall from the bones. The hair was gone from the head, the countenance so disfigured that there was no identifying it, and the bones of one hand were as clean and white as chalk. Every particle of clothing was gone, and down in the ashes below Gloveson found a number of boot nails.

Although not certain that the roasted body was that of his relative, Gloveson secured no other evidences of the man's death, and yet fails to find that he is living. It seems that the victim, whoever he was, had been caught in that vicinity by the fire, and having no other resort, crawled into the log, hoping that the flames would sweep over it. The dry end caught fire, and he was roasted alive, enduring the most horrible death imaginable. There was an excavation close at hand, made by the uprooting of a tree, and into this place the skeleton was dragged and the bank caved in on it as a covering. Returning to Forrestville, Gloveson made such inquiries as led him to believe that the skeleton was not that of any resident of that locality, and he then ended his search.

CHAPTER XLII.

HON. WM. A. HOWARD, writing to Hon. Alexander H. Rice, of Boston, concerning the needs of Michigan, says:—

If you will now look at Port Huron, at the foot of Lake Huron, and follow the shore northerly clear around and up Saginaw Bay, and extending back from the shore into the country for miles, you will find a country where the horror and suffering is almost past belief. The loss of life is very great. Fifty dead bodies were found in the rear of our town alone. Women and children saved their lives by rushing into the Lake, with their hair burned off. These people are being cared for by Detroit and all the eastern portion of this State, nobly aided by Toledo, Cleveland, and, indeed, all over Ohio, New York, and last, but not least, glorious old New England. Excuse the contradiction, but she seems to me not only the best but the oldest country in the world, perhaps because I first saw the light there.

If you would now find the field of the western branch of our State Committee, you will please place yourself at Grand Haven, at the mouth of Grand River and the western terminus of the Detroit and Milwaukee Railroad. If you go north along the coast about ninety miles, you will find the place where Manistee was. If you go south from Grand Haven about twenty miles, and turn inland and cross a little lane, you will find the ruins of Holland. Of the villagers you will find three hundred families, in all perhaps two thousand persons, who were utterly destitute of food and shelter, and of clothing, except what they had on their persons when they fled for their lives. Instead of finding succor from the surrounding country, the country itself is devastated. One hundred and thirty farms were stripped of buildings, fences, crops, and their inhabitants were driven before the fire to the village.

But I need not describe the scenes at Holland or Manistee. They simply needed everything there. But the liberality of the

American people, and especially of all the railroads in carrying free, and giving preference to cars loaded with supplies, has enabled the committee to accomplish more than could have been expected. I inclose a printed slip that has some suggestions deemed valuable for Holland. The people of Manistee will more readily find employment in getting out lumber. When once provided with shelter, bedding, and the men set to work, we shall be relieved to a great extent. The fire extended across the whole State. I have tried to point you to the shores. It was less severe in the interior, although there are many cases that do and must secure attention.

At Manistee and Holland we have depots, and earnest, faithful men and women working to distribute to the necessities of the people. Our great depot is at Grand Haven, from which point we readily replenish smaller ones. We seek local contributions in kind, food, second-hand clothing, cooking stoves, etc. I hope $1,000 of your bounty has reached Peshtigo or vicinity. Some of it is at Holland, in the shape of ticking for straw beds, or cheap prints and wadding being made into warm comfortables; hammers, nails, putty, glass, flour, pork, etc.—a few dollars in a shanty here and barracks there. When a people finds the wolf at the door, and is fighting him for life, it can spread a little money very thin. Our operations must continue a long time. We feel as though we had entered on a winter's campaign. We are very thankful for your people's liberality. We invoke your sympathy and aid in the future. I beg leave to say that if some of your wealthy men would loan to those Holland farmers some money to aid them in building, they would combine business with charity. Probably prompt payment of interest could hardly be expected the first year, but the whole principal and interest could be made very safe. They are the most industrious people I ever saw. They settled twenty-five years ago on that flat land, densely covered with large timber, and, with nothing but their hands, they converted nine townships into a garden.

This was written Oct. 30th, and on Nov. 15th the committee of Boston relief reported upon Michigan as follows:

On the 16th of October, Governor Baldwin appointed two State relief committees, one for the eastern shore, and another for the western shore.

(1) The committee for the eastern shore is located in Detroit and consists of four gentlemen, of whom Charles M. Garrison is chairman. This committee has charge in general of the peninsula between Saginaw Bay and Lake Huron, the burnt region of which comprises twenty-three townships severely, and eighteen partially burned, and embraces an area of more than 1,400 square miles. In the parts severely burned the committee say that nine-tenths of the houses were consumed. Extreme drought had prevailed throughout the West for many weeks, and there had not been a rainy day since the beginning of June. During this time fires were raging in the woods in many localities. The same gale which blew upon Chicago on Sunday night, October 8, swept over the burning woods of Michigan and Wisconsin, and in places increasing to tornadoes, fanned the scattered fires on the east side of Michigan into a general conflagration.

Its fearful power may be illustrated by the case of White Rock, on the coast. Here the population, that had fought the fire for weeks, were aroused at one o'clock at night by the roar of the tornado, and fled before it. They waded out into the Lake up to their necks and remained there until seven in the morning, when, exhausted, they returned to the beach and slept till noon. Boats which went to relieve the sufferers were unable to go within miles of the shore for nearly two days, on account of the dense smoke and fiery cinders. And yet we were told that not more than twenty lives were lost in this eastern division, although from three thousand to four thousand people were rendered utterly destitute.

The Detroit committee has thoroughly canvassed this whole district. Lists of all the needy inhabitants are in the hands of the Supervisors of each township, through which individual wants are ascertained. This committee has established seven stations along the shore, at the most convenient points for distribution. The region is almost inaccessible for supplies during the winter; and the committee is endeavoring to accumulate, before the closing of water communication, sufficient stores to last until spring. The clothing which we saw was much of it poor and unassorted, but the committee believed that there was a sufficient supply both for present and prospective use. There was not, however, money enough to meet the requisitions of the agents on the ground, and further contributions are especially desired for the purchase of such articles as can best be obtained there.

(2) The committee for the western shore, whose headquarters are at Grand Rapids, consists of five gentlemen, with Hon. Thomas D. Gilbert as chairman. The territory under its charge lies in two distinct sections, and embraces the region around Holland and the region around Manistee.

Manistee, which is situated about one hundred and fifty miles north of Grand Haven, is a lumber settlement. About one-half of the mills and one-half of the houses of this town were burned. Of these burned houses one-half were owned by the wealthier mill-owners, and the remainder by the inhabitants of Manistee. These latter were stripped of everything. Nevertheless, as regular fall supplies were on their way, and the mill-owners were giving employment to the laboring population, we did not deem it necessary to visit this place.

Holland, a fine town of about 3,000 inhabitants, was settled twenty-five years ago by a poor Dutch colony under the lead of their religious teacher, Rev. A. C. Van Raalte. In a quarter of a century, by their thrift and industry, they had changed the wilderness into nine prosperous townships. The fire had been raging for several days in the immediate vicinity previous to Sunday, October 8. That night it struck the town and raged from a little after midnight until six o'clock the next afternoon, laying waste about two-thirds of the whole town, including the business portion of the city. There were destroyed five churches, three hotels, sixty-eight stores, and more than three hundred dwelling-houses; all of which were a total loss in consequence of a religious prejudice of the people against insurance. Although the inhabitants had to flee suddenly in the night, but a single life was lost.

Besides, more than one hundred and forty farms were swept of everything, buildings, fences, and trees. These sufferers, both in town and country, were left almost entirely destitute, and what is worse, without work for the immediate future. The committee at Grand Rapids informed us that sufficient clothing had been received, except some special articles, but that contributions of money will be very acceptable.

The Grand Rapids committee agreed with the principal citizens of Holland that the most efficient aid that could be rendered to that section is a loan to the burnt-out tradesmen and farmers, of from fifty thousand to one hundred thousand dollars, in sums of from five hundred to one thousand dollars each, on good security, with interest.

We have made arrangements with reliable persons for extending relief to special cases which would not otherwise be reached.

It should be added that isolated fires throughout the northern portion of Michigan have caused no inconsiderable loss and suffering, which must needs depend for relief chiefly on local charity."

Through the liberality so universal and divine, the pressing necessities of this immense army of sufferers have been supplied, and now that winter is upon them with its snow and icy blasts, they must be remembered still in their poverty and loneliness. If it be asked what their neighbors and fellow-citizens are doing, we could answer by telling this story:—

A well-known Detroit lawyer, while conversing with a gentleman, incidentally learned that the latter had been driven out of

Chicago by the great fire, with nothing in the way cf clothing except a single suit. The lawyer, without waiting to hear anything more, hastily pulled off a new beaver-cloth overcoat and gave it to the Chicago man, with the remark that he had another, and could buy more if need be. It was a spontaneous act of charity, and one which the recipient seemed to fully appreciate.

The Western people are generally free-hearted and sympathetic, and will share the last loaf with the unfortunate. Everything that can be done they will do for the victims of these awful calamities.

CHAPTER XLIII.

OUR sketch of Western fires we bring to a close, with notices of the remoter districts where the population was sparse. A party from Dakota says:—

For some days previous to leaving Cheyenne River, in Dakota, at a point seventy-five miles west of the crossing of the Northern Pacific Railroad at Red River, a dense smoky atmosphere prevailed, which each day grew more dense, warning us that immense "prairie fires" were approaching our quarters rapidly, and our party deemed it prudent to move eastward as fast as possible. We made immediate preparations, but found that we were in the saddle none too soon. The intense heat and weight of smoke affected us very much, and soon after starting we were forced to ride as rapidly as it was possible for our beasts to carry us. All through that long day we toiled along, our eyes nearly blinded; with parched throat and cracked lips and intense thirst we rode on and on, till at nightfall we came in sight of Red River, having ridden seventy-five miles without rest or halt but once. Glad were the hearts of our party, and much rejoicing

ENTERPRISING YOUNG MERCHANT DISPOSING OF RELICS OPPOSITE THE RUINS OF THE SHERMAN HOUSE.

was there at our escape from great danger, if not from loss of life. At points along the route the wall of flame would be quite near us. Its roar could be heard many miles, and its rapid motion was surprising. The line of fire seemed to be a solid wall of flame of about twenty to thirty feet in height, and moved as rapidly as a fleet horse could run. Occasionally a portion of the line would break away in bodies of forty or more feet square, and be carried with almost electric rapidity a distance of fifty or a hundred rods ahead, and then strike the high, dry grass, which would immediately ignite and add its destroying force to the already gigantic conflagration.

After resting at Red River, our party, reduced to three persons, moved on eastward and southward, passing over a district but lately burned. We could not distinguish any object fifty yards away, great, heavy clouds of smoke hanging like a pall through all the distance of 250 miles we travelled before reaching the Mississippi River, and even there the smoke was very oppressive. We deviated somewhat from a usual route travelled, and found at different points the charred remains of three human beings, nothing left but the bodies, and those burned to a crisp. The sight was horrible in all particulars, and not a thing could be found that would in any way identify the burned corpses.

We heard of one case that showed great presence of mind and much calmness. A man who had been with Sherman in his "march to the sea," was caught in the midst of a fire which was approaching him from all sides. Having no matches to create what is called "setting a back fire," and death staring him in the face, his wit suggested a "gopher hole." Setting at work with the will that a man would use who was working for life, he attacked the sod with a large hunting-knife. Cutting a large piece away, he rolled it back, and at once commenced throwing the soft, dry earth upward and outward, and soon had a hole dug of sufficient size to admit his body. Carefully drawing the sod toward him, he succeeded in drawing it over his body, and then filled up the "chinks" with dirt from within. He lay there until the fire passed over him and was speeding furiously on its way miles distant, then slowly he crawled out of his living grave, heated fearfully, but injured in no way whatever. His soldier experience had saved his life.

No one who has not witnessed this besom of destruction on the "plains," can form any adequate idea of its magnitude, its velocity, its fiendish-like cruelty, its thundering roar, and its vast destruction.

The prairie fires in the section of country above and below Yankton, Dakota, on Wednesday, were terrible.

In the afternoon of that day the flames swept into the village of Bon Homme with resistless fury, and the terror-stricken and helpless populace saw a mill and three or four dwellings disappear in the fiery blast. Among the latter was a house occupied by a widow woman as a boarding-house, and while she was expressing to our informant her fear that the fire would reach the town, a wave of flame came whirling on like a frightened steed, and before an effort could be made to save anything in or about the premises, the house was wrapped in flames and everything was lost.

The down coach found the country pretty well burned over to within a mile or two of Yankton, and the fire is still burning in various directions. The ruins of four smouldering houses were seen, grain and hay stacks were blazing on all sides, and burning fences swept across the country in all directions. At one point a little girl, some ten years of age, appeared at the roadside and piteously petitioned the people in the coach for help, saying that her father and mother were away from home, and that she had two sisters and a little brother, all younger than herself, in the house, and the premises were in immediate danger of destruction. Leaving one of the passengers to watch the horses, the rest ran to the house, and by starting a contrary fire succeeded in saving the place.

Another terrible fire raged the same day this side of Yankton, and within a few miles of that city. The flames swept toward Yankton, and in their course devoured several houses, besides numerous barns, sheds, and stacks of grain. The coach due in Yankton on Wednesday evening had an exciting time of it. It was discovered that the fire was coming, and a race was instituted. The driver plied his whip, and away the horses went on a gallop. Nearer and nearer came the fire. The red glare filled the sky; the forked tongue shot out; the terrible hissings of the demon were in the ears of the affrighted passengers. The driver gathered his lines, drew the leaders from the road, the horses gathered, jumped, a rail fence was beneath the wheels of the coach, the coach was on a piece of ploughed ground, and the fire went by with a roar like a cataract.

The particulars of a miraculous escape from death from a prairie fire were related to us by two gentlemen who arrived here last evening from Grant County, Wisconsin, Messrs. J. L. Finley and W. Kinney. They came the entire distance with a mule team and wagon across the country. Their trip was without any incident of note until they reached a point upon the prairie about six miles east of Le Mars, where they were surrounded by a prairie fire on Wednesday, the 4th inst. They were encamped upon a small bottom, preparing their noon meal, when the fire made its appearance. As soon as the fire was discovered approaching toward them Mr. Kinney ran from it and attempted to start a fire and follow after it, while Mr. Finley set himself to work hitching the team to the wagon, but owing to the high wind Kinney was unsuccessful in his effort. The fire was now fast approaching them, and they abandoned the attempt of setting another fire, and ran for their lives. Mr. Kinney struck out on foot, and his companion put the whip upon the mules and attempted to make his escape with them, but finding this impossible he reined his team around toward the fire, and, after repeated attempts, succeeded in running them through it, with no damage to himself and but a little to the mules. Mr. Kinney, who had become separated from his companion, ra.. into the bed of a dry stream, on the banks of which the fire was raging. He was nearly suffocated with the smoke, and wild with terror, and in his attempt to get out of the stream he fell into the fire, burning his hands seriously; after which he lay close to the bed of the stream until the fire passed over, when he came out and found his companion, who was some distance from him, uninjured.

They had two dogs with them, one of which was burned to death, and the other one took refuge in a deep sink hole or well, near where they had taken their dinner, and he was saved.

In view of all the miseries and calamities, crimes and casualties of the past twelve months, *The Chicago Tribune* christens this *The Black Year*.

The year 1871 will hardly be considered in history a year of grace. In point of fatality to human life, and destruction to material values by extraordinary natural causes, no year in the history of the world can equal it. Overwhelmed as we are by our own disaster, we have given little attention to what has been transpiring abroad, and have almost come to consider ourselves the only sufferers. The retrospect, however, is a terrible one. War, famine, pestilence, fire, wind, water, and ice, have been let loose, and done their worst, and with such appalling results, and with such remarkable phenomena accompanying them, that it is not to be wondered at men have sometimes thought the end of the world had come. We have seen our own fair city laid in ashes, throughout almost its entire business limits, and seventy thousand people left homeless. On that same night the conflagration swept through Northern Wisconsin and Michigan, sweeping village after village with horrible loss of life, and ruining thousands of acres of timber, the cutting and milling of which formed the main industry of that region. Illinois, Minnesota, Indiana, New York, Pennsylvania, Kansas, Missouri, and California, the Alleghanies, the Sierras, and the Rocky Mountains have been ravaged by fire, destroying immense amounts of property and entailing wide-spread suffering. Chicago is not the only city which has suffered. Peshtigo, Manistee, Cacheville, and Vallejo, Cal., Urbana, Darmstadt, and Geneva, under the Alps, have all been visited by terrible fires; and the torch of the incendiary has been applied successively to Louisville, St. Louis, Toronto, Montreal, and Syracuse.

The pestilence has walked at noonday. The cholera has steadily travelled from Asia westward through Europe, and our despatches of yesterday announced its arrival at New York Quarantine. One of the most appalling plagues of modern times, arising from yellow fever, has swept over portions of South America, and in Buenos Ayres alone 28,000 bodies were buried in one cemetery. Persia has been almost depopulated by the plague, which has been rendered all the more terrible by the added horrors of famine; and now, in our own country, small-pox has appeared as an epidemic in nearly every large city.

Storms, in their various manifestations, have never been so destructive before. In one night, a river in India suddenly rises, swollen by a storm, and sweeps away an entire city, destroying 3,000 houses, and utterly prostrating the crops. The little French seaport town of Pornic has been almost utterly destroyed by a tidal wave. The icebergs of the Arctic have caught and imprisoned within their impassable walls thirty-three whalers, inflicting a loss of a million and a half of dollars upon the city of New Bedford, and seriously crippling an important branch of industry. St. Thomas has been devastated by a hurricane which left 6,000 people homeless, and strewed its coasts with wrecks. A typhoon of terrible power has swept along the Chinese coast, destroying everything in its course,—towns, shipping, and life. A hurricane at Halifax has inflicted a severe blow upon English shipping. The storms on the English coast have never been so severe before, nor so fruitful in maritime disasters. A tidal wave at Galveston swept off all the shipping in port. A tornado has swept through Canada, doing serious damage in Toronto, Montreal, and Quebec. The Island of Formosa has been nearly destroyed by an earthquake.

Add to these the unusual crop of murders and suicides in this country, the alarming increase of railroad and steamboat disasters, the monstrous villanies which have been brought to light in public offices and private corporations, the Franco-German war with its attendant horrors, and the statement of the astronomers that there has been an explosion in the sun, and that two or three comets are just now in danger of losing their tails by their proximity to that orb,—and we may be justified in assuming that the year 1871 will be known in future calendars as the Black Year.

HISTORY OF THE GREAT FIRES IN THE PAST.

CHAPTER XLIV.

Fire is a good servant, but a bad master, says the old proverb, and this has been the experience of men in all ages. Virgil, the Latin poet, in his poem the Æneid, introduces his hero Æneas as the narrator of the Siege of Troy by the Greeks, and its final fall, and destruction by the torch of the incendiary. The reader will observe the abundance of supernatural machinery which accompanies the description, for in those days, as Paul observed in Athens, men were very religious, and had as many gods as occasion required, to account for the events which transpired, and the conduct of men.

The Greeks had long besieged Troy in vain. They contrived at last a huge wooden horse and placed it near the gates, filled with armed men, their bravest and best. To induce the Trojans to admit the monster, they persuaded one Sinon to throw himself in the way of capture, and enter Troy and explain the object of this horse in such a way as to secure its admission. The treacherous scheme succeeds, and here we let the hero tell his own tale of woe We premise that Priam is king of Troy, Calchas is a soothsayer of the Greeks, and Laocoon is a Trojan warrior, who has hurled his spear against the horse in disdain. Laocoon and his two sons were sacrificing at the altars, when lo! two serpents, with orbs immense, bear along the sea, and with equal motion shoot forward to the shore. They, with resolute motion, advance towards Laocoon; and first both serpents, with close embraces, twine around the little bodies of his two sons, and with their fangs mangle their wretched limbs. Next they seize himself, as he is coming up with weapons to their relief, and bind him fast in their mighty folds; and now grasping him twice about the middle, twice winding their scaly backs around his neck, they overtop him by the head and lofty neck. He strains at once with his hands to tear asunder their knotted spires: at the same time he raises hideous shrieks to heaven. Meanwhile, the two serpents glide off to the high temple. All urge with general voice to convey the statue into the city. The fatal machine passes over our walls, pregnant with arms. Four times it stopped in the very threshold of the gate, and four times the arms resounded in its womb; yet we, heedless, and blind with frantic zeal, urge on, and plant the baneful monster in the sacred citadel. The Trojans, dispersed about the walls, were hushed: deep sleep fast binds them weary in his embraces. And now the Grecian host, in their equipped vessels, set out from Tenedos, making towards the well-known shore, by the friendly silence of the quiet moonshine, as soon as the royal galley had exhibited the signal fire; and Sinon, preserved by the will of the adverse gods, in a stolen hour unlocks the wooden prison to the Greeks shut up in its womb: the horse, from his expanded caverns, pours them forth to the open air, and with joy issue from the hollow wood Thessandrus and Sthenelus the chiefs, and dire Ulysses, sliding down by a suspended rope, with Athamas and Thoas, Neoptolemus, the grandson of Peleus, and Machaon who led the way, with Menelaus, and Epeus the very contriver of the trick. They assault the city buried in sleep and wine. The sentinels are beaten down; and with opened gates they receive all their friends, and join the conscious bands.

Meanwhile, the city is filled with mingled scenes of woe; and though my father Anchises' house stood retired, and inclosed with trees, louder and louder the sounds rise on the ear, and the horrid din of arms assails. I start from sleep, and, by hasty steps, gain the highest battlement of the palace, and stand with erect ears. Then, indeed, the truth is confirmed, and the treachery of the Greeks disclosed. Now Deiphobus' spacious house tumbles down, overpowered by the conflagration; now, next to him, Ucalegon blazes; the Straits of Sigæum shine far and wide with the flames. The shout of men and clangor of trumpets arise. My arms I snatch in mad haste. I hurry away into flames and arms. Who can describe in words the havoc, who the deaths of that night? or who can furnish tears equal to the disasters? Our ancient city, having borne sway for many years, falls to the ground; great numbers of sluggish carcasses are strewn up and down, both in the streets, in the houses, and the sacred thresholds of the gods. Nor do the Trojans alone pay the penalty with their blood: the vanquished, too, at times, resume courage in their hearts, and the victorious Grecians fall; everywhere is cruel sorrow, everywhere terror and death in thousand shapes. I stood aghast; the image of my dear father arose to my mind, when I saw the king, of equal age, breathing out his soul by a cruel wound; Creüsa,* forsaken, came into mind, my rifled house, and the fate of the little Iülus. I look about, and survey what troops were to stand by me. All had left me through despair, and flung their fainting bodies to the ground, or gave them to the flames. And thus now I remained all alone.

Then, indeed, all Ilium seemed to me at once to sink in the flames, and Troy, built by Neptune, to be overturned from its lowest foundation. Down I come, and under the conduct of the god, clear my way amidst flames and foes: the darts give place, and the flames retire. And now, when arrived at the gates of

---

* Creüsa, daughter of Priam, and the wife of Æneas, who was lost in the streets of Troy, when Æneas made his escape with his father Anchises and his son Ascanius.

my paternal seat and ancient house, my father, whom I was desirous first to remove to the high mountains, and whom I first sought, obstinately refuses to prolong his life after the ruin of Troy, and to suffer exile.

Such purpose declaring, he persisted, and remained unalterable. On the other hand, I, my wife Creüsa, Ascanius, and the whole family, bursting forth into tears, besought my father not to involve all with himself, nor hasten our impending fate.

"Now, son, I resign myself indeed, nor refuse to accompany you in your expedition," he said; and now throughout the city the flames are more distinctly heard, and the conflagration rolls the torrents of fire nearer. "Come then, dearest father, place yourself on my neck; with these shoulders will I support you, nor shall that burden oppress me. However things fall out, we both shall share either one common danger or one preservation: let the boy Iülus be my companion, and my wife may trace my steps at some distance." This said, I spread a garment and a tawny lion's hide over my broad shoulders and submissive neck, and stoop to the burden: little Iülus is linked in my right hand, and trips after his father with unequal steps; my spouse comes up behind. We haste away through the gloomy paths; and I, whom lately no showers of darts could move, nor Greeks inclosing me in a hostile band, am now terrified with every breath of wind; every sound alarms me anxious, and equally in dread for my companion and my burden. By this time I approached the gates, and thought I had overpassed all the way, when suddenly a thick sound of feet seems to invade my ears just at hand; and my father, stretching his eyes through the gloom, calls aloud, "Fly, fly, my son, they are upon you; I see the burnished shields and glittering brass." Here, in my consternation, some unfriendly deity or other confounded and bereaved me of my reason; for while in my journey I trace the by-paths, and forsake the known beaten tracks, alas! I know not whether my wife Creüsa was snatched from wretched me by cruel fate, or lost her way, or through fatigue stopped short; nor did these eyes ever see her more. Nor did I observe that she was lost, or reflect with myself, till we were come to the rising ground, and the sacred seat of ancient Ceres: here, at length, when all were convened, she alone was wanting, and gave disappointment to all our retinue, especially to her son and husband. To my friends I commend Ascanius, my father Anchises, with the gods of Troy, and lodge them secretly in a winding valley. I myself repair back to the city, and brace on my shining armor. I am resolved to renew every adventure, revisit all the quarters of Troy, and expose my life once more to all dangers. First of all, I return to the walls and the dark entry of the gate by which I had set out, and backward unravel my steps with care amidst the darkness, and run them over with my eye. Horror on all sides, and at the same time the very silence affrights my soul. Thence homeward I bent my way, lest by chance, by any chance, she had moved thither; the Greeks had now rushed in, and were masters of the whole house. In a moment the devouring conflagration is rolled up in sheets by the wind to the lofty roof; the flames mount above; the fiery whirlwind rages to the skies. Now adventuring even to dart my voice through the shades, I filled the streets with outcry, and in anguish, with vain repetition, again and again, called on Creüsa. While I was in this search, and with incessant fury ranging through all quarters of the town, the mournful ghost and shade of my Creüsa's self appeared before my eyes, her figure larger than I had known it. I stood aghast! my hair rose on end, and my voice clung to my jaws. Then thus she bespeaks me, and relieves my cares with these words: "My darling spouse, what pleasure have you thus to indulge a grief which is but madness? These events do not occur without the will of the gods. It is not allowed you to carry Creüsa hence to accompany you, nor is it permitted by the great ruler of heaven supreme. In long banishment you must roam, and plough the vast expanse of the ocean: to the land of Hesperia you shall come, where the Lydian Tiber, with his gentle current, glides through a rich land of heroes. There, prosperous state, a crown, and royal spouse, await you: dry up your tears for your beloved Creüsa. And now farewell, and preserve your affection to our common son."

With these words she left me in tears, ready to say many things, and vanished into thin air. There thrice I attempted to throw my arms around her neck; thrice the phantom, grasped in vain, escaped my hold, swift as the winged winds, and resembling most a fleeting dream. Thus having spent the night, I at length revisit my associates. And here, to my surprise, I found a great confluence of new companions: matrons, and men, and youths, drawn together to share our exile, a piteous throng!

CHAPTER XLV.

SENATOR THURMAN, of Ohio, in a speech made in behalf of the contributions for Chicago, thus alluded to the burning of Ancient Rome:—

The memorable fire at Rome, in the reign of Nero, destroyed nearly five-sevenths of the city, and included within the ruins were her most stately temples and public buildings, and her rarest and most valuable collections of literature, science, and the arts. But it is not probable that the pecuniary loss was as great as that at Chicago. And the individual suffering, dreadful as it was, was mitigated by the warmth of a summer sun, for the fire occurred in July. Yet it was horrible; and, in order to recall the scene to your memories, and because in many points it resembles that at Chicago, I cannot refrain from reading the vivid description of it by Tacitus:—

"A dreadful calamity," says he, "followed in a short time after, by some ascribed to chance, and by others to the execrable wickedness of Nero. The authority of historians is on both sides, and which predominates it is not easy to determine. It is, however, certain, that of all the disasters that ever befell the city of Rome from the rage of fire, this was the worst, the most violent and destructive. The flames broke out in that part of the circus which adjoins on one side to Mount Palatine, and on the other to Mount Cælius. It caught a number of shops stored with combustible goods, and, gathering force from the wind, spread with rapidity from one end of the circus to the other. Neither the thick walls of houses, nor the inclosure of temples, nor any other building, could check the rapid progress of the flames. A dreadful conflagration followed. The level parts of the city were destroyed. The fire communicated to the higher buildings, and again, laying hold of interior places, spread with a degree of velocity that nothing could resist. The form of the streets, long and narrow, with frequent windings and no regular opening, according to the plan of ancient Rome, contributed to increase the mischief. The shrieks and lamentations of women, the infirmities of age, and the weakness of the young and tender, added misery to the dreadful scene. Some endeavored to provide for themselves, others to save their friends, in one part dragging along the lame and impotent, in another waiting to receive the tardy or expecting relief themselves; they lingered, they obstructed one another; they looked behind, and the fire broke out in front; they escaped from the flames, and in their place of refuge found no safety; the fire raged in every quarter; all were involved in one general conflagration. The unhappy wretches fled to places remote, and thought themselves secure, but soon perceived the flames raging round them. Which way to turn, what to avoid, or what to seek, no one could tell. They crowded the streets; they fell prostrate on the ground; they lay stretched in the fields, in consternation and dismay, resigned to their fate. Numbers lost their whole substance, even the tools and implements by which they gained their livelihood, and, in that distress, did not wish to survive. Others, wild with affliction for their friends and relations whom they could not save, embraced a voluntary death, and perished in the flames.

"During the whole of this dismal scene no man dared to attempt anything that might check the violence of the dreadful calamity. A crew of incendiaries stood near at hand denouncing vengeance on all who offered to interfere. Some were so abandoned as to heap fuel on the flames. They threw in fire-brands and flaming torches, proclaiming aloud that they had authority for what they did. Whether, in fact, they had received such horrible orders, or, under that device, meant to plunder with greater licentiousness, cannot now be known."

I make no apology for reading to you this lengthy extract, written nearly 1,800 years ago, to describe what befell Rome in the sixty-fourth year of the Christian era. It presents to your imaginations a more lively picture of what happened to an American city, within the present month, than anything I could say. When it was written the American continent was unknown. More than seventeen centuries after it was written, the city of Chicago sprang into existence; and yet, so similar are mankind in all ages, and so invariable at all times are the laws of Nature; that the words of the Roman historian describe, with almost equal fidelity, the destruction of the ancient mistress of the world and of the commercial mistress of an American State.

CHAPTER XLVI.

THE BURNING OF MOSCOW.

We reproduce from the pages of Sir Archibald Alison, whose history is deservedly ranked as standard authority, the following description of the burning of the ancient capital of Muscovy, an event which, more than all others combined, broke the power of the first Napoleon:—

At eleven o'clock on the 14th September, 1812, the advanced guard of the French army, from an eminence on the road, descried the long-wished-for minarets of Moscow. The domes of above two hundred churches, and the massy summits of a hundred palaces, glittered in the rays of the sun—the form of the cupolas gave an Oriental aspect to the scene; but, high above all, the cross indicated the ascendancy of the European faith. The scene which presented itself to the eye resembled rather a province adorned with palaces, domes, woods, and buildings, than a single city; a boundless accumulation of houses, churches, public edifices, rivers, and parks, stretched out over swelling eminences and gentle vales as far as the eye could reach. The mixture of architectural decoration and pillared scenery, with the bright-green foliage, was peculiarly fascinating to European eyes. Everything announced its Oriental character, but yet without losing the features of the West. Asia and Europe met in that extraordinary city.

. . . .

Struck by the magnificence of the spectacle, the leading squadrons halted, and exclaimed: "Moscow! Moscow!" and the cry, repeated from rank to rank, at length reached the emperor's guard. The soldiers, breaking their array, rushed tumultuously forward, and Napoleon, hastening in the midst of them, gazed impatiently on the splendid scene. His first words were: "Behold at last that famous city!" the next, "It was full time!" Intoxicated with joy, the army descended from the heights. The fatigues and dangers of the campaign were forgotten in the triumph of the moment, and eternal glory was anticipated in the conquest which they were about to complete. Murat at the head of the cavalry speedily advanced to the gates, and concluded a truce with Miladowitch for the evacuation of the capital. But the entry of the French troops speedily dispelled the illusions in which the army had indulged. Moscow was found to be deserted. Its long streets and splendid palaces resounded only with the clang of the hoofs of the invaders' horses. Not a sound was to be heard in its vast circumference; the dwellings of three hundred thousand persons seemed as silent as the wilderness. Napoleon waited in vain until evening for a deputation from the magistrates or chief nobility. Not a human being came forward to deprecate his hostility, and the mournful truth could at length be no longer concealed, that Moscow, as if struck by enchantment, was bereft of its inhabitants. Wearied of fruitless delay, the emperor, on the morning of the 15th, advanced into the city, and entered the ancient palace of the czars, amidst no other concourse than that of his own soldiers.

The Russians, however, in abandoning their capital, had resolved upon a sacrifice greater than the patriotism of the world had yet exhibited. The Governor, Count Rostopchin, set the example of devotion by preparing the means of destruction for his country palace, which was splendidly furnished, and adorned with the finest works of art, which he set fire to by applying the torch with his own hands to his nuptial chamber; and to the gates of the palace he had affixed the following inscription: "During eight years I have embellished this country-house and lived happily in it in the bosom of my family. The inhabitants of this estate, to the number of seventeen hundred, quit it at your approach, in order that it may not be sullied by your presence. Frenchmen! at Moscow I have abandoned to you my two houses, with their furniture, worth half a million roubles; here you will find nothing but ashes."

The nobles were prepared, in a public assembly, to have imitated the example of the Numantians, and destroy the city they could no longer defend, and Kutosoff had promised to give Rostopchin three days' notice before he evacuated the city, in order that it might be held. But owing to the advance of the French being more rapid than had been anticipated, the notice was not given or the meeting held, and the governor was left to act on his own responsibility. Everything, however, had been prepared for that noble sacrifice. The authorities, when they retired, carried with them the fire-engines, and everything capable of arresting a conflagration, and combustibles were disposed in the principal edifices to favor the progress of the flames. The persons intrusted with the duty of firing the city only awaited the retreat of their countrymen to commence the work of destruction. Rostopchin was the author of this sublime effort of patriotic devotion, but it involved a responsibility greater than either government or any individual could support, and he was afterward disgraced for the heroic deed.

The sight of the grotesque towers and venerable walls of the Kremlin first revived the emperor's imagination, and rekindled those dreams of Oriental conquest which, from his earliest years, had floated through his mind. His followers, dispersed over the vast extent of the city, gazed with astonishment on the sumptuous palaces of the nobles, and the gilded domes of the churches. Evening came on, and with increasing wonder the French troops traversed the central parts of the metropolis, recently so crowded with passengers, but not a living creature was to be seen to explain the universal desolation. It seemed like a city of the dead. Night approached; an unclouded moon illuminated those beautiful palaces, those vast hotels, those deserted streets—all was still; the silence of the tomb. The officers broke open the doors of some of the principal mansions in search of sleeping quarters. The found everything in perfect order; the bed-rooms were fully furnished as if guests were expected; the drawing-rooms bore the marks of having been recently inhabited; even the work of the ladies was on the tables, the keys in the wardrobes; but not an inmate was to be seen. By degrees a few of the lower class of slaves emerged, pale and trembling, from the cellars, showed the way to the sleeping apartments, and laid open everything which these sumptuous mansions contained; but the only account they could give was that the whole of the inhabitants had fled, and that they alone were left in the deserted city. But the terrible catastrophe soon commenced. On the night of the 14th a fire broke out in the Bourse, behind the Bazaar, which soon consumed that noble edifice, and spread to a considerable part of the crowded streets in the vicinity. This, however, was but the prelude to more extended calamities. At midnight on the 15th, a bright light was seen too illuminate the northern and western parts of the city; and the sentinels on watch at the Kremlin soon discovered the splendid edifices in that quarter to be in flames. The wind changed repeatedly during the night, but to whatever quarter it veered the conflagration extended itself; fresh fires were every instant seen breaking out in all directions, and Moscow soon exhibited the spectacle of a sea of flame agitated by the wind. The soldiers, drowned in sleep or overcome by intoxication, were incapable of arresting its progress; and the burning fragments, floating through the hot air, began to fall on the roofs and courts of the Kremlin. The fury of an autumnal tempest added to the horrors of the scene; it seemed as if the wrath of heaven had combined with the vengeance of man to consume the invaders of the city they had conquered.

But it was chiefly during the nights of the 18th and 19th that the conflagration attained its greatest violence. At that time the whole city was wrapped in flames, and volumes of fire of various colors ascended to the heavens in many places, diffusing a prodigious light on all sides, and attended by an intolerable heat. These balloons of flame were accompanied in their ascent by a frightful hissing noise and loud explosions, the effect of the vast stores of oil, resin, tar, spirits, and other combustible materials with which the greater part of the shops were filled. Large pieces of painted canvas, unrolled from the outside of the buildings by the violence of the heat, floated on fire in the atmosphere, and sent down on all sides a flaming shower, which spread the conflagration in quarters even the most removed from where it originated. The wind, naturally high, was raised by the sudden rarefaction of the air produced by the heat, to a perfect hurricane. The howling of the tempest drowned even the roar of the conflagration; the whole heavens were filled with the whirl of the volumes of smoke and flame which rose on all sides, and made midnight as bright as day; while even the bravest hearts, subdued by the sublimity of the scene, and the feeling of human impotence in the midst of such elemental strife, sank and trembled in silence.

The return of day did not diminish the terrors of the conflagration. An immense crowd of hitherto unseen people, who had taken refuge in the cellars and vaults of their buildings, issued forth as the flames reached their dwellings; the streets were speedily filled with multitudes flying in every direction with their most precious articles; while the French army, whose discipline this fatal event had entirely dissolved, assembled in drunken crowds, and loaded themselves with the spoils of the city. Never in modern times had such a scene been witnessed. The men were loaded with packages, charged with their most precious effects, which often took fire as they were carried along, and which they were obliged to throw down to save themselves. The women had often two or three children on their backs, and as many led by the hand, which, with trembling steps and piteous cries, sought their devious way through the labyrinth of flame. Many old men, unable to walk, were drawn on hurdles or w el-barrows by their children and grandchildren, while their burnt beards and smoking garments showed with what difficulty they had been rescued from the flames. Often the French soldiers, tormented by hunger and thirst, and loosened from all discipline by the horrors which surrounded them, not contented with the booty in the streets, rushed headlong into the burning edifices, to ransack their cellars for the stores of wine and spirits which they contained, and beneath the ruins great numbers perished miserably, the victims of intemperance and the surrounding fire. Meanwhile the flames, fanned by the tempestuous gale, advanced with frightful rapidity, devouring alike in their course the palaces of the great, the temples of religion, and the cottages of the poor. For thirty-six hours the conflagration continued at its height, and during that time above nine-tenths of the city was destroyed. The remainder, abandoned to pillage and deserted by its inhabitants, offered no resources to the army. Moscow had been conquered; but the victors had gained only a heap of ruins. It is estimated that 30,800 houses were consumed, and the total value of property destroyed amounted to £30,000,000.

CHAPTER XLVII.

THE GREAT FIRE IN LONDON.

We must go back more than a couple of centuries to find a parallel to the terrible fire which has wrapped the city of Chicago in a sea of resistless flame. On the 2d of September, 1666, the city of London was almost entirely destroyed by what has since been known as the Great Fire. This awful conflagration gained headway with the same terrible rapidity as that of Sunday night, and in five dreadful days of ruin and terror and panic laid two-thirds of the English metropolis in ashes. Like the fire at Chicago, it broke out upon a Sunday, though at a different hour—two o'clock in the morning. It originated in a bakehouse, kept by a man with the quaint name of Farryner, at Pudding lane, near the Tower. At that period the buildings in the English capital were chiefly constructed of wood, with pitched roofs, and in this particular locality, which was immediately adjacent to the water side, the stores were mainly filled with materials employed in the equipment of shipping, mostly of course of a highly combustible nature. To add to the conspiring causes of the immense mischief in which the fire ultimately resulted, the pipes from the New River—the source of the water supply of the city—were found to be empty, and the engine which raised water from the Thames was among the first property destroyed. The vacillation and indecision of the lord mayor aggravated the confusion. For several hours he refused to listen to the counsel given him to call in the aid of the military, and when the probable proportions of the fire were plainly apparent, and when it was clear that the destruction of a block of houses was absolutely necessary to the preservation of the city, he declined to accept the responsibility of destroying them until he could obtain the consent of their owners. All through Sunday the wind increased in violence, and the fire sped with incredible rapidity from house to house, from street to street, on its work of havoc. We cannot now do better than transcribe the account of the further mischief caused by the fire, given by Mr. John Evelyn, in his "diary." It reads as follows:—

"*Sept.* 3. The fire continuing, after dinner I took coach with my wife and son, and went to the Bankside, in Southwark, where we beheld that dreadful spectacle—the whole city in dreadful flames near ye water side: all the houses from the bridge, all Thames street, and upwards towards Cheapside down to the Three Cranes, were now consumed.

"The fire having continued all this night (if I may call that night which was as light as day for ten miles round about after a dreadful manner) when conspiring with a fierce eastern wind in a very drie season; I went on foot to the same place, and saw the whole south part of the city burning, from Cheapside to the Thames, and all along Cornehill (for it kindled back against the wind as well as forwards), Tower street, Fenchurch street, Gracious street, and so along to Bainard's castle, and was now taking hold of St. Paul's church, to which the scaffolds contributed exceedingly. The conflagration was so universal and the people so astonished, that, from the beginning—I know not from what, despondency or fate—they hardly strived to quench it, so that there was nothing hearde or seene but crying out and lamentations, running about like distracted creatures, without at all attempting to save even their goods, such a strange consternation there was upon them—so, as it burned both in length and breadth, the churches, public halls, Exchange, hospitals, monuments and ornaments, leaping after a prodigious manner from house to house and streete to streete, at greate distance one from ye other; for ye heate, with a long set of fair and warme weather had even ignited the air, and prepared the materials to conceive the fire which devoured after an incredible manner houses, furniture and everything. Here we saw the Thames covered with goods floating, all the barges and boats laden with what some had time and courage to save, as on ye other, ye carts, &c., carrying out to the fields, which for many miles were strewed with moveables of all sorts, and tents erecting to shelter both people and what goods they could get away. Oh, the miserable and calamitous spectacle such as haply the world had not seene the like since the foundation of it, nor to be outdone till the universal conflagration. All the sky was of a fiery aspect, like the top of a burning oven, the light seene above forty miles round about for many nights. God grant my eyes may never behold the like, now seeing above ten thousand houses all in one flame; the noise and crackling and thunder of the impetuous flames; ye shrieking of women and children, the hurry of people, the fall of towers, houses and churches was like an hideous storme, and the fire all about so hot and inflamed that at last one was not able to approach it, so that they were forced to stand stille and let the flames burn on, which they did for neere two miles in length and one in breadth. The clouds of smoke were dismall, and reached upon computation near fifty miles in length. Thus I left it in the afternoone burning—a resemblance to Sodom or the last day. London was, but is no more!

"*Sept.* 4. The burning still rages, and it was now gotten so far as the Inner temple, olde Fleete streete, the Olde Bailey, Ludgate Hill, Warwick lane, Newgate, Paule's Chain, Watling streete, now flaming and most of it reduced to ashes; the stones of Paule's flew like grenades, ye melting lead running downe the streetes in a streame, and the very pavements glowing with fiery rednesse, so as no horse or man was able to tread on them, and the demolition had stopped all the passages, so that no help could be applied. The eastern wind still more impetuously drove the flames forward. Nothing but ye almighty powers of God was able to stay them, for vaine was ye helpe of man.

"*Sept.* 5. It crossed towards Whitehalle; oh, the confusion there was then at that court! It pleased his majesty to command me among the rest to looke after the quenching of Fetter lane, and to preserve if possible that part of Holborne, while the rest of ye gentlemen tooke their several posts and began to consider that nothing was so likely to put a stop but the blowing up of so many houses as might make a wider gap than any that had yet been made by the ordinary method of pulling them down by engines."

. . . .

Then after a description of the abating of the wind, and the gradual dying out of the fire, the quaint old diarist continues:—

"The poore inhabitants were dispersed about St. George'

Fields and Moorfields, as far as Highgate, and several myles in circle; some under tents, some under miserable hutts and hovels, many without a rag or any necessary utensils, bed or board, who from delicatenesse, riches and easy accommodation in stately and well furnished houses, were reduced now to extreamest misery and poverty."

. . . .

And again:—

"I then went towards Islington and Highgate, where one might have seene 200,000 people of ranks and degrees dispersed and lying along by their heapes of what they could save from the fire, deploring their losse, and though ready to perish from hunger and destitution, yet not asking one penny for relief, which to me appeared a stranger sight than I had yet beheld."

How vivid an idea of the suffering and misery entailed by this terrible visitation we find in this simple but expressive narrative! Nearly two-thirds of the entire city was destroyed. Thirteen thousand houses, eighty-nine churches, and many public buildings were reduced to charred wood and ashes. Three hundred and seventy-three acres within, and sixty-three acres without the walls were utterly devastated. Well might Mr. Evelyn compare the fire to that which overwhelmed Sodom and Gomorrah, or that other and yet more awful one which will engulf the entire world at the Day of Doom.

John Howe preached a sermon on the rebuilding of London, taking for his text these words, "The street shall be built again, and the wall even in troublous times." In a note to that discourse the editor thus describes the fire and the restoration of the city: "The dreadful fire so often alluded to began on September 2d, 1666, near the place where the monument now stands, by which one of the noblest and most magnificent cities in the world was turned into ashes in a few days. A raging east wind, we are told, fomented it to an incredible degree, which in a moment raised the fire from the bottom to the tops of the houses, and scattered prodigious flakes in all places, which were mounted so vastly high into the air as if heaven and earth were threatened with the same conflagration. The fury soon became insupportable against all the arts of men and power of engines; and besides the dreadful scenes of flames, ruins, and desolations, there appeared the most killing sight under the sun—the distracted looks of so many citizens, the wailings of miserable women, and the cries of poor children and decrepit old people, with all the marks of confusion and despair."

The inscription on the pillar erected by that famous architect, Sir Christopher Wren, in memory of this calamity, tells us: "The fire, with incredible fury and noise, destroyed eighty-nine churches, among which was the Cathedral of St. Paul; many public hospitals, schools, libraries, a vast number of stately edifices, thirteen thousand two hundred dwelling-houses, four hundred streets, etc. The destruction was sudden; for in a short time the same city which was seen in a flourishing condition, was reduced to nothing; and in a few days, when the fatal fire had, in appearance, overcome all means of resistance and human counsels, by the will of Heaven it stopped and was extinguished. All persons were indefatigable in the work of rebuilding, and making provision for the resurrection of this city; and Sir Jonas Moor, after having raised Fleet street according to the model appointed, from that beginning the city advanced so hastily towards a general perfection that, within the compass of a few years, it far transcended its former splendor."

CHAPTER XLVIII.

NEW YORK'S GREAT FIRE.

THAT great event in the history of New York, the "Great Fire," occurred on the night of the 16th of December, 1835. It was declared by the croakers of the time a damper upon the city's prosperity and a clog to the wheels of its progress towards its present position. But though the people lost a great part of their capital, they did not lose their strength, energy, and enterprise, and the proper application of those qualities caused their city to rise Phœnix-like from its ashes, more beautiful, stronger, and fuller of life than before.

At between eight and nine o'clock of the evening above stated, the fire was discovered in the store No. 25 Merchant street, a narrow street that led from Pearl into Exchange street, near where the Post-office then was. The flames spread rapidly, and at ten o'clock forty of the most valuable dry-goods stores in the city were burned down or on fire. The narrowness of Merchant street, and the gale which was blowing, aided the spread of the destructive element. It passed from building to building, leaped across the street, between the blocks, urged by the gale and in nowise deterred by the feeble forces opposing it. The night was bitterly cold, and, though the firemen were most energetic, the freezing of the hose and the water in their defective engines, combined with their sufferings from the weather, made their efforts of little avail. The flames spread north and south, east and west, until almost every building on the area bounded by Wall, South, and Broad streets, and Coenties' slip, was burning, gutted, or levelled to the ground. There was not a building destroyed on Broad street, nor on the block on Wall street from William to Broad street, the fire taking an almost circular course just at the rear of the buildings on the streets named. The scene in the night was one of indescribable grandeur, the glare from the three hundred buildings that were at one time burning brightly lighting up the whole city. In all five hundred and thirty buildings were destroyed; they were of the largest and most costly description, and were filled with the most valuable goods. The total loss, estimated at about $20,000,000, was afterwards found to be about $15,000,000. Of the buildings destroyed the most important were the Merchants' Exchange, the Post-office, the offices of the celebrated bankers the Josephs, the Allens, and the Livingstons, the Phœnix Bank, and the building owned and occupied by Arthur Tappan, then much despised for his anti-slavery sympathies. The business portion of the city was alone that burned over, so that few poor were rendered otherwise than without employment.

NEW YORK, 1845.

The greatest fire since that of December, 1835, that has devas tated property in New York, began on the morning of the 20th of July, 1845. The fire originated in the sperm oil store in New street, near the corner of Exchange place, about three o'clock on the morning named, and spread over a great part of the territory which had been the scene of the conflagration of 1835. The flames were communicated to a chair factory adjoining and nearer to the corner of Exchange place, whence they passed along Exchange place to Broad street. There they enwrapped a building in which was a quantity of saltpetre, or gunpowder, on storage. When the building had been burning for about fifteen minutes a most awful explosion took place, which shook the city like an earthquake. The building was blown up, and with it some other buildings. Immediately after the explosion fire was discovered in four different places, and shortly the rear of the entire block was blazing. Soon the fire leaped to the south side of Broak street, passing at the same time to Broadway. All this time the firemen, although making the most strenuous efforts, had effected but little toward suppressing the flames. On Broadway they spread downward toward the Bowling Green; and on Broad street north toward Wall street and south to Beaver street, along which they passed to New street, both sides of which had been devastated. The fire was checked ere it had reached the magnificent Merchants' Exchange on its way to Wall street. Both sides of Exchange place, from Broadway to Broad street and half way down to William, were burned. Every building on Broadway from Exchange place down was levelled, and then the flames turned into Marketfield street, where they were checked. Altogether about three hundred buildings were destroyed, among which were the costly shrines of commerce and finance, and the abodes of the poverty-stricken. A liberal estimate of the total loss is made at $6,000,000, but this is belittled when the lamentable loss of life of which the explosion was the occasion is thought of. The number of persons whose lives were destroyed never was accurately ascertained; but it was generally believed at the time that about six persons perished.

A REMINISCENCE OF THE GREAT FIRE OF 1835.

On the night of the 16th of December, 1835, I was sitting with a literary friend, about nine o'clock, in one of the private boxes of Hamblin's magnificent Bowery Theatre. Suddenly the big bell of the City Hall boomed loud and long over the metropolis, and "Fire! fire!" echoed around and within the theatre. We were off in an instant, rushing out of the slamming doors, and onward toward the scene of the conflagration, which was "glaring on Night's startled eye" away down town.

When we reached Wall street, near Water, the Tontine Coffee House had caught, and dark smoke in huge masses, tinged with flickering flashes of bright flame, was bursting from all the upper windows. The night, as all who were out in it will well remember, was intensely cold. There was but little wind, but as the fire advanced there was plainly perceptible the "food of fire" in the air, as I firmly believe there always is in all great conflagrations; something mysterious as yet, and unexplainable. It was so in our great fire, for I saw its evidences myself, and I see that reports of the same evidences are mentioned as features of the still more terrible and vastly greater conflagration in Chicago, which has "roused the world." Science, there is little doubt, will find out, by and by, what this mysterious power is, and tell us how it is worked and how it may be guarded against, if not conquered. Whether it is atmospheric or electric, or whatever else it may be, is yet to be determined. A word or two more concerning this a little further on.

Our great fire travelled south and west faster than a man could walk. Water froze in all the gutters; thick ice coated the hydrants; crunched in the hose-pipes that encumbered the streets, and lay in "floes" where there was a shadow from the heat and the flame. But in a little while no water was wanted. Engines were soon useless, and no energetic "Sykesy" was required to "take the butt." Clouds of smoke, like dark mountains suddenly rising into the air, were succeeded by long banners of flame, rushing to the zenith, and roaring for their prey. Street after street caught the terrible torrent, until over acre after acre there was rolling and booming an ocean of flame! "All of this I saw, and part of it I was." The printing-office of the *Knickerbocker*, at first in South William street, was moved three times far beyond the prevailing fire, but was gradually followed by the raging enemy, and finally devoured.

As we were standing upon the roof of the Exchange, looking down upon the scene when in mid-progress, buildings far beyond the line of fire, and in no contact with it, burst in flames from the interior. The same thing, I observe, happened in Chicago, and was attributed to incendiaries; but there were no incendiaries suspected in our great fire. What latent power enkindled the inside of these advanced buildings, while externally they were un touched? A scientific writer at the time contended, I think in the old *Daily Advertiser*, that at a certain period there is what he called an "inflammable vacuum" in the air, which is self-igniting and irresistible. Perhaps, a hundred years or so from now, some safeguard against this mysterious element, now lying latent and sleeping in nature, may be discovered. It is not so very long since the old tea-kettle first lifted its lid to the science of steam, and talking round the world under water is a much younger wonder.

CHAPTER XLIX.

PITTSBURGH, 1845.

PITTSBURG, Pa., was visited by a most destructive conflagration the 10th of April, 1845. By it a very lage portion of the city was laid waste, and a greater number of houses destroyed than by all the fires that had occurred previously to it. Twenty squares, containing about 1,100 buildings were burned over. Of these buildings the greater part were business houses containing goods of immense value—grocery, dry-goods, and commission houses—and the spring stocks of the latter had just been laid in. The fire commenced in a frame building at the corner of Second and Ferry streets, and the prevailing strong wind urged it with fearful rapidity through the city. So short was the time between the discovery of the flames and their spread through the city, that many persons were unable to save any of their household goods, while others, having got theirs to the walk, were compelled to flee and leave them to be seized and destroyed by the element.

The merchants were equally unsuccessful in saving anything from their warehouses. The loss was estimated at $10,000,000

PHILADELPHIA, 1850.

A conflagration, by which an immense amount of property was destroyed, took place in Philadelphia, on the 9th of July, 1850. It began about four o'clock on the afternoon of that day, in a store at 78 North Delaware avenue. The fire was beyond control when discovered, and soon spread, despite the most strenuous efforts to prevent it, to the store-houses adjoining. When the fire had reached the cellar of the building in which it had originated, two explosions occurred, which rent the walls of the building and threw flakes of combustible matter in all directions, setting fire to many other buildings. Delaware avenue and Water street were covered with persons who exhibited little fear at these evidences of dangerous substances being stored in the building. Suddenly a third and most terrific explosion occurred, by which a number of men, women, and children were killed, and several buildings demolished. This disaster caused a panic among the firemen and spectators, and in the efforts of all to escape from danger many were trampled upon and injured. Some were thrown into the Delaware, and others jumped in to get away from the falling bricks and beams sent up from the burning building by the explosion. The number of persons who lost their lives by the explosion was about thirty—nine persons who jumped into the river in a fright were drowned—and about one hundred persons injured. The area over which the fire spread contained about four hundred buildings. Its locality was one of the most densely populated in the city, and a large number of the residents, having been poor people, the suffering caused was immense. The loss was about $1,000,000, and the fire would be a comparatively small one had there been no loss of life.

PHILADELPHIA, 1865.

The most terrible conflagration of which Philadelphia was the theatre, after that of July, 1850, occurred there on the morning of February 8, 1865. Like its predecessor, it brought death to many, and in the most horrible and painful manner. The fire originated among several thousand barrels of coal-oil that was stored upon an open lot on Washington street near Ninth. The flames spread through the oil as if it had been gunpowder, and in a very short time, 2,000 barrels were ablaze, and sending a huge volume of flame and smoke upward. The residents of the vicinity, awakened by the noise of the bells and firemen, and affrighted by the glare and nearness of the fire, rushed in their night garments into the streets that were covered with snow and slush. The most prompt to leave their homes got off with their lives, but those near the spot where the fire commenced, and not prompt to escape, were met by a terrible scene.

The blazing oil poured into Ninth street and down to Federal, making the entire street a lake of fire that ignited the houses on both sides of the street for two blocks. The flames also passed up and down the cross streets, and destroyed a number of houses. The fiery torch was whirled back and forth along the street at the pleasure of the wind, and as it passed destroyed everything in or near its course. People leaving their blazing homes, hoping to reach a place of safety, were roasted to death by it. Altogether, about twenty persons were roasted in the streets or houses. Firemen making vain endeavors to save the poor creatures from their horrible fate were fearfully burned. The loss of property amounted to about $500,000, and fifty buildings were destroyed. From Washington street to Federal, on Ninth, every building was burned.

SAN FRANCISCO.

The city of San Francisco was retarded in its progress toward its present proud position by many causes, but by nothing more than fire. The most destructive of the many conflagrations which have occurred in that city began on the 3d of May, 1851, at eleven o'clock P.M., and was not overmastered until the 5th. The loss that was caused by it amounted to $3,500,000, and it destroyed 2,500 buildings. The fire began in a paint-shop on the west side of Portsmouth Square, adjoining the American House. Although but a slight blaze when discovered, the building was within five minutes enwrapped with flames; and before the fire-engines could be got to work, the American House and the building on the other side of the paint-shop were also burning. The buildings being all of wood and extremely combustible, the fire spread up Clay street, back to Sacramento, and down Clay street towards Kearney, with fearful rapidity. Soon the fire department was compelled to give up every attempt to extinguish it, and to confine their work to making its advance less rapid.

Pursuing this plan, they checked the flames on the north side at Dupont street. But in every other direction it took its own course, and was only arrested at the water's edge and the ruins of the houses that had been blown up. The shipping in the harbor was only protected by the breaking up of the wharves. Thousands of persons were made homeless, and for a long time after lived in tents. The custom-house, seven hotels, the post-office, the offices of the steamship company, and the banking-house of Page, Bacon & Co. were destroyed. During the continuance of the fire a number of persons were burned, and others died from their exertions toward subduing it.

Another large fire devastated a great portion of San Francisco in June, 1851. It occurred on the 22d of that month, and 500 buildings were destroyed by it. The loss was estimated at $3,000,000.

PORTLAND (ME.), 1866.

The terrible fire which laid in ruins more than half of the city of Portland, Me., commenced at five o'clock on the afternoon of the 4th of July, 1866. Beginning in a cooper's shop, at the foot of High street, caused by a fire-cracker being thrown among some wood shavings, it swept through the city with frightful rapidity. With difficulty did the inhabitants of the houses in its path escape with their lives. Little effort was made to save household goods when this saving involved a possibility of death. Everything in the track of the flames was destroyed; and so completely, that when they had been overcome even the streets could hardly be traced. For a space of one mile and a half long by a quarter of a mile wide, there seemed a straggling forest of chimneys, with parts of their walls attached. From the place of beginning the fire was swept by a violent gale in a devious way, sparing nothing in its passage until it was checked by the ruins of the houses which had been blown up. The utmost endeavors of the firemen of the city, aided by those from other cities and towns, were of little avail until the plan of blowing up had been carried out, and then only to prevent the fire from spreading, and cause it for want of fuel to burn out. One-half of the city, and the one which included its business portion, was destroyed. Every bank and all the newspaper offices were burned; and it is somewhat singular to note that all the lawyers' offices in the city were swept away. The splendid city and county building on Congress street was considered fire-proof and safe, and was filled with furniture from the neighboring houses, and then the flames catching it laid it in ruins. All the jewelry establishments, the wholesale dry-goods houses, several churches, the telegraph offices, and the majority of other business places were destroyed. The custom-house, though badly burned, was not destroyed. Most singularly a building on Middle street, occupied by a hardware firm, was left unscathed by the sea of flame which surged and devastated all around it.

Two thousand persons were rendered houseless, and were sheltered in churches and tents erected for them.

In all, the loss was estimated at $10,000,000, which was but in small part covered by insurance.

CHARLESTON, 1838.

Charleston, S. C., was, on the 27th of April, 1838, visited by one of the most destructive fires that have ever occurred in any city in this country. A territory equal to almost one-half of the entire city was made desolate. The fire broke out at a quarter past eight o'clock on the morning of the day mentioned, in a paint shop on King street, corner of Beresford, and raged until about twelve A.M. of the following day. It was then arrested by the blowing up of buildings in its path. There were 1,158 buildings destroyed, and the loss occasioned was about $3,000,000. The worst feature of the catastrophe was the loss of life which occurred while the houses were being blown up. Through the careless manner in which the gunpowder was used, four of the most prominent citizens of the city were killed and a number injured.

CHICAGO, 1857, 1859, 1866, 1868.

On the morning of the 10th of October, 1857, a fire occurred in Chicago, which, though notable from the amount of property destroyed by it, was made awful by the loss of human life which it caused. The fire broke out in a large double store in South Water street, and spread east and west to the buildings adjoining, and across an alley in the rear to a block of new buildings. All these were completely destroyed. When the flames were threatening one of the buildings, a number of persons ascended to its roof to there fight against them. Wholly occupied with their work, they did not notice that the wall of the burning building tottered, and, when warned of their danger, they could not escape ere it fell, crushing through the house on which they were, and carrying them into its cellar. Of the number fourteen were killed and more injured. The loss in property caused by the fire amounted to over half a million of dollars.

A fire, the most disastrous after that of October, 1857, took place on September 15, 1859. It broke out in a stable, and, spreading in different directions, consumed the block bounded by Clinton, North Canal, West Lake, and Fulton streets, on which the stable was situated. From this block the fire was communicated to Blatchford's lead works and to the hydraulic mills, whence it passed to another block of buildings, all of which were destroyed. The total loss was about five hundred thousand dollars.

Property to the amount of $500,000 was destroyed by fire on the 10th of August, 1866. The fire originated in a wholesale tobacco establishment on South Water street, and passed to the adjoining buildings, occupied by wholesale grocery and drug firms. The first two buildings and contents were utterly, while the other was but partially, destroyed.

A fire, which destroyed several large business houses on Lake and South Water streets, took place November 18, 1866. It originated in the tobacco warehouse of Banker & Co., and the loss caused by it was about $500,000.

The fire which occurred on the 28th of January, 1868, was the most destructive by which Chicago had ever been visited. It broke out in a large boot and shoe factory on Lake street, and destroyed the entire block on which that building was situated. The sparks from those buildings set fire to others distant from them on the same street, and caused their destruction. In all the loss was about $3,000,000.

CHAPTER L.

TABLE OF FORMER GREAT FIRES.

NORFOLK, Va., destroyed by fire and the cannon-balls of the British. Property to the amount of $1,500,000 destroyed. January 1, 1776.

City of New York, soon after passing into possession of the British; 500 buildings consumed. September 20–21, 1776.

Theatre at Richmond, Va. The governor of the State and a large number of the leading inhabitants perished. December 26, 1811.

City of New York; 530 buildings destroyed; loss, $20,000,000. December 16, 1835.

Washington City; General Post Office and Patent Office, with over ten thousand valuable models, drawings, etc., destroyed. December 15, 1836.

Philadelphia; fifty-two buildings destroyed; loss, $500,000. October 5, 1839.

Quebec, Canada; 1,500 buildings and many lives destroyed. May 28, 1845.

Quebec, Canada; 1,300 buildings destroyed. June 28, 1845.

City of New York; 300 buildings destroyed; loss, $6,000,000. June 20, 1845.

St. John's, N. F., nearly destroyed; 6,000 people made homeless. June 12, 1846.

Quebec, Canada; Theatre Royal; 47 persons burned to death June 14, 1846.

Nantucket; 3(0 buildings and other property destroyed; value, $800,000. July 13, 1846.

At Albany; 600 buildings, steamboats, piers, etc., destroyed; loss, $3,000,000. August 17, 1848.

Brooklyn; 300 buildings destroyed. September 9, 1848.

At St. Louis, 15 blocks of houses and 23 steamboats; loss estimated at $3,000,000. May 17, 1849.

Fredericton, N. B.; about 300 buildings destroyed. November 11, 1850.

Nevada, Cal.; 200 buildings destroyed; loss, $1,300,000. March 12, 1851.

At Stockton, Cal.; loss, $1,500,000. May 14, 1851.

Concord, N. H.; greater part of the business portion of the town destroyed. August 24, 1850.

Congressional Library at Washington, 35,000 volumes, with works of art, destroyed. December 24, 1851.

At Montreal, Canada, 1,000 houses destroyed; loss, $5,000,000. July 8, 1852.

Harper Brothers' establishment, in New York; loss over $1,000,000. December 10, 1853.

Metropolitan Hall and Lafarge House, in this city. January 8, 1854.

At Jersey City, 30 factories and houses destroyed. July 30, 1854.

More than 100 houses and factories in Troy, N. Y.; on the same day a large part of Milwaukee, Wis., destroyed. August 25, 1854.

At Syracuse, N. Y., about 100 buildings destroyed; loss, $1,000,000. November 8, 1856.

New York Crystal Palace destroyed. October 5, 1858.

City of Charleston, S. C., almost destroyed. February 17, 1856.

At Quebec, Canada, 2,500 houses destroyed; loss, $2,500,000.

CONCLUSION.

The tales of horror with which this book is filled are relieved by the deeds of heroism and mercy which have been faithfully rehearsed. Human experience repeats itself from age to age

The thing that is, is that which hath been and shall be, so that there is nothing new under the sun. Lessons of human insufficiency, and weakness and vanity, mingle with lessons of man's greatness and nobleness. The heart turns to God in the midst of all this confusion and unrest, and finds unchangeable perfection, excellence, beauty, and joy. Though vanity of vanities is written on all that is earthly, there is in God pure satisfaction, absolute rest for every soul that seeks Him with a sincere purpose.

APPENDIX.

SPECIAL REPORT OF THE RELIEF AND AID SOCIETY.

IT will be a matter of interest to the generous donors to know how the money which they so kindly gave Chicago in her hour of need has been expended. The following report gives the fullest reliable information on all points, and we append it to this work as a memorial to the Christian charity of the world, to stand through all ages.

The Committee who gave their services to the herculean task of providing for the wants of 80,000 homeless people are as follows:—

Directors—Henry W. King, President; Wirt Dexter, E. C. Larned, T. M. Avery, T. W. Harvey, Marshall Field, John V. Farwell, N. S. Bouton, Murry Nelson, J. T. Ryerson, N. K. Fairbank, George M. Pullman, Dr. H. A. Johnson, H. E. Sargent, Julius Rosenthal, C. H. S. Mixer, A. B. Meeker, B. G. Caulfield, J. McGregor Adams, R. B. Mason, C. G. Hammond, Joseph Medill, *ex-officio.* [Elected Mayor Nov. 7.]

Executive Committee—Wirt Dexter, Chairman; Colonel C. G. Hammond, Henry W. King, T. M. Avery, T. W. Harvey, N. K. Fairbank, George M. Pullman, Dr. H. A. Johnson, E. C. Larned, N. S. Bouton, J. McGregor Adams.

SHELTER.

Concerning the work of providing the first requisite—shelter—for the houseless thousands, the Executive Committee says in its report: The Bureau of the Shelter Committee is very thoroughly organized, with an efficient corps of clerks and examiners, through whom the claim of the applicant goes for a careful and thorough examination, with all possible checks to detect imposition, while all are listened to with the utmost sympathy and patience. The houses given are of two sizes; one of 20x16 feet for families of more than three persons; the other of 12x16 feet for families of three only. The floor joists are of 2x6 inches timber, covered with a flooring of planed and matched boards; the studding is of 2x4 inches, covered with inch boards and battened on the outside; the inside walls are lined with thick felt paper; and each house has a double iron chimney, two panelled doors, three windows, and a partition to be put up where the occupant pleases. The establishment is completed in a simple but sufficient way for comfortable living, by the addition of a cooking-stove and utensils, several chairs, a table, bedstead, bedding, and sufficient crockery for the use of the family; the total cost of the house, when thus furnished, is $125. The majority of those who receive prepared material for these houses are mechanics enough to put them together for themselves, or have the means to hire builders; but for the large class of widows, infirm or otherwise helpless persons, the house is built and put in complete readiness for the proposed tenant by the Committee. There were on Saturday, the 18th inst., 5,497 of these houses put up, or in process of erection, most of which are completed and occupied. The applications for them, at the same date, numbered 7,246; and it is calculated that the demand for them which it will be prudent for the society to meet, with the means at their disposal, will be about eight thousand. This will provide, at the usual estimate of five to a family, and, as the houses chosen are almost entirely of the larger size, respectable and comfortable homes for from thirty to thirty-five thousand persons. Where the Committee think that the circumstances justify it, the house and its furnishings are an outright gift. In the majority of applications this is the case. But where the Committee have reason to know or to believe that the applicant has means that will become available, or that he will soon be able to command, he is requested to give an obligation to repay in one year, but without interest, three-fourths of the value advanced him. So far is this from being considered a hardship, that the applicant, in most cases, prefers to accept the obligation to return the money, that it may again be used to aid others who may be in need; as it frees him from being the recipient of public bounty, and allows him to retain an honorable feeling of independence. He may refund the amount before the year expires, if it shall suit his convenience; but if it shall appear at the end of that time, upon a reinvestigation of the case, that he is evidently unable to refund it, he is simply considered by the Committee as belonging to that class from which no return could be expected for bounty bestowed.

The actual rental of these houses may be estimated as worth $10 per month, based upon what the society is paying, in many instances, for similar accommodations to keep people from being turned out of doors. This rental for six months would amount to $60, and as the cost of the "shelter houses," exclusive of furniture, is nearly $100, they will have paid, by the 1st of May next, 60 per cent. of their cost. It must not be understood, however, that this is a rental charged, but only a rental estimated, and which is saved to the owner of the house in six months. In no case is any rent taken from the occupants of these houses.

The stock of lumber destroyed in Chicago by the fire was not less than sixty-five millions of feet, and the supply destroyed in the lumber regions, ready for shipment to this market, was also immense. The price of lumber, consequently, has rapidly enhanced, and since the 26th of October has been $20 per thousand. By the wise forethought and activity of the Shelter Committee this rise in value was anticipated, and all their purchases have been made at an average price of $16.50 per thousand. They have used, thus far, nearly twenty-seven millions of feet with this large saving in cost.

The Barracks.

Besides the isolated houses, there are in different sections of the city four barracks, in which are lodged one thousand families. They are mainly of the class who have not hitherto lived in houses of their own, but in rooms in tenement houses. Each family in these barracks has two separate rooms to itself, and they are furnished in precisely the same way with the isolated houses. Their occupants are undoubtedly very nearly, if not quite, as comfortable as they were before the fire; and as only one thousand two hundred and fifty people are gathered together in one community, and those are under the constant and careful supervision of Medical and Police Superintendents, their moral and sanitary condition is unquestionably better than that which has heretofore obtained in that class. There has been among them but a single death up to the 25th of November.

Distribution of Supplies.

The checks upon waste or fraud in the issue of supplies of food, etc., are thus detailed: The first step was to district the city under the direction of O. C. Gibbs, who for years had been Superintendent of the Relief and Aid Society, and it was accordingly divided into five large districts, made as nearly equal as possible with regard to population. These were subdivided, at first, into thirteen smaller sub-districts, but which are now, as rearranged from time to time, and as rations are given out at longer intervals, six only. The whole are under the General Superintendent; but to each district is given a Superintendent, with supervision over his whole district, and to each sub-district a Sub-superintendent, with supervision over his immediate depot of supplies. Sufficient assistance is given to each Superintendent, averaging about ninety men and women to each district, the duties of a part of whom are to administer to the wants of applicants for food and clothing. Another part of this force is made up of a corps of visitors who are constantly busy in visiting all whose names are registered in the books at the offices of the relief stations, and in searching for sufferers who need aid, but do not know where to find it. It is the business of the visitor to keep himself constantly informed as to all the persons who are thus entered in his district, and to make periodical returns at the office. He is to learn by observation and inquiry the exact condition of the registered; whether they are well or ill; whether they are idle or industrious; whether they are voluntarily idle, in which case they are peremptorily cut off from aid; whether they are entitled to entire or only partial support; whether they have other means of support than public bounty; and in short, any circumstances in relation to their condition or habits, or character, which will be a guide as to the care which should be given them at the stations. There a ledger account is opened with each of them, in which appear the returns of the visitors, the supplies given, with their dates, and when they were cut off, if discontinued, and the reasons why.

The Superintendents are required to keep a strict account of all their requisitions of supplies, as well as of their distribution: and as they are accountable for a judicious and energetic discharge of their duties to the General Superintendent, so they hold their own subordinates strictly accountable for all their actions. Full and careful reports are made daily from each district, and the Superintendents meet one evening in the week with the Executive Committee to make or hear suggestions, to answer criticism or complaints, to report progress and suggest improvements, if possible, in the working machinery. The districts are frequently visited by a general inspector, to examine into their condition and management; and a Committee on Complaints is always ready at headquarters to listen to any complaint of neglect or improper treatment, and to provide for their immediate correction, if found, on inquiry, to be well founded.

Number of Families Aided.

The total number of families receiving aid from the Society on the 11th of November was, 12,765; on the 18th, 14,137; on the 25th, 15,122; total number since full reports were had, 18,478. The number increased as the cold weather came on, notwithstanding many families, having been enabled to earn their livelihood, dropped off the rolls of the Society.

Rations.

Food was given at first not only indiscriminately, but in uncertain quantities, for want of conveniences in measuring and weighing. As soon as possible, however, it was reduced to fixed rations, and as the system of distribution was perfected, these were given out at intervals of two or three days, and now of a week. The following ration for a family of five persons has been found to be sufficient for one week. At first bread was given instead of flour, as the people had few conveniences for cooking, at an increased cost of forty-two cents to the ration. This is now almost wholly saved, as most of the applicants are supplied with stoves, and can bake their own bread. Coffee or tea is given, as the applicant prefers; but tea, which is the cheaper, is the more usually chosen. The cost of the ordinary weekly ration given for a family of five is $1.98, as shown by the following exhibit:

Exhibit of the Amount and Cost of One Week's Ration for Two Adults and Three Children.

Three pounds pork, at 5½ cents	.16½
Six pounds beef, at 5 cents	.30
Fourteen pounds flour, at 3 cents	.42
One and a quarter pecks potatoes, at 20 cents	.25
One-quarter pound tea, at 80 cents	.20
One and one-half pounds sugar, at 11 cents	.16½
One and one-half pound rice, at 8 cents, or 3½ pounds beans, at 3¾ cents	.12
One and one-quarter pounds soap, at 7 cents	.09
One and one-half pounds of dried apples, at 8 cents	.12
Three pounds fresh beef, at 5 cents	.15
Total	$1.98
If bread, at 4 cents per pound, is used instead of flour, the cost is increased	.42
If crackers, at 7 cents per pound	1.05
If 1½ pounds of coffee instead of tea	.17

To the cost of the weekly ration of food for a family of five should be added the allowance of one ton of coal a month, or a quarter of a ton a week. Fortunately for such an exigency as this, the supply of bituminous coal for Chicago is ample, through the Wilmington Coal Company, which owns and works extensive coal mines in Will County, Illinois, with sufficient means of transportation at their command over the Alton, St. Louis and Chicago Road. With this company the committee has made a contract for the delivery of coal by the ton or half ton, at the door, for $4.50 per ton. This brings the weekly cost of coal for the family at $1.12½, which, added to the cost of the weekly ration, brings the cost of food and fuel at $3.10½. As the demand for fuel is as constant and next in importance to that of food, a large depot of coal from other sources is kept in reserve for emergencies, as in case of interruption to railroad transportation by snow-storms, or other causes, during the winter.

CLOTHING.

The demand for clothing has been, and continues to be, incessant and immense. The larger proportion of those who were sufferers by the fire lost all their personal apparel and their household goods. Immediate and urgent need was only very partially met by the bountiful supplies which were sent forward from all quarters. Much of this supply was of second-hand summer clothing, which was all that people could lay their hands on in the first emergency. It answered a good, though only temporary purpose, and the necessity of substituting for it better and warmer garments is constant and imperative. The markets of this country cannot supply the demand for blankets alone. Where the supply of ready-made clothing has been insufficient, piece goods are given out in measured quantity to applicants to make up for themselves. In this work great assistance is rendered by associations of ladies, as the Ladies' Relief and Aid Society, the Ladies' Industrial Aid Society of St. John's Church, the Ladies' Christian Union, Ladies' Society of Park Avenue Church, and Ladies' Society of the Home of the Friendless—all of whom employ a large number of sewing women, thrown out of employment by the fire, in making up garments, bed comforters, bed-ticks, and other articles, from piece goods supplied by the Relief Committee, and returned thus manufactured, to the several depots for distribution.

But a comparatively small portion of those in need of warm and sufficient clothing for the winter is as yet supplied, and the labor and expenditure to meet this want must be very large for some time to come. Of the actual quantity received by gift from abroad and distributed, it is impossible to make a detailed statement, as much of that received was given out in the first calamitous days of destitution to all comers and without count. The United States Government, through the active efforts of General Sheridan, has furnished us seven thousand blankets, and has also on the way for our use five thousand each of undershirts, drawers, and socks. We are promised by President Grant, through the Hon. W. W. Belknap, Secretary of War, who has recently visited Chicago, such further supplies as we may need, so far as the government may have them in store. This branch of the work, however, is being reduced to a system, like the rest, and the following table is condensed from the reports of the several districts for the week ending November 25, giving the number distributed of several articles of prime necessity. To it is added the number previously reported since an accurate record was kept:—

DISTRIBUTION OF ARTICLES FOR WEEK ENDING NOVEMBER 25.

Articles.		*Previously reported.*	*Total.*
Mattresses	2,131	8,606	10,737
Blankets	4,615	20,724	25,339
Coal, tons	1,522	2,131	4,653
Stoves	644	3,795	4,459
Shoes	22,531	....	22,539
Men's wear	8,846	45,883	54,729
Women's wear	8,895	57,091	65,986
Children's wear	8,984	35,953	44,937

The above table does not include the stoves and mattresses given out by the Shelter Committee, who furnished both articles to the large proportion of their houses and the barracks. Neither does it include the furniture and crockery, both large items of expenditure, the aggregate of which is not yet reached.

PAY-ROLLS FOR THE WEEK ENDING NOVEMBER 18.

Districts.	*Persons employed.*	*Amount.*
No. 1	142	$1.549 00
No. 2	116	1,190 17
No. 3	106	1,347 75
No. 4	50	531 82
No. 5	75	855 66
Special Bureau	9	118 50
Superintendents' salaries		225 50
Warehousemen, receiving, storing, and delivering supplies	111	1,259 87
Transportation		2,148 87
Total for distribution		$9,226 64

Employment Bureau	2	$36 00
Clerks in offices of Treasurer, Auditor, Transportation Committee, Purchasing Committee, and Executive Committee	32	496 34
Total general business		$9,758 91

The Bureau of Special Relief.

No branch of the work has given the committee so much anxiety and perplexity as that which has come to be known as "The Bureau of Special Relief." Among the sufferers by the fire is a large class of persons who, it was soon apparent, would not be reached by the established method of relief, but who were the least accustomed to deprivation and hardship. They shrank from an exposure of their poverty, though it was no fault of their own, and, though sufferers in common with tens of thousands of others, from a great public calamity, they would perish rather than appear as the recipients of public bounty. If they were to be helped at all, they must be helped in some special way. It was no time to stop and consider whether the feeling was altogether reasonable or not. It was painfully evident that a want existed, growing out of previous conditions in life of the sufferers, and public opinion, as well as private feeling, made it necessary to devise some way to meet it. It was believed that the personal and confidential relations between pastor and people, and between the officers and members of benevolent and fraternal societies, would reveal a great many cases of this sort; and many, it was thought, would ask aid for themselves, if encouraged to do so by being permitted to seek relief where publicity could be avoided, and the shock be lessened to their sensitiveness and reserve.

Moved by these considerations, the committee invoked the aid and counsel of clergymen and the representatives of the societies just referred to. By the establishment of a bureau, to be devoted exclusively to special relief, and to be under the control of a committee appointed by this body of our fellow-citizens, it was proposed to bring into line all effort on behalf of the unfortunate, that none should be left to perish for the want of sympathy and help. Most of those whose aid was invoked entered heartily into the work, and with a sincere desire to lighten the general labors of the Society. Perhaps it was too much to expect, even in a cause involving only the single purpose of feeding the hungry and clothing the naked, that the plan should succeed in satisfying all those who sought to make use of the means in the hands of the Society. It is by no means easy to say always where the obligations of those entrusted with the delicate task of deciding between the claims of different classes begin or where they end; and the most careful judgment and the most even justice will not save their decisions from sometimes seeming invidious. But nevertheless, since the Special Bureau, E. C. Larned, Chairman, was opened at the Church of the Messiah, its usefulness has become daily more and more manifest, and more and more appreciated. Mr. Larned has been assisted by Rev. Laird Collier, B. G. Caulfield, Rev. E. P. Goodwin, Louis Wahl, G. R. Chittenden, Orrington Lunt, Mrs. Joseph Medill, Mrs. David A. Gage, and Mrs. J. E. Tyler, all of whom give their time without compensation.

Up to the 25th instant, aid has been given to one thousand five hundred and twenty-five families. Very few, if any, of these had previously sought for relief through the ordinary channels, and would, no doubt, some from pride and some from inability through sickness or infirmity, have suffered the very extremity of distress before they would or could have looked for succor in that direction. While great care is taken that there shall be from its stores no duplication of supplies from other distributing points, all applications are received and considered with all the delicacy and reserve that the nature of the business admits of; and there can be no doubt that it has relieved the wants of thousands who would otherwise have been left uncared for, or dependent upon the chance charity of those who should happen to know of their condition. One of its methods of relief, especially, has saved many worthy women from penury and despair, by putting into their hands the means of immediate and comfortable subsistence. Arrangements have been made with all the principal manufacturers of sewing machines, by which they generously agreed to accept a large discount on the usual price for a single machine. A payment of twenty dollars is made as the first instalment on that price, and one hundred and thirty-two machines have thus been purchased and given to that number of deserving women who brought satisfactory evidence that they had been sufferers by the fire. So many are reinstated, and many more will be in the same way, to their former means of earning a livelihood.

Charitable Institutions.

The Committee on Charitable Institutions, N. S. Bouton, Chairman, have extended aid to the following institutions:—The Home for the Friendless, the Protestant Orphan Home, the St. Joseph

Asylum for Orphans, the Old Ladies' Home, the House of the Good Shepherd, the Foundlings' Home, the Half-Orphans' Home. To all of these institutions a monthly allowance in money is given, and those which have been burned out have been supplied, in addition, with food, clothing, bedding, and stores sufficient for their immediate necessities. Still further aid will be extended to them all, if it should be found requisite to carry them through the winter. There were other institutions of a similar character which were destroyed and their inmates dispersed. These have not yet provided themselves with permanent residences, and the Committee do not feel justified, by the means at their disposal, to advance the large sums that would be required for their re-establishment. They can only undertake to supplement, in some measure, to those whose responsibilities are still existent, the resources of which they are deprived by the general disaster. The Sailors' Home only was made an exception to this rule. Though that institution lost its house by the fire, the inmates were kept together, and the Shelter Committee has offered to put up a temporary building for them, at a cost not to exceed $5,000.

Employment Bureau.

There has been no lack of employment, particularly of unskilled labor, since the fire; but as that could not be foreseen, it was thought prudent to establish an employment bureau in connection with the general work. An employment committee, N. K. Fairbank, Chairman, was therefore appointed, with headquarters in a temporary building in the Court-House yard. This has been a sort of labor exchange in the very heart of the burnt district, where those wanting mechanics or laborers could find them, and where those in need of work were provided with it. The Superintendents at all the points of distribution are instructed to send every able-bodied man or boy who applies to them for aid to the bureau of the Employment Committee, and the ticket he takes becomes a certificate of character. If labor is found for him—as is almost invariably the case—he surrenders the ticket, and it is returned to the Superintendent who issued it. If the ticket is not presented at the Employment Bureau, and not returned, therefore, to the Superintendent, it is presumptive evidence that the bearer prefers to eat the bread of idleness rather than work for his own subsistence, and if he again presents himself at the distributing station, his claim for relief is rejected. If, having obtained work—of which the returned ticket is evidence—he asks again for relief, the proper inquiry decides whether his labor is not sufficient to sustain himself and his family, if he has one, or whether he has asked for bounty of which he is not in need.

Most of the mechanics who apply at the Employment Bureau for work are in want of tools, without which they can do nothing at their trades. This want the committee has supplied, and by giving the applicant from ten to twenty dollars' worth of tools, he is at once made self-supporting, and ceases to be dependent upon the Relief Society. A large number of carpenters have thus been effectively and permanently helped, as the demand for their labor is greater than for that of any other class. Bricklayers, gas-fitters, shoemakers, and other mechanics have also been aided in the same way. The Bureau has not undertaken to find employment for women, but has turned that class over to the organizations that have hitherto made its care their special business. Excepting seamstresses—who are received and cared for by the Bureau of Special Relief—women seeking employment have been left under the direction of such societies, and especially of the Ladies' Christian Union, which in this part of the work has been a valuable coadjutor of the Relief and Aid Society.

Nationalities of Beneficiaries.

The following table is an accurate return from the books kept at all the distributing stations:—

Nationality.	*No. families.*	*Nationality.*	*No. families.*	*Nationality.*	*No. families.*
Americans	1,724	Swiss	30	Welsh	10
English	599	Danish	14	Belgian	23
Scotch	195	Spanish	2	Holland	5
Irish	5,512	Polish	90	Greek	1
German	7,280	Russian	2	Scandinavian	2,104
French	185	Jewish	43	African	241
Italian	112	Hungarian	4		
Canadians	94	Bohemian	203	Total	18,478

Future Expenses.

The Committee makes the following estimate of expenditures for six months, from October 9, 1871, to April 9, 1872:—

Food and fuel rations for 15,122 families of five persons each, at $3.10½ per week	$46,953 81	$1,220,799 06
Shelter Committee, 8,000 furnished houses, at $125 each	1,000,000 00	
Barracks and furniture for 2,000 families, at $80 each	160,000 00	
Hospital and store-houses	83,000 00	1,243,000 00
Stoves, in addition to those used in new houses and barracks		75,000 00
Special Bureau		250,000 00
Charitable institutions		25,000 00
Deducting contributions of clothing and mattresses and furniture furnished by the Shelter Committee, already charged, it is estimated that 10,081 of the 15,122 families must be supplied with clothing, shoes, furniture, beds and bedding, at a cost of		866,966 00
Current expenses, at $9,758.98 per week, twenty-six weeks	253,733 48	
General expenses Shelter Committee	42,000 00	295,733 48
		$3,976,498 54
Total of contributions		3,418,188 20
Deficit		$558,310 34

How Certain Articles were Procured.

If, say the Committee, the fund at our disposal were sufficient to buy all needed things, it is simply impossible to purchase in so short a time any considerable portion of the necessary articles that were to be found in the houses of a hundred thousand people. Take as an illustration the article of stoves. No stoves were saved from the fire. To replace them a medium-sized soft coal cook-stove was needed. No other would answer. To some extent wood stoves were sent, but we had no wood. Under the active direction of A. B. Meeker, the Society has managed to buy, between the Atlantic and the Missouri River, eight thousand five hundred stoves of the kind demanded, and here and elsewhere has obtained pipe and furniture for them. But with the stoves purchased, our difficulties were by no means surmounted. We have been subjected to delays in railway transportation and movement of goods about the city, in common with our merchants, many of whom had merchandise lying three weeks within a few miles of Chicago, which could neither be received nor stored. Every possible facility and courtesy has been extended to this Society by railway and express companies connected with Chicago, and also by the Western Union Telegraph Company and the Atlantic Cable Company, in transmitting free our answers to despatches and orders for goods. Yet, with all these advantages, we have been able to actually deliver, up to November 25, but six thousand six hundred stoves. There is precisely the same difficulty in the delivery of mattresses, blankets, and many articles furnished by the Society which have been drawn from Buffalo, Detroit, Montreal, St. Louis, Indianapolis, Louisville, Cincinnati, and other points. With this delay in the arrival of goods, the best we could do was to give applicants orders to be filled in turn, informing them at the same time of the situation. Some of these orders, we regret to say, were many times presented at our depot without being filled; but this was not the result of any defect in our system, but simply because a supply of the article asked for was not to be had. Nor was there any delay in contracting for these things. Within three days from the time the Society assumed the work of relief, large engagements were made both in the United States and Canada for supplies which it was plain would be needed. We refer to these facts in order that applicants as well as the public at large may form some idea, if possible, of the difficulties surrounding the transaction of all business in Chicago during the last few weeks. Yet we may be allowed to call attention to the fact that we have just passed through two weeks of unusually severe winter weather, accompanied by snow, but that the needy are provided for and the sick and infirm attended to.

Treasurer's Report of Receipts and Expenditures.

From the Report of George M. Pullman, Treasurer, we make the following extracts, which embrace the receipts and expenditures of the Society passing through his books since the 14th of October, when the duty of Treasurer devolved upon him, to November 18th, the date of the Report. Space permits us only to give the total amounts received from the States and Countries that donated, without entering into detail, and only cash receipts are embraced:—

UNITED STATES.

State	Amount	State	Amount	State	Amount
New York	$401,556 96	Delaware	$8,070 70	West Virginia	$15,526 40
Massachusetts	297,305 12	Kentucky	5,108 90	Nebraska	12,689 00
Pennsylvania	228,363 37	Iowa	4,596 25	Colorado	12,653 03
California	148,281 60	Arkansas	2,536 55	New Mexico	1,495 50
Maryland	179,327 84	Georgia	2,065 75	Florida	1,049 23
New Jersey	124,464 32	Vermont	3,692 20	South Carolina	1,001 90
Connecticut	65,968 68	Nevada	1,505 83	Michigan	632 25
Ohio	50,714 71	Dist. Columbia	34,065 65	Washington Ter.	600 00
Rhode Island	45,179 70	Indiana	24,750 84	Wisconsin	856 00
Illinois	45,950 96	Minnesota	24,108 40	North Carolina	105 00
Maine	11,616 26	Tennesee	23,655 20	Mississippi	48 50
Louisiana	11.604 80	New Hampshire	21,834 85	Alabama	5 00
Virginia	11,228 41	Kansas	21,257 85		
Oregon	10,000 00	Missouri	16,974 70	Total	$1,875,062 80

FOREIGN.

England	$358,410 72	Scotland	$31.000 00	Miscellaneous	327 20
Canada	82,695 57	Austria	12,500 00		
France	42,200 00	Cuba	10,000 00	Total	$610,821 75
Ireland	36,948 18	Italy	271 38	United States	1,875,062 80
Germany	36,256 70	Holland	112 00		
				Grand total	$2,485,884 55

LEDGER STATEMENT.

The following is the ledger statement to November 18, inclusive:—

Donation fund	$2,485,884 55
Cash on hand	189,864 93
Pay rolls	34,009 20
Insurance	697 50
Expense	809 22
Office furniture	1,503 31
Postage	261 15
Transportation (city teaming)	2,927 34
Freight	800 91
Stationery, printing, and advertising	2,557 22
Supplies—of clothing, food, stoves, bedding, etc., etc	130,870 71
Shelter—lumber and building materials	311,162 46
Hospital expenses	1,984 80
Rents—of supply depots and offices	2,174 83
Buildings	85 64
Cash advances	3,033 00
Employment Committee	5 80
Bureau of special relief	2,555 27
Passenger fares	22 00
Deposits in banks	1,800,559 26
Total	$2,485,884 55

BANK BALANCES.

Chicago—	
Third National Bank	$103,087 25
Union National Bank	103,060 24
Northwestern National Bank	33,017 15
First National Bank	87,667 92
Merchants' National Bank	58,340 97
Mechanics' National Bank	33,100 00
German National Bank	33,257 74
Second National Bank	19,394 33
Fifth National Bank	46.752 31
Manufacturers' National Bank	6,285 12
Total	$523,963 03
New York—	
National City Bank	$275,109 78
Bank of Commerce	200,000 00
Drexel, Morgan & Co	493,982 22
Total	$969,092 00
Boston—	
Kidder, Peabody & Co	$128,461 84
Second National Bank	82,239 30
Total	$210,701 14
Bank of Montreal	68,803 09
Providence, R. I.	28,000 00
Total	$1,800,559 26

All the above bank balances are drawing 5 per cent interest.